VOLUME SEVENTY NINE

Current Topics in MEMBRANES

Piezo Channels

CURRENT TOPICS IN MEMBRANES, VOLUME 79

Series Editors

ROBERT BALABAN
National Heart, Lung and Blood Institute
National Institutes of Health
Bethesda, Maryland, USA

SIDNEY A. SIMON
Department of Neurobiology
Duke University Medical Centre
Durham, North Carolina, USA

VOLUME SEVENTY NINE

CURRENT TOPICS IN MEMBRANES

Piezo Channels

Edited by

PHILIP A. GOTTLIEB
State University of New York at Buffalo, Buffalo, NY, United States

ELSEVIER

ACADEMIC PRESS
An imprint of Elsevier

Academic Press is an imprint of Elsevier
50 Hampshire Street, 5th Floor, Cambridge, MA 02139, United States
525 B Street, Suite 1800, San Diego, CA 92101-4495, United States
125 London Wall, London EC2Y 5AS, United Kingdom
The Boulevard, Langford Lane, Kidlington, Oxford OX5 1GB, United Kingdom

First edition 2017

ISBN: 978-0-12-809389-4
ISSN: 1063-5823

For information on all Academic Press publications visit our website at https://www.elsevier.com/books-and-journals

Publisher: Zoe Kruze
Acquisition Editor: Ashlie M. Jackman
Editorial Project Manager: Alina Cleju
Production Project Manager: Vignesh Tamil
Designer: Greg Harris

Typeset by TNQ Books and Journals

CONTENTS

CONTRIBUTORS

C. Alcaino
Mayo Clinic College of Medicine, Rochester, MN, United States

S.L. Alper
Beth Israel Deaconess Medical Center and Harvard Medical School, Boston, MA, United States

E.O. Anderson
Yale University, New Haven, CT, United States

S.N. Bagriantsev
Yale University, New Haven, CT, United States

N. Bavi
Victor Chang Cardiac Research Institute, Darlinghurst, NSW, Australia; University of New South Wales, Darlinghurst, NSW, Australia

D.J. Beech
University of Leeds, Leeds, United Kingdom

A. Beyder
Mayo Clinic College of Medicine, Rochester, MN, United States

I. Borbiro
Rutgers, New Jersey Medical School, Newark, NJ, United States

C.D. Cox
Victor Chang Cardiac Research Institute, Darlinghurst, NSW, Australia; University of New South Wales, Darlinghurst, NSW, Australia

G. Farrugia
Mayo Clinic College of Medicine, Rochester, MN, United States

N. Gao
Tsinghua University, Beijing, China

J. Geng
Tsinghua University, Beijing, China

R. Gnanasambandam
State University of New York at Buffalo, Buffalo, NY, United States

P.A. Gottlieb
State University of New York at Buffalo, Buffalo, NY, United States

F. Guilak
Washington University in St Louis and Shriners Hospitals for Children, St Louis, MO, United States

S.Z. Hua
State University of New York at Buffalo, Buffalo, NY, United States

A.J. Hyman
University of Leeds, Leeds, United Kingdom

W. Lee
Duke University Medical Center, Durham, NC, United States

W. Li
Tsinghua University, Beijing, China

W. Liedtke
Duke University Medical Center, Durham, NC, United States

M.M. Maneshi
State University of New York at Buffalo, Buffalo, NY, United States

B. Martinac
Victor Chang Cardiac Research Institute, Darlinghurst, NSW, Australia; University of New South Wales, Darlinghurst, NSW, Australia

T. Rohacs
Rutgers, New Jersey Medical School, Newark, NJ, United States

F. Sachs
State University of New York at Buffalo, Buffalo, NY, United States

E.R. Schneider
Yale University, New Haven, CT, United States

S. Tumova
University of Leeds, Leeds, United Kingdom

B. Xiao
Tsinghua University, Beijing, China

M. Yang
Tsinghua University, Beijing, China

T. Zhang
Tsinghua University, Beijing, China

Q. Zhao
Tsinghua University, Beijing, China

PREFACE

The ability of cells to sense and adapt to mechanical stress in their environment is critical for survival, and they have developed specialized sensors for this purpose. Transmembrane protein enzymes, called mechanical sensitive ion channels (MSCs), convert mechanical stress into chemical and electrical potentials that elicit an appropriate cellular response. A family of eukaryotic MSCs, Piezo1 and Piezo2, was identified and cloned in 2010, and these are involved in a variety of mechanical processes ranging from proprioception to blood pressure regulation, as well as disease. The 12 chapters in this volume describe properties of the Piezo channels and their role as mechanical transducers in a variety of biophysical, biochemical, and biological processes. Chapter 1 provides an overview of the properties of the Piezo channels and the important work that led to their discovery.

Chapter 2, written by the Beech group, summarizes the effect of fluid shear stress on the developing vascular system and the central role that is played by Piezo1. The related Chapter 12 by Hua and colleagues maps the biophysical response of PIEZO1 on cells during fluid shear stress using microfluidic devices and precision controlled stimuli. This approach is useful for rapidly measuring the effect of inhibitors or activators on Piezo channels.

A central question regarding Piezo channels is the mechanism of activation. Chapter 3 by Cox and Martinac summarizes our current understanding of the forces that allow the channel to open (gate) and describes the Piezo properties in the context of previous work related to known MSCs, such as the prokaryotic mechanical channels and the eukaryotic potassium selective MSCs. The structural basis for gating the channel comes from the cryo-electron microscopic structure described in Chapter 5 by Yang and colleagues. Although we are in the early stages of structural analysis of the channel, it is clear that it is unique in relation to other known mechanical channels. Complementary work by the Xiao group in Chapter 6 describes work with mutagenesis and chimera channels to define important regions of the protein such as the pore and the roles of different protein domains. The channel has three states (closed to open to inactivated), and the model described in Chapter 11 by Sachs and colleagues provides an overview of channel kinetics and how mutations can change those properties. Permeation properties of cations through the PIEZO1 channel is thoroughly summarized as well.

There are two chapters dedicated exclusively to Piezo2 channels. Chapter 7 by Bagriantsev et al. shows that Piezo2 in somatosensory neurons provides a selective advantage for tactile-foraging duck. The work described by Beyder in Chapter 8 shows that Piezo2 is present on enterochromaffin cells found in the small bowel and is involved in the release of serotonin with increased mucosal secretion. This mechanical response has implications for diseases like irritable bowel syndrome.

Mutations to the Piezo channels have paved the way for a new understanding of diseases that lead to an aberrant response of cells to mechanical forces. For example, hereditary mutations identified in Piezo1 causes xerocytosis, a condition related to hemolytic anemia of red blood cells. The chapter by Seth Alper (Chapter 4) catalogs and describes our current understanding of diseases caused by mutations to Piezo1 or Piezo2. Chapter 10 by Liedtke discusses the importance of Piezo channels in chondrocytes involved in joint health with obvious implications for aging. This work also describes the cooperation between Piezo1 and Piezo2 necessary for the mechanical response of cells. Piezo2 is linked to mechanical pain, an effect potentially mediated by a cellular signaling pathway described by the Rohacs group in Chapter 9.

New advances in our understanding of cell mechanics have been made through this work on Piezo channels. There remain many questions that need to be answered. At the molecular level, we have the mystery of why the Piezo channels are so large in comparison to all other mechanical channels. Are there other properties yet to be discovered? Our understanding of the cell's pathological response related to mechanics requires a deeper understanding of the cellular environment in which these channels reside and has implication for understanding diseases.

The rapid expansion of this field has so far given us only a hint of their significance in disease and their overall impact on life itself. We anticipate further fruitful discoveries and collaborations in the near and distant future.

Philip A. Gottlieb
SUNY at Buffalo

CHAPTER ONE

A Tour de Force: The Discovery, Properties, and Function of Piezo Channels

P.A. Gottlieb[1]
SUNY at Buffalo, Buffalo, NY, United States
[1]Corresponding author: E-mail: philgott@buffalo.edu

Contents

Current Topics in Membranes, Volume 79
ISSN 1063-5823
http://dx.doi.org/10.1016/bs.ctm.2016.11.007

Abstract

Mechanical transducers appear throughout cell biology and are used to convert mechanical stress into chemical or electrical signals that allow the cell to respond to environmental changes. In the past six years, a eukaryotic mechanical channel family with two members, Piezo1 and Piezo2, has been identified. Piezo1 was shown to be a cation-selective channel that does not require ancillary proteins for activity. Mouse Piezo1 is large, with over 2500 amino acids, and is not homologous to other ion channels. Both piezo channels have rapid voltage-dependent inactivation with a reversal potential near 0 mV. The CryoEm structure of Piezo1 at 4.8 Å shows trimer stoichiometry.

Since the discovery of the piezo channels, their roles in the physiological response of cells have started to emerge. Significant progress has been made in understanding the intrinsic properties of the channels and how these properties are modulated by cytoskeletal elements. Specific diseases, such as hereditary xerocytosis affecting red blood cells, have mutations in Piezo1 that alter the cell's response to force, typically slowing inactivation and introducing a latency for activation. A number of physiological functions for piezo channels have been identified. These range from sensing the stiffness of surrounding substrate, to the response to light touch, to serotonin release from the gut. This review provides a general overview of the properties and roles of Piezo1 and Piezo2 in eukaryotic mechanotransduction.

1. BACKGROUND

Cells that experience mechanically induced stress in their environment have developed mechanisms that allow them to adapt. For bacteria, the response to mechanical cues is well established. Their mechanosensitive channels (MSCs) open (gate) with increased membrane tension that occurs with hypoosmotic shock (Hamill & Martinac, 2001; Martinac, 2011; Sukharev, Blount, Martinac, Blattner, & Kung, 1994). This lets excess solutes leave the cell, reducing osmotic stress (Sukharev & Sachs, 2012).

For eukaryotic cells, MSCs are needed for a variety of functions including hearing, touch, balance, and blood pressure sensing, and the MSCs are sensors for organs that undergo mechanical stress. Our understanding of the structure and function of these channels has been limited, although ion channels that respond to force were shown to exist about 35 years ago (Corey & Hudspeth, 1979; Guharay & Sachs, 1984). Even in the absence of cloned channels, endogenous cation selective ion channels have been studied extensively and provided important information about channel properties (Suchyna, Besch, & Sachs, 2004). For example, the single channel recordings of mechanically activated (MA) endogenous channels of astrocytes revealed a rapid voltage-dependent inactivation (Suchyna et al., 2004). Studies also revealed the relationship between cytoskeletal stress and channel function (Suchyna & Sachs, 2007). The impairment of a cell's mechanical response to stress can arise not only from mutagenic changes to properties of the channel, but from changes to the structural elements that surround and support the channel (Ingber, 2006; Suchyna & Sachs, 2007). For example, defects in the cytoskeletal protein dystrophin, responsible for muscular dystrophy, affect the cell's ultrastructure and changes the open probability of mechanical channels by altering local environment around the channel (Gailly, 2002; Lansman, 2007; Lansman & Franco-Obregon, 2006; Yeung & Allen, 2004).

The search for a eukaryotic MSC tended to focus on the TRP family (Clapham, Runnels, & Strubing, 2001). Data suggested their involvement in mechanotransduction, but there was never clear evidence of their direct participation in the response to force (Christensen & Corey, 2007; Eijkelkamp, Quick, & Wood, 2013). The question of what constituted a mechanical channel led to the specific criteria for assessing whether a target ion channel is involved in a cell's mechanical response (Arnadottir & Chalfie, 2010; Christensen & Corey, 2007). There is the obvious condition that cells that respond to mechanical stress have a gene that codes for proteins, so when they are removed the response is lost. It is equally important to express the gene in mechanically nonresponsive cells, thus imparting to the cell the ability to respond to mechanical stimuli. Signature features, such as the intrinsic kinetics of the channel, should be at least roughly preserved in the new cellular environment. Mutations to the gene that alter channel properties, such as conductance and selectivity, can be used to verify the channel. In general, it is rewarding to have reconstitution of purified channels to allow for a direct measure of mechanosensitivity, if the MSC is activated by bilayer tension. Many channels, such as the voltage-dependent channels, are modulated by bilayer tension (Beyder et al., 2012; Morris, 2011; Morris & Juranka, 2007), but we are restricting our definition of MSCs to those that can be driven over the whole dynamic range by mechanical stress alone. The mechanism of force from lipid activation of mechanical channels is extensively addressed in chapter "Origin of the Force: The Force-From-Lipids Principle Applied to Piezo Channels" by Cox et al. (2016).

The family of K_{2P} potassium selective channels is an obvious example that satisfies all the aforementioned criteria. They have been shown to directly sense mechanical force in the membrane (Honore, 2007). Heterologous expression of these channels, for example TREK-1, confers mechanical properties to the cell that were not observed in untransfected cell (Patel & Honore, 2001; Patel & Honore, 2002). Because K_{2P} channels are potassium selective, the biophysical properties are easy to measure above background. In general, TRP channels, thought to possess mechanosensitive properties when expressed in a neutral background, did not generally fulfill the criteria of transferable mechanical properties (Gottlieb et al., 2008).

A key issue for MSCs is the mechanism of activation. Whereas TRAAK and TREK channels (Brohawn, Su, & MacKinnon, 2014; Honore, 2007) behave much like bacterial channels because they gate via membrane tension, some channels use tethers, where forces are transmitted to the channel through the cytoskeleton and/or the extracellular matrix (Sukharev & Corey, 2004). An example is the TRP channel, NOMPC, from *Drosophila*, that is involved in touch sensation and hearing, and operates through ankyrin repeats connected to tubulin (Zhang et al., 2015). To achieve heterologous expression of this channel would require that the channel in the new environment be able to associate with tubulin to transmit the forces. NOMPC, which is highly expressed in Class III neurons but not Class IV, when introduced into class IV neurons confers mechanical properties to this population. NOMPC is also responsive when expressed in the *Drosophila* S2 cell line (Yan et al., 2013). Another example of tethered channels is the DEG/ENaC family that is an amiloride-sensitive sodium channel. For the nematode *Caenorhabditis elegans,* this channel is involved in touch (O'Hagan, Chalfie, & Goodman, 2005) and is thought to be gated by forces applied through extracellular and intracellular proteins (Gu, Caldwell, & Chalfie, 1996; Sukharev & Corey, 2004).

2. DISCOVERY OF PIEZO CHANNELS

2.1 Identification of *Piezo* genes

2.1.1 Neuro2A cell line produces mechanical current

For many years, our understanding of the details of mechanical transduction for MSCs in depolarizing eukaryotic cells was hampered by the absence of a cloned target. The effort of the Patapoutian lab, in particular Coste, eventually led to the discovery of a bona fide mechanical channel they called piezo, a name based on the Greek word πίεση which means pressure (Coste et al., 2010). The work was initiated by first identifying a cell line that produced large MA currents when the cells were indented with a pipette. The Neuro2A (N2A) line from mouse was ideal because mechanical currents were observed using different stimuli [although there appears to be variability in this cell line (Lee et al., 2014)]. The first technique recorded whole-cell currents while using a fire-polished pipette to press on the cell. The second assay was the patch clamp where suction was applied to a patch using a high speed pressure servo to precisely control hydrostatic pressure (Besch, Suchyna, & Sachs, 2002).

Fig. 1 shows the current traces of whole-cell recordings and patch recordings from N2A cells. Whole-cell recordings demonstrated that increasing depth of penetration increased the current to many hundreds of pA (Fig. 1A). In keeping with the properties expected of cation-selective channels, the reversal potential was near 0 mv (Fig. 1B). The single channel recordings (Fig. 1C) clearly showed voltage-dependent inactivation (compare rates +80 and −80 mV) and a reversal potential around 0 mV (Fig. 1D). In patch clamp mode, a series of increasing suction pulses gave rise to increasing currents (Fig. 1E), which allowed for a Boltzmann fit of normalized current versus pressure:

$$P_{open} = 1\Big/\Big(1 + \exp\Big(\alpha\Big(P - P_{1/2}\Big)\Big)\Big)$$

where P_{open} is the open probability, alpha, α, is the slope sensitivity, P is pressure, and $P_{1/2}$ (pressure for half activation) which was around −30 mmHg (Fig. 1F).

2.1.2 Identification of Piezo1 (FAM38a)

Having identified a cell line that produced robust mechanical currents, Coste et al. (2010) used microarray analysis to look for the gene and narrowed their search to proteins that spanned the membrane at least twice. They generated

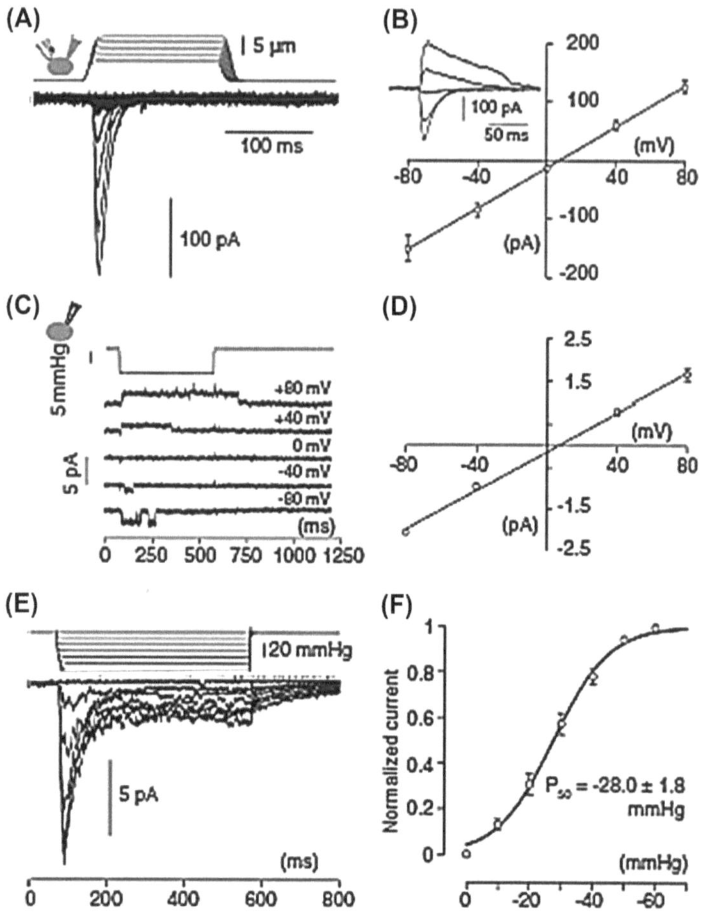

Figure 1 Mechanically activated currents from Neuro2A cells. (A) Whole cell–inward currents were generated by pressing on cells with a polished glass pipette. Above the current traces are the pulses used; 1 μm movements at a holding potential of −80 mV. (B) The current–voltage relationships of mechanically activated currents shows a reversible potential around 0 mV (see Inset). (C) Single-channel currents with suction pulses at holding potentials ranging from −80 to +80 mV. (D) Average current–voltage relationships of stretch-activated single channels. Single channel conductance was calculated from the slope of the linear regression line of each cell giving $\gamma = 22.9 \pm 1.4$ pS (mean ± SEM). Note the voltage-dependent inactivation of the channels (compare—80 to +80 mV). (E) Representative currents induced by suction pulses of 0 to −60 mmHg, Δ 10 mmHg. (F) Normalized current–pressure relationship of stretch-activated currents at −80 mV fitted with a Boltzmann equation. *Reproduced with permission from Coste, B., Mathur, J., Schmidt, M., Earley, T. J., Ranade, S., Petrus, M. J., … Patapoutian, A. (2010). Piezo1 and Piezo2 are essential components of distinct mechanically activated cation channels.* Science, 330*(6000), 55–60. http://dx.doi.org/10.1126/science.1193270.*

multiple silencing RNAs (siRNAs) for each gene, introduced them into cells, and tested the cells for whole-cell currents. Over 70 different genes were targeted before there was a noticeable reduction in current, not evident in controls using randomized siRNA sequences. The sequence identified with the channel was the gene *Fam38a* now called *Piezo1*. The protein product of *Fam38a* was first identified in activated astrocytes in response to senile plaques where it was upregulated (Satoh et al., 2006).

Once they had demonstrated that a specific gene in N2A was responsible for the MA currents, they cloned and sequenced the mouse gene. The protein is over 2500 amino acids long and there is little homology to other known channels (Coste et al., 2010). The initial predication was that there were 24–32 transmembrane domains (TMDs). This has been adjusted to ~18 TMDs based on the recent CryoEM structure (Ge et al., 2015) and biochemical mapping of the intracellular and extracellular domains (Coste et al., 2015).

2.1.3 Piezo1 in the plasma membrane

The channel was shown associated with the plasma membrane using piezo antibodies and TRPA1 as a control surface marker (Coste et al., 2010). We confirmed the presence of the channel on the plasma membrane using a human PIEZO1 protein that was fluorescently labeled at an internal site that was functional (Cox et al., 2016). We transfected HEK293 cells with both PIEZO1 tagged with mCherry and TREK-1 labeled with GFP (Cox et al., 2016).

Using super illumination microscopy of those cells, we showed that both PIEZO1 and TREK-1 are at the surface and the image clearly shows that the channels do not overlap (Fig. 2). Another distinctive anatomical feature was the presence of large clusters, domains, of PIEZO1 channels. The existence of domains of piezo channels was predicted from electrophysiological data (Bae, Gnanasambandam, Nicolai, Sachs, & Gottlieb, 2013). The TREK-1 channels were organized also in clusters and in lines as if they were attached to the underlying cytoskeleton, consistent with earlier suggestions (Lauritzen et al., 2005).

2.1.4 Heterologous expression

The properties of piezo channels were verified by expression in N2A cells and then HEK293 cells by Coste et al. (2010). For HEK cells, Fig. 3A and B shows the robust whole-cell currents whose reversal potential is near 0 mV. Cell-attached patch (Fig. 3C and D) currents plotted as a function of increasing suction gave rise to a Boltzmann distribution with $P_{1/2}$

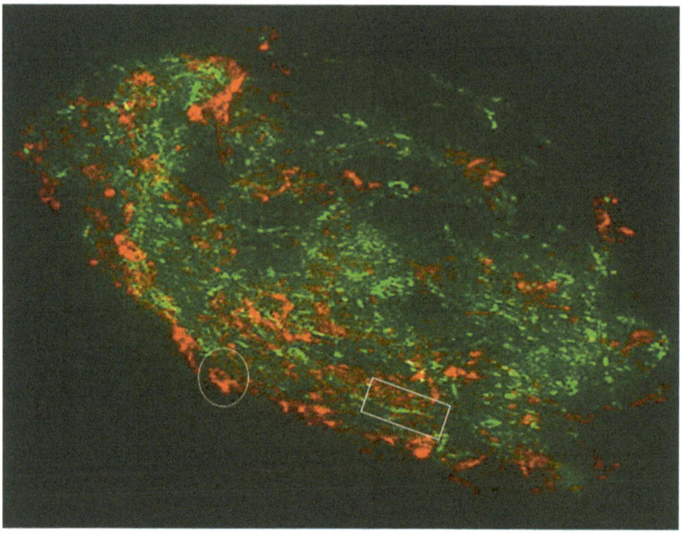

Figure 2 PIEZO1 channels form clusters. Image of fluorescent proteins PIEZO1 (red) and TREK-1 channels (green) on the plasma membrane. PIEZO1 is tagged with internal mCherry and TREK-1 is linked to GFP through its C-terminus. Twenty-four Z-stacks were imaged by SIM and reconstructed into a 3-D object. TREK-1 forms beadlike structures that appear to track an underlying cytoskeletal structure (rectangle). In comparison, PIEZO1 forms larger cluster (circle) and there is no significant overlap of the two channels.

around −30 mmHg. Heterologous expression of Piezo1 imparted a mechanical response in HEK cells that are normally unresponsive to mechanical stimulation. The kinetic features of the channels observed in the N2A cells were similar to those in transfected HEK293 cells. Coste et al. showed that Piezo1 carries monovalent cations as expected, and significantly, is somewhat permeable to Ca^{2+}.

A thorough analysis of channel selectivity and ion permeation has been published recently (Gnanasambandam, Bae, Gottlieb, & Sachs, 2015; Li et al., 2015; Zhao et al., 2016) and is addressed in detail in chapter "The Kinetics and the Permeation Properties of Piezo Channels" by Gnanasambandam et al. (2016)

2.1.5 Inhibitors of Piezo channels

2.1.5.1 Small molecule

Piezo channel currents are inhibited by common nonspecific inhibitor ruthenium red (RR) and Gd^{3+} (Coste et al., 2010); RR blocks inward but not outward currents (Coste et al., 2012). Using a channel chimera approach RR was shown to interact with the C-terminal and attempts

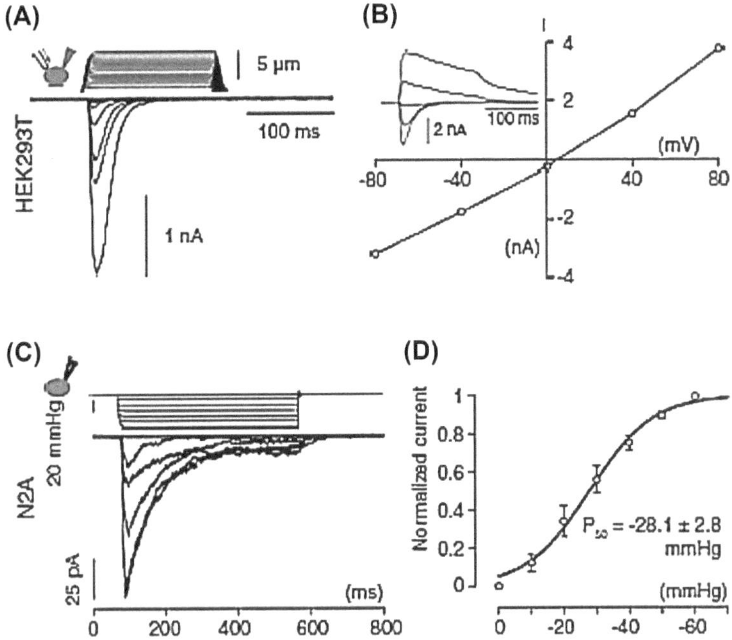

Figure 3 Robust mechanically activated currents from cells overexpressing Piezo1. (A) Mechanically activated (MA) currents of Piezo1-expressing HEK293 T cells recorded in the whole-cell configuration. (A) Cells were subjected to a series of mechanical steps in 1 μm increments using a glass probe at a holding potential of −80 mV. (B) Representative current–voltage relationships of MA currents expressed in cells transfected with Piezo1. Inset, MA currents evoked at holding potentials ranging from −80 to +40 mV. (C) HEK293 T cells in cell-attached configuration. Representative averaged currents induced by negative pipette pressure (0 to −60 mmHg, Δ 10 mmHg). (D) Normalized peak currents elicited at −80 mV in Piezo1 HEK293 T cells are plotted as a function of suction and fitted with a Boltzmann equation. P_{50} is the average value for all individual cells. *Reproduced with permission and modified Coste, B., Mathur, J., Schmidt, M., Earley, T. J., Ranade, S., Petrus, M. J., … Patapoutian, A. (2010). Piezo1 and Piezo2 are essential components of distinct mechanically activated cation channels.* Science, 330*(6000), 55–60. http://dx.doi.org/10.1126/science.1193270.*

were made to locate the specific regions involved in RR binding that blocks conduction. A more extensive discussion on RR inhibition can found in chapter "In Touch With the Mechanosensitive Piezo Channels: Structure, Ion Permeation and Mechanotransduction" by Geng et al. (2016).

2.1.5.2 Peptide

Many years ago we isolated a peptide, called GsMTx4, from the venom of tarantula that was shown to inhibit endogenous stretch channels

(Suchyna et al., 2000). The peptide is 34 amino acids long and can be chemically synthesized and folded (Ostrow et al., 2003). It has three disulfides and the NMR solution structure is globular (Oswald, Suchyna, McFeeters, Gottlieb, & Sachs, 2002). We demonstrated that both enantiomeric forms of the peptide inhibit endogenous mechanically activated currents in the patch indicating that the inhibitor is positioned at the interface between the protein and lipids (Suchyna et al., 2004). We then showed that the peptide in both enantiomeric forms was able to inhibit mouse Piezo1 mechanical currents from patches and whole cell (Bae, Sachs, & Gottlieb, 2011). Similarly we demonstrated that both human PIEZO1 and their mutations that slowed inactivation were inhibited by the peptide (Bae, Gnanasambandam, et al., 2013).

2.1.5.3 Protein

Recent work indicates that inhibition of stretch-activated currents can be modulated by other proteins. An example of this is MA currents that were detected on the basolateral side of renal proximal convoluted tubules that are associated with Piezo1 since siRNA-targeting *Piezo1* diminishes these currents (Peyronnet et al., 2013). The expression of Polycystine-2 (PC2) or its pathogenic mutant PC2-740X reduces currents from Piezo1 channels. The interaction between Piezo1 and PC2 is supported by co-immunoprecipitation of the two proteins.

2.1.6 Distribution of Piezo1 and Piezo2

A second piezo channel, called *Piezo2* (*Fam38b*), was isolated and cloned by Coste et al. (2010). Whole-cell recordings of Piezo2 channels expressed in HEK293 showed fast inactivation ($\tau = 7.3$ ms). Like Piezo1, Piezo2 is a cation-selective channel with a reversal potential near 0 mV. The distribution of the channels in various tissues shows that Piezo1 is highly expressed in the lung, bladder, and skin and lower expression in the colon, stomach, and kidney, whereas Piezo2 was found in the lung and dorsal root ganglion (DRG) neurons (Coste et al., 2010). The presence of Piezo2 channels in DRG neurons was analyzed with siRNA-targeting *Piezo2* resulting in a depletion of neurons with rapidly inactivating currents.

These two channels are not mutually exclusive for cell function. Recent work by Liedtke and collaborators (see chapter: Role of Piezo Channels in Joint Health and Injury by Lee et al., 2016) suggest that these two channels can work synergistically in cartilage cells that are related to high-strain mechanical stress in joints (Lee et al., 2014).

2.1.7 Bilayer reconstitution of Piezo1

To determine whether the Piezo1 protein was sufficient to reconstitute activity of the channel, the purified protein was inserted into *droplet interface* lipid bilayers (DIB) assembled from two monolayers to mimic eukaryotic cell bilayers (Coste et al., 2012). The second method used proteoliposomes (Coste et al., 2012). Recordings from the DIB preparation showed single channel openings characteristic of Piezo1 and they were inhibited by ruthenium red when applied to the correct side. Similar findings were observed for the azolectin liposomes. One of the most important findings from reconstitution is that the Piezo1 channel will function without any ancillary proteins. The reconstituted channel I/V is ohmic over the range of −100 to 200 mV with a conductance of ~100 pS. This is higher than that measured in the cellular environment (Coste et al., 2010; Gottlieb, Bae, & Sachs, 2012), possibly due to the different ionic solutions used, and the cellular I/V plot showed mild outward rectification (Coste et al., 2012). The reconstituted bilayer channels are constitutively active if the bilayers have asymmetric lipid composition. This is consistent with the highly asymmetric structure of the channel (Ge et al., 2015). Significantly, the reconstituted channels do not inactivate showing their kinetics are environmentally sensitive. Later measurements using symmetric bilayers did not show spontaneous activity (Syeda et al., 2015). Recent work has shown that the Piezo1 channels in cells are opened by bilayer tension (Cox et al., 2016; Lewis & Grandl, 2015; see chapter Origin of the Force: the Force-From-Lipids Principle Applied to Piezo Channels by Cox et al., 2016).

Direct evidence that the Piezo1 channel is inherently mechanosensitive has now been reported (Syeda et al., 2016). Purified Piezo1 channels were reconstituted in lipid bilayers in the absence of other proteins. Two methods were used to activate the channels mechanically: osmotic gradient with mannitol and direct expansion of a droplet monolayer by solvent injection. In both cases channel activity was observed confirming that Piezo1 is activated mechanically.

2.2 Piezo1 functional stoichiometry

The Piezo1 monomer is similar in size to that of a sodium channel (Noda et al., 1984). Sodium channels have internal repeat units that form the channel pore. An examination of the piezo sequence shows no obvious sequence element that would indicate a channel could be formed from a single subunit. Initial work on channel pore formation revealed that the channel was comprised of a tetramer (Coste et al., 2012) based on cross-linking with paraformaldehyde. These data were confirmed using a fluorescently labeled Piezo1 molecule and monitoring the discrete loss of fluorescence with bleaching of the probe. However, the cryoEM structure of mouse Piezo1 has been recently solved to 4.8 Å resolution and shows the structure as a trimer (Ge et al., 2015; see chapter: The Structural Basis for Sensing by the Piezo1 Protein by Li et al., 2016).

Based on the cryoEM structure and biochemical studies, the C-terminal segment was shown to house the pore (Coste et al., 2015; Ge et al., 2015). This suggested that the N-terminal portion of the channel lowers the energetic transition allowing the C-terminus pore to open with mechanical stimuli. Experiments that replaced the C-terminal domain of the Piezo1 channel with an ASIC channel created chimera channels in which mechanosensitivity was conferred on the ASIC channels (Zhao et al., 2016; see chapter: In Touch With the Mechanosensitive Piezo Channels: Structure, Ion Permeation and Mechanotransduction by Geng et al., 2016). These results show that the channel is modular. The ability of the chimera to assemble into a functional ASIC channel is consistent with the formation of a trimer. The domain nature of the channel was also seen when we coexpressed two segments of human PIEZO1 as a split protein; they assembled into an active channel (Bae, Suchyna, Ziegler, Sachs, & Gottlieb, 2016). Each protein domain, when expressed alone, does not produce current.

Direct evidence for trimer stoichiometry comes from electrophysiological data. Mutations near the pore (position E2117) of human PIEZO1 can alter the channel's conductance (WT ~50 pS at −100 mV). Substitution of an alanine at this site lowers conductance (~28 pS at −100 mV), and introduction of an aspartic increases conductance (~75 pS at −100 mV). Similar substitutions were first made in homologous sites of mouse Piezo1 (Coste et al., 2015) and our changes produced similarly altered conductance. These sites are conserved in piezo channels (Prole & Taylor, 2013). By coexpressing high and low conducting subunits, we observed intermediate unitary conductances which we explained by the association of different subunits (Gnanasambandam, Bae, Gottlieb, & Sachs, 2016). Single channel recordings showed two new conductances (39 and 58 pS at −100 mV) and this is consistent with a PIEZO1 trimer. These results do not preclude the possibility of another stoichiometry.

2.3 Biophysical properties of mouse Piezo1

2.3.1 Mouse Piezo1 inactivation

During stimulation (hyperpolarization), mouse Piezo1 can rapidly lose current with time. There are two possible mechanisms that accounts for this. It can arise by relaxing the forces associated with viscoelastic coupling of the system, a property that is extrinsic to the channel and referred to as adaptation. Alternatively, it can occur as a channel-dependent property and in this case it is termed inactivation. To distinguish between the two models, we used a pulse train applied to mouse Pieoz1–transfected HEK293 cells in cell-attached mode (Gottlieb et al., 2012). The first pulse (−40 mmHg) was designed to activate all the channels. If the channels were desensitized by adaptation, a second pulse of −80 mmHg would reactivate Piezo1 channels. We did not observe such channel reactivation. Rather, the second pulse gave rise to small currents, indicating the channels remained inactivated (Gottlieb et al., 2012).

We were able to determine the rate constants for a three-state model (closed to open to inactivated) using the MAC routine in the QuB (Milescu, Akk, & Sachs, 2005). The model requires only a single pressure-dependent step from closed to open to explain the data. A more thorough description of the kinetics of piezo channels is found in chapter "The Kinetics and the Permeation Properties of Piezo Channels" by Gnanasambandam et al. (2016).

2.3.2 Piezo channel clusters

For mouse Piezo1 channels, we observed a sudden shift in a large number of channels to a noninactivating form. This transition from rapid inactivation to a sustained current occurs randomly after repeated stimuli. The remarkable feature of this transition is that it can apply to a population of tens of channels simultaneously, and it is irreversible (Gottlieb et al., 2012) (Fig. 4).

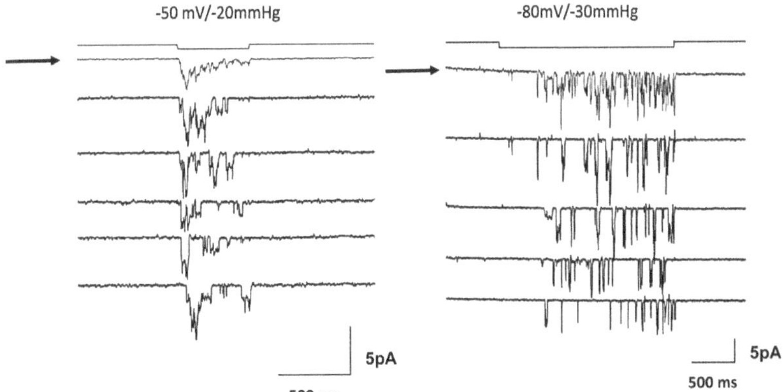

Figure 4 Transition of the inactivating Piezo1 to a noninactivating form. (Panel A) The initial response in cell-attached mode showing a transient and a steady-state response. The arrow indicates the average response of the pulses. (Panel B) After repeated stimulation at high pressure and voltage, Piezo1 loses inactivation irreversibly.

We found almost no cases where we observed the two populations (activating and noninactivating) in a given patch. This kind of binary channel behavior has been observed before by Hamill's lab when studying endogenous channels of *Xenopus oocytes* (Hamill & McBride, 1997). One explanation for this behavior is that the channels are clustered (Fig. 2), and they are shielded from the mean tension of the cell's membrane by a line tension around the cluster. Repeated stimuli appears capable of fracturing the clusters so that the channels are now exposed to the mean external tension. The exact composition of these clusters and the forces that hold them together are unknown at this time.

2.3.3 Effects of modulating the cytoskeleton

Cell-attached patches of mouse Piezo1-transfected cells treated with cytochalasin D did not change the inactivation rate, but it decreased the steady state—resting current (Gottlieb et al., 2012). The effect on whole-cell currents was different; we observed a dramatic reduction in the current as though actin fibers were in series with the domain boundary. In this model, actin fibers "tug" on the membrane increasing the coupled bilayer tension (Akinlaja & Sachs, 1998; Cox et al., 2016) activating the channel.

To further explore the properties of Piezo1, we used hypotonic solutions to swell the cell and ask what the channel's response was in a prestressed environment. Treatment of transfected HEK293 cells with a 50% hypotonic solution increased currents by about threefold for whole-cell currents via an increased sensitivity to the depth of penetration. Similar properties were observed for Piezo2 channels (Jia, Ikeda, Ling, Viatchenko-Karpinski, & Gu, 2016).

2.3.4 Protonation of PIEZO1

PIEZO1 currents were measured at different pHs and we observed that at lower pH there was a reduction in current (Bae, Sachs, & Gottlieb, 2015). The effect was use dependent, meaning that currents fell off with sequential stimuli, but this was reversible when the pH returned to normal. Both cell-attached and whole-cell recordings showed these properties.

We asked whether this pH effect is related to titration of the closed state or the inactivated state by using a mutant channel that did not inactivate (Bae, Gottlieb, & Sachs, 2013). These channels were pH insensitive. We did, however, observe a slow decay when channels were subjected to repeated mechanical stimulation that we attribute domain stability. Our results support a model where protonation stabilizes the inactivated state.

2.3.5 Diseases of piezo mechanical channels

Two human *PIEZO1* channel mutations that were associated with hereditary xerocytosis (HX), a hemolytic anemia of red blood cells (RBCs), provided us with a basis for understanding the pathology (Zarychanski et al., 2012). The two mutations were R2456H and M2225R, both of them in the C-terminal, the former on the intracellular side and the latter in the extracellular side. Other sites causing similar pathology were identified (Albuisson et al., 2013; Andolfo et al., 2013), as well as mutations in *PIEZO2* that lead to distal arthrogryposis type 5, that causes limited eye movements, restrictive lung disease, and reduction of cruciate knee ligament size (Coste et al., 2013). A common element of many of these mutations is a slowing of inactivation, often referred to as a gain-of-function mutation since the effect is to keep the channel open for longer times and increasing the cation influx. One Piezo2 mutation (I820F) speeds the return from the inactivated state.

When we made human PIEZO1 with the mutations involved in HX, both mutants caused a slowing of inactivation in both cell-attached patch and whole-cell modes (Albuisson et al., 2013; Bae, Gnanasambandam, et al., 2013). The patch recordings revealed that, in addition to changes of inactivation rate, the mutations introduced a random latency for activation that could be as long as hundreds of milliseconds (Bae, Gnanasambandam, et al., 2013). The combined effect of activation and inactivation to the channel kinetics needs to be taken into account when modeling the pathological response of the channel and its role in the cell. Since RBCs in the blood stream are exposed both to shear stress when flowing through large vessels and compressive stress squeezing through capillaries, these mutations will alter channel properties so that they are no longer in sync with the cell's ability to respond to mechanical cues.

2.4 Chemical activation of Piezo1

Piezo1 channels are traditionally activated by mechanical force, and Patapoutian's lab developed a high throughput assay to monitor both Piezo1 and Piezo2 activity using calcium indicators (Syeda et al., 2015). Syeda et al. (2015) then set out to look for chemical activators of Piezo channels using this assay. They screened a library of low molecular weight compounds for their ability to activate Piezo1 or Piezo2 and identified a compound they called Yoda1 that activated only Piezo1. The calcium response did not result from internal calcium release but depended on Ca^{2+} influx. Mutant channels associated with xerocytosis were similarly activated by Yoda1, and they observed higher whole-cell currents than typical of WT channels, consistent with slowing of the inactivation rate (Albuisson et al., 2013; Bae, Gnanasambandam, et al., 2013; Bae, Gottlieb, et al., 2013).

The effect of Yoda1 was examined in more detail at the single channel level. In the cell-attached configuration and Yoda1 in the pipette, there were a number of changes to the biophysical responses. First, there was a slowing of inactivation. Second there was a leftward shift of the activation curve to lower pressures of about 15 mmHg. Lastly, there were spontaneous channel openings without applied pressure where the channels seemed to be activated by the significant resting tension in patches (Slavchov, Nomura, Martinac, Sokabe, & Sachs, 2014; Suchyna, Markin, & Sachs, 2009). Whole-cell currents showed a pronounced slowing of inactivation with Yoda1.

These channel properties are similar to what we observed when we introduced two mutations into the C-terminus of human PIEZO1 (Bae, Gnanasambandam, et al., 2013; Bae, Gottlieb, et al., 2013). This includes a loss of inactivation at the single channel level, a leftward shift in the activation curve, and spontaneous channel currents, presumably activated by tension in the patch (Slavchov et al., 2014). Similarly, we observed a change in the steady state of inactivation at the whole-cell level. There was no change in the conductance. The common biophysical kinetic properties of Yoda1 and the C-terminal mutations suggest a common feature. Because the position of the HX mutations are near the interface of the Piezo1 protein (Ge et al., 2015) that forms the pore, perhaps Yoda1 works by binding to the C-terminus. Interestingly, Yoda1 has no effect on Piezo2.

To better define the activation mechanism of Yoda1, its effect was examined with reconstituted channels in a lipid bilayer (Syeda et al., 2015). Purified Piezo1 was inserted into a symmetric droplet interface lipid bilayer composed of diphytanoyl-sn-glycero-3-phosphocoline. Yoda1 (1 μM) induced single channel currents. Higher concentrations produced multiple channel activation. The effect of Yoda1 is thought to be mediated by a direct interaction with Piezo1 because Yoda1-like molecules modified by removing substituent groups, do not activate the channel. Yoda1 can serve as an agonist for Piezo1. Syeda et al. (2015) suggest that there may be a Yoda1-like soluble factor that modulates channel activity. This is not an unreasonable assumption since PIP2, another well-characterized diffusible factor, inhibits Piezo1 when it is depleted (Borbiro, Badheka, &

Rohacs, 2015; see chapter: Regulation of Piezo Channels by Cellular Signaling Pathways by Borbiro and Rohacs, 2016).

Having identified an activator of Piezo1 channels, Cahalan et al. (2015) used this compound to analyze the mechanism of RBC dehydration associated with xerocytosis. When Piezo1 was removed from RBCs by a conditional knockout, overhydration and swelling of RBCs resulted. Using Yoda1 on wild-type RBCs, the calcium influx through Piezo1 activated a calcium-activated potassium channel. They proposed that potassium and water are prompted to leave the cell resulting in dehydration.

2.5 Cytoskeleton modulates Piezo1 function

2.5.1 Tuning Piezo1 activity

The STOML3 protein is an example of the influence of structural domains on piezo channel function. Sensory neurons were grown on flexible pillar arrays allowing piezo activation by deflecting pillars about 13 nm (Poole, Herget, Lapatsina, Ngo, & Lewin, 2014). Neurons that were harvested from mice with ablated STOML3 proteins were tested but they required an order of magnitude increase in pillar deflections to activate the channels.

To understand the relationship between STOML3 and MA currents, Poole et al. (2014) focused on how STOML3 modulates Piezo1 channel currents when N2A cells are grown on pillars. By moving the pillars with a probe, induced currents were observed. They showed that robust MA currents were observed when the pillars were moved ~200 nm. The source of these currents was Piezo1 because miRNA for Piezo1 inhibited the MA currents. Overexpression STOML3 in these cells generated larger currents with even smaller displacements of ~10 nm. Reduction of endogenous STOML3 by miRNA significantly reduced currents with even larger displacements up to 1000 nm. The effect of STOML3 was observed for both Piezo1 and Piezo2 in HEK293. Other STOML3 proteins did not influence Piezo1 activity. The conclusion is that STOML3 is able to tune piezo sensitivity like other cytoskeletal proteins, perhaps by altering bilayer tension or local stresses near the channel.

A recent study on neurons suggests that STOML3 is associated with cholesterol and is critical for the formation of lipid rafts, and controls piezo channels (Qi et al., 2015). The presence of both STOML3 and cholesterol appears to alter membrane mechanics. Ablation of STOML3 ($STOML3^{-/-}$) or removal of cholesterol by beta cyclodextran resulted in softer cells attenuating the mechanical response of piezo channels. These data in conjunction with imaging of STOML3 and one of its mutants suggest that STOML3 is associated with lipid rafts and these specialized structures shield the molecules inside from the mean tension of the cell. Anishkin and Kung have previously suggested lipid rafts can serve as stiff structures that may be critical for redirecting and focusing forces to specialized proteins and enzymes within the platform domain (Anishkin & Kung, 2013).

Unlike the effect of cholesterol depletion described earlier, mechanical currents were increased in myotubes when beta cyclodextran was used to deplete cholesterol, or when Caveolin-3 was depleted (Huang, Bae, Sachs, & Suchyna, 2013). These currents are thought to be associated with Piezo1 because their overexpression in myotubes produced the same channel kinetics as the endogenous channel. These experiments clearly demonstrate that ancillary proteins regulate mechanically induced currents in the patch and in whole-cell recordings. The mechanism by which the presence of ancillary proteins can potentiate channel activity remains to be determined. Three potential mechanisms have been suggested (Huang et al., 2013). Caveolin-3 may interact directly with the channel altering the channel's activity; another possibility is that local stress is modulated by association with cytoskeletal elements; a third suggestion is that microdomains are disrupted.

Why STOML3 and Caveolin-3 give rise to opposite responses is unclear. The cytoskeleton and the extracellular matrix of cells have many parallel pathways to distribute forces experienced by cells and how individual cell types choose to respond to mechanical forces may be a matter of the types of proteins expressed. Overall these results indicated that while an examination of channel's intrinsic properties is invaluable for understanding channel activity, the channel's environment and interactions with other molecules are essential for describing the overall channel response in the cell.

2.5.2 Filamin-A modulates Piezo1 function

The work by Retailleau et al. (2015) has linked the role of Piezo1 and Filamin-A (FlnA) with small diameter arterial remodeling during hypertension. This study centers on the caudal artery where Piezo1 expression levels are high in relation to most other arteries. Using a Piezo1 conditional knockout mouse, MA currents were removed from smooth muscle. However, there was no observable change in the myogenic tone of caudal artery as well as the cerebral artery that also expresses Piezo1.

To understand the function of Piezo1 in these arteries, Retailleau and colleagues looked at remodeling of the arteries during hypertension. By inducing hypertension in Piezo1-ablated arteries, remodeling was *not* evident. They examined the contribution of FlnA, a cortical cytoskeletal element, to myogenic tone by creating a KO mouse (Retailleau, Arhatte, Demolombe, Jodar, et al., 2016; Retailleau, Arhatte, Demolombe, Peyronnet, et al., 2016) and showed that this mutation elevated cytosolic calcium. No increase in cytosolic calcium is observed when Piezo1 and FlnA are ablated, meaning the influx of cytosolic calcium is linked to Piezo1 activity.

On the other hand, MA currents from Piezo1 are potentiated in the *FlnA* conditional knockout mouse along with an increase in cytosolic calcium. Perhaps FlnA is in parallel with the channel domain reducing the stress in the domain. Under these circumstances, remodeling of caudal artery occurs even in the absence of induced hypertension. The Piezo1-mediated rise in calcium is correlated with a greater activity of transglutaminases, a cross-linking enzyme in small artery remodeling. Interestingly, the removal of a single Piezo1 allele was sufficient to prevent remodeling. This detailed work has established a connection between Piezo1 MA currents and a clinically relevant arterial remodeling effect.

2.6 Cell motility and Piezo1

2.6.1 Integrin activation and Piezo1 (Fam38a)

Even before the Piezo1 was shown to be a mechanical channel, it (Fam38a) was linked to integrin activation indicating its involvement in cell adhesion of epithelial cells to the extracellular matrix (McHugh et al., 2010). The mechanism for activation was not thought to be at the plasma membrane but rather in the endoplasmic reticulum where it is proposed to recruit R-Ras leading to calcium release from internal stores and calpain activation. Of interest is that a C-terminal fragment of Fam38a, which does not form a functional channel (Bae et al., 2016), is sufficient to induce R-Ras signaling as well. The need for Fam38a was confirmed by siRNA to reduce β1 integrin and cell adhesion. These knockdown experiments also showed that cells were able to increase migration which has obvious implications for cancer and metastatic cells (McHugh, Murdoch, Haslett, & Sethi, 2012).

2.6.2 Breast cancer

An examination of the role of Piezo1 channels in breast cancer underscores the complex relationship between the mechanical response of piezo channels and its potential role in cell migration. Two breast cancer cell lines were compared, MCF-7 and MCF-10a, and they showed that the malignancy of the cells correlates with the presence of Piezo1, the more malignant cell (MCF-7) expressed a higher level of Piezo1 that was easily detected by patch recordings (Li et al., 2015). The results appear to be at odds with the depletion of Piezo1 involved in increased migration and metastasis for Piezo1 integrin activation discussed earlier.

2.6.3 Prostate cancer

Understanding the cell's mechanical response in cancer is further complicated by transitions of channel properties as a function of use. The mechanically induced currents of two prostate cell lines, PC-3 and LNCaP, have been compared (Maroto, Kurosky, & Hamill, 2012). The former is associated with an aggressive form of prostate cancer because it acquires the ability to migrate, while the latter is benign. Although the MSCs were not established as Piezo1 yet, the mechanical responses were similar and PC-3 MA currents are blocked by GsMTx4. The main distinguishing feature between them was dynamic, the PC-3 cells do not inactivate while LNCaP cells inactivate, but with applied stimuli the latter can become a noninactivating channel (Maroto et al., 2012). This type of transition from inactivating to noninactivating has been observed for mouse Piezo1 (Gottlieb et al., 2012).

2.7 Piezo1's role in cell contact

There is now considerable evidence to suggest that extracellular surfaces have an impact on cell function whether it involves cell—cell or ECM interactions (Sun, Chen, & Fu, 2012). The cell produces traction forces so there is undoubtedly a mechanosensory component. Following, are three examples where Piezo1 is thought to be involved in processing these external signals allowing cells to respond to environmental cues.

2.7.1 Lineage choice in neural stem cells

Neural stem cells (hNSPC) differentiate into neurons, astrocytes, and oligodendrocytes and the surrounding surfaces play a critical role in the lineage choice of the stem cells. To define components involved in cell differentiation, Pathak and colleagues looked at mechanically currents in neural stem cells and demonstrated currents in both cell-attached and whole-cell mode (Pathak et al., 2014). These currents were inhibited by the peptide GsMTx4 suggesting a piezo-type channel, and they were able to show that there were high expression levels of *Piezo1*. Using siRNAs directed against *Piezo1*, they effectively inhibited MA currents confirming the involvement of Piezo1.

They then focused on the interface between the substrate and the cell and monitored spontaneous calcium currents by TIRF microscopy, visualizing cells loaded with calcium indicators. These calcium fluxes are expected to be related to the traction forces experienced by the cell, and by introducing siRNA for *Piezo1*, they were able to reduce the calcium influx, confirming that the spontaneous channel openings were associated with the presence of Piezo1.

Traction forces are known to be involved in a number of cellular processes, including differentiation. They are generated by an ATP-dependent nonmuscle myosin II with actin fibers and transmitted to focal adhesions. To demonstrate the involvement of myosin II, they used the myosin II inhibitor blebbistatin and it blocked the spontaneous currents, supporting the idea of traction forces as the source of spontaneous events. Because tractions forces are dependent on substrate stiffness, the question was whether the frequency of spontaneous events changed with different surfaces. Cells were grown on silicone elastomers of varying stiffness and they found that the frequency of spontaneous calcium release events scaled with substrate stiffness. Minimal numbers of transients were observed on the soft substrates and more with stiffer substrates. This is consistent with the idea that Piezo1 channel activity is related to traction forces. Interestingly, a recent work examined confinement sensing of cells and showed that both Piezo1 and myosin activation enables the cell to sense and adapt to different physical environments (Hung et al., 2016).

For neural cell lines, differentiation is dependent on substrate stiffness (Leipzig & Shoichet, 2009; Saha et al., 2008). In these experiments, two transcriptional factors were monitored: Yap and Taz. In epithelial cells, higher substrate stiffness sends these transcriptional factors to the nucleus, while low stiffness keeps them in the cytoplasm (Dupont et al., 2011). For hNSPC cells a similar pattern was observed. An interesting observation is that even on stiffer substrates, knock down by siRNA of *Piezo1* resulted in exclusion of Yap from the nucleus, suggesting that Piezo1 is involved in sensing substrate stiffness.

Pathak and collaborators analyzed the differentiation of hNSPC by growing cells on different substrates and analyzing the neuron/astrocyte composition (Pathak et al., 2014). There was a greater propensity of neurons with stiffer substrates (in contrast, rodent neural stem cells had an opposite response). Two approaches were used to confirm that Piezo1 is central to differentiation. First there was a pharmacological approach where GsMTx4 was incubated with the hNSPC cell line, and using cell markers for neurons versus astrocytes typing, they showed an increase in the astrocyte population. A second approach was the use siRNA to lower *Piezo1* expression and that resulted in an increase in the astrocyte population. These data indicated that the Piezo1 ion channel is critical in the lineage choice of neural stem cells.

2.7.2 Neuron—astrocyte interaction

Interactions between astrocytes and neurons are essential for a variety of functions such as maintenance of neurotransmission. There are undoubtedly mechanical forces between the two cell types that can convey information in the absence of chemical factors. To study this process, Blumenthal et al. used a cell line called PC-12 that is neuron-like and can be used to follow cell differentiation (Blumenthal, Hermanson, Heimrich, & Shastri, 2014). Cells were grown on different surfaces that were modified by colloidal particles altering the nanoroughness of the surface from an Rq (small values indicate a smoother surface while large values indicate a rougher surface) of 3.5 nm (glass surface) to 80 nm. In the presence of nerve growth factor, they observed that with an Rq of 32 nm, the cells became highly polarized with fewer neurite extensions compared to growth on glass (Blumenthal et al., 2014).

They next isolated hippocampal cells that are a mixture of neurons and astrocytes. When these cells were plated on various surfaces, they observed highly polarized neurons at Rq of 32 nm similar to what was observed for PC12 cells. A surprising observation is that the neurons dissociated from the astrocytes, since the neuron—astrocyte interaction is known to be essential for neuron survival (Cui, Allen, Skynner, Gusterson, & Clark, 2001). But, under these conditions, neurons were able to survive many weeks.

Because these results suggested mechanical sensing, Blumenthal et al. (2014) asked whether piezo channels were involved. Using antibodies to Piezo1 they were able to determine that the distribution of Piezo1 channels was different for different surfaces. The changes were not associated with other soluble factors or expression levels of the channel. Next GsMTx4 was used to inhibit Piezo1 channels and showed that even on smooth glass surfaces, the neurons and the astrocytes can dissociate, suggesting that piezo channels are needed for surface sensing. The surface roughness of an astrocyte was determined to be around 26—28 nm which is close to that found for the surface that allows the dissociation of the neurons from astrocytes. The role of Piezo1 in cell–cell contact was examined by introducing siRNA for Piezo1 into a mixed population of neuron and astrocytes. Neurons were unaffected but the siRNA appeared lethal for astrocytes. A more systematic study on the isolated populations is needed to define Piezo1 involvement.

Consistent with these results is the recent report that Piezo1 is critical for sensing substrate stiffness (Koser et al., 2016). Tissue stiffness regulates the axon length and their degree of spreading for retinal ganglion cells, where high stiffness allows axons to elongate and low stiffness is associated with spreading. An interesting observation is that when cells are grown on high stiffness, inhibition of Piezo1 by GsMTx4, produces an axon response that is normally observed on low stiffness substrates.

2.7.3 Cell homeostasis

Epithelial cells maintain homeostasis by matching the number of dividing cells with lost cells in the epithelial layer. Rosenblatt and colleagues examined this process, called cell extrusion, in zebrafish epithelial cells and MDCK cells, an epithelial cell line, to identify the processes used to remove excess cells from the epithelial layer (Eisenhoffer et al., 2012). Extruded cells were assayed for an apoptotic maker, caspase-3, and surprisingly, there were cells that did not show apoptosis indicating that there is a live cell pathway for homeostasis.

Extrusion of cells occurs where there is high density of cells. To mimic this process, MDCK cells were seeded on a flexible silicone membrane that was stretched by about 30%. Once cells reached confluency, the membrane was released, increasing cell density; but after 6 h, the overcrowded cells returned to confluency and cells extruded from the epithelial layer were shown to be outside the apoptotic pathway.

To investigate what regulates live cell extrusion, they examined two candidate molecules, JUN kinase and MSCs. Inhibitors for JUN kinase did not block the extrusion process, but the generic MSC inhibitor Gd^{+3} blocked the homeostatic response and overcrowding remained high. Similar experiments in zebrafish showed cell proliferation and migration of epithelial cells blocked extrusion and formed masses in regions where cells failed to extrude (Eisenhoffer et al., 2012).

In support of MSC involvement in extrusion, photocleavable morpholino probes were used to knock down expression of Piezo1. Similar to the Gd^{+3} results, extrusion was inhibited and cell masses formed at sites of tissue strain. This indicated that Piezo1 is essential for homeostatic control of cell extrusion. Rosenblatt and collaborators propose a cell-death pathway involving sphinosine-1-phosphate leading to detachment from the ECM.

2.8 Functional role of Piezo

2.8.1 Nociception of Drosophila *larvae*

The in vivo role of Piezo1 was first demonstrated in the avoidance response of *Drosophila* larvae. Piezo1 from *Drosophila melanogaster* were shown first to produce currents in a heterologous system, and channels were found in all types of sensory neurons as well as some nonneuronal cells in both adults and larvae (Kim, Coste, Chadha, Cook, & Patapoutian, 2012).

Piezo1 knockout flies were made by genomic recombination and both the adult and larvae were viable with no overt changes in behavior such as coordination. Piezo1-ablated larvae were tested with a von Frey filaments and showed a significant reduction in their rollover response compared to wild type (~50%). Other noxious stimuli such as heat, which also can potentially produce a response, were unaffected. Knocking down *Piezo1* in the neurons involved in nociception reduced the rollover response and could be rescued by expression of *Piezo1*. These neurons also express the genes for *pickpocket* (DEG/ENAC), and the ablation of both *pickpocket* and *Piezo1* genes abolished the rollover response indicating that these two pathways are functionally in parallel.

2.8.2 Shear stress response of Piezo1

Piezo1 channels have been shown to respond to shear stress forces and are determinants of vascular structure (Li et al., 2014; Ranade, Qiu, et al., 2014). The physiological role of shear stress is described by Beech and colleagues in chapter "Piezo1 Channels in Vascular Development and the Sensing of Shear Stress" by Hyman et al. (2016), and Meneshi et al. discuss the biophysical properties of shear stress activation of Piezo1 channels in chapter "A Microfluidic Approach for Studying Piezo Channels"by Maneshi et al. (2016).

2.8.3 Calcium influx and ATP release

Stretch-induced calcium influx through Piezo1 ion channels has been linked with the release of ATP in at least two cases. In the mouse urothelium, there is a high level of Piezo1 expression as well as in urothelial cell cultures (Miyamoto et al., 2014). Cell stretch–activated mechanical channels allow internal calcium to increase as well as the release of ATP. siRNAs targeting

Piezo1 diminished the calcium response and lowered amounts of ATP released. When GsMTx4 was used to inhibit Piezo1 function a similar reduction in Calcium and ATP levels was observed.

A second example of ATP release associated with Piezo1 was observed for RBCs (Cinar et al., 2015). In this case, shear stress was used to activate calcium influx and ATP release. An interesting observation of this work is that mutations associated with HX that slow inactivation would be expected increase calcium influx and presumably ATP release. On the contrary, calcium influx was significantly lower and this is attributed to the long latency (>100 ms) to activation observed for mutated channels (Bae, Gnanasambandam, et al., 2013). Cinar and colleagues using 10 ms pulses, appear to be below the threshold of activation for a mutant channel.

2.8.4 Piezo2 and light touch

Part of our ability to experience our environment is dependent on the touch receptors in our skin. Merkel cells, which exists in fingertips and specialized regions in hairy skin called touch domes, have long been associated with touch sensation (Iggo & Muir, 1969). An interesting feature of Merkel cells is their close association with Aβ sensory neurons similar to a synaptic contact. Pressure applied to touch domes having Merkel cell–neurite complexes have slowly adapting discharges characterized by a high frequency dynamic phase followed by a sustained discharge phase (Woo, Lumpkin, & Patapoutian, 2015).

One of the first questions asked was whether Merkel cells are touch sensitive. Three different groups have examined this question and, while the approaches are different, all three groups show that Merkel cells respond to mechanical force dependent on the presence of Piezo2 (Ikeda & Gu, 2014; Maksimovic et al., 2014; Woo et al., 2014). Interestingly, a variant of Piezo2 is thought to be involved in the light touch response of zebrafish (Faucherre, Nargeot, Mangoni, & Jopling, 2013).

2.8.4.1 Optogenetic approach

Maksimovic et al. (2014) examined Merkel cell activation and sustained firing of action potentials in tactile afferent neurons. The basic approach was to preferentially express channel rhodopsin-2 in Merkel cells. This allowed selective activation of Merkel cells without directly stimulating associated afferents. Conversely, sustained firing was attenuated by silencing Merkel cell currents by expressing the light-activated proton pump AchT. This work showed that depolarization of Merkel cell is necessary and sufficient for sustained action potential firing in SAI afferents establishing a connection between Merkel cells and afferents. They then ablated Merkel cells and measured SAI response using ex vivo skin—nerve recordings. Firing patterns were converted to intermediary adapting (IA) responses where the dynamic phase was reduced but not lost.

2.8.4.2 Knockout approach

This work directly linked Piezo2 in Merkel cells to the cell's response to mechanical force (Woo et al., 2014). Using conditional knock out technology, Piezo2 was ablated from Merkel cells. Whole-cell currents from dissociated Merkel cells had no MA currents as compared to normal Merkel cells. Skin-nerve recordings from *Piezo2*^{cko} showed that dynamic firing was normal and that sustained firing was truncated.

2.8.4.3 Biochemical and knockdown approach

A third group looked at isolated rat whisker follicles containing Merkel cells (Ikeda & Gu, 2014), a slightly different approach to the earlier experiments which used isolated Merkel cells. Nonetheless, the currents were similar to currents observed for Merkel cells, being sensitive to mechanical stimulation with rapidly inactivating currents. To demonstrate Piezo2 involvement, antibodies for Piezo2 or shRNA directed against *Piezo2* were shown to diminish the MA currents of Merkel cells. Ikeda et al. observed compound action potentials from sensory afferent bundles of the whisker follicle bundles during follicle movement. They were reduced using shRNA directed against *Piezo2*, but were not abolished.

The studies have revealed that central role of Piezo2 in the touch response expressed both in Merkel cells and afferents. This is consistent with the study where ablation of Piezo2 from Merkel cells and sensory neurons abolished the light touch response (Ranade, Woo, et al., 2014). Recently, a two receptor model has been proposed for the Merkel cell—neurite complex that can account for the dynamic and static phases of SAI firing patterns and this has been reviewed extensively (Nakatani, Maksimovic, Baba, & Lumpkin, 2015; Ranade, Syeda, & Patapoutian, 2015; Vasquez, Scherrer, & Goodman, 2014; Woo, Lumpkin, et al., 2015).

2.9 Piezo2 channels in other roles

2.9.1 Piezo2 in foraging duck

Functional Piezo2 channels serve as mechanosensors in other venues. For example, mechanical stimulation of duck trigeminal neurons evokes high-amplitude depolarizing current with a low threshold of activation, high signal amplification gain, and slow kinetics of inactivation that are thought to be related to the presence of Piezo2 (Schneider et al., 2014) although Piezo2 channels are normally rapidly inactivating. A discussion of the Piezo2 channel and their adaptation to duck requirements are discussed in chapter "Piezo2 in Cutaneous and Proprioceptive Mechanotransduction in Vertebrates" by Anderson et al. (2016).

2.9.2 Piezo2 in proprioception

Proprioception is the perception of body and limb position. Recent work by the Patapoutian lab showed that Piezo2 is the principal MA channel of proprioceptors in mice (Woo, Lukacs, et al., 2015). They first showed that there is high expression of *Piezo2* in sensory endings of proprioceptors innervating muscle spindles and golgi tendon organ. Ablation of *Piezo2* in proprioceptive neurons produced mice that are severely uncoordinated. In addition, the parvalbumin-expressing neurons that identify proprioceptors required Piezo2 expression for mechanically induced currents.

2.9.3 Piezo2 channels in the gut

Piezo channels in the gastrointestinal system in mouse and human is starting to emerge. In the gut epithelium, a cell with substantial functional and developmental similarities to the Merkel cell (see earlier) is the enterochromaffin (EC) cell. Like the Merkel cell, it is a primary mechanosensor of the epithelial forces. EC cells are known to synthesize and release the vast majority of circulating serotonin (5-HT) in response to chemical and mechanical forces. The primary EC cell mechanotransducer appears to be Piezo2 that is distributed exclusively within the EC, and these cells are important for mucosal secretion in response to pressure. Inhibitors targeting Piezo2 in EC cells significantly decreased mechanically induced secretion, showing that Piezo2 is critical for EC cell mechanical sensitivity, 5-HT release, and downstream physiological processes (see Chapter: Mechanosensitive Piezo Channels in the Gastrointestinal Tract by Alcaino et al., 2016).

2.9.4 Piezo2 and pain

The Piezo2 channel has been shown to be involved in hyperalgesia and mechanical allodynia. These two types of responses appear to have different intracellular pathways. Elevated mechanosensitive currents associated with Piezo2 were demonstrated in the inflammatory response associated with the bradykinin receptor beta 2 (BDKRB2) (Dubin et al., 2012).

Coexpression of Piezo2 and BDKRB2 in HEK293 cells produce normal Piezo2 currents until challenged with bradykinin (BK). A sixfold increase in current was observed when the cells were challenged with 300 nM BK. This effect is in part due to a slowing of inactivation. Direct activation of PKC and PKA resulted in enhanced Piezo2 currents while inhibitors to PKC and PKA abolished BDKRB2 effects. A similar effect was demonstrated on a subclass of sensory neurons expressing Piezo2.

A second example of Piezo2-related hyperalgesia is associated with the inflammation response, by endothelin-1, a pronociceptive mediator that is vasoactive (Ferrari, Bogen, Green, & Levine, 2015). The authors propose that Piezo2 is the mechanotransducer for endothelial cells because application of Piezo2 antisense oligonucleotide reduced endothelin-1-induced hyperalgesia for these cells.

Mechanical allodynia (pain) can result from innocuous stimuli such as light touch (Lolignier, Eijkelkamp, & Wood, 2015). Since mechanical allodynia is associated with the cAMP messenger pathway mediated by the protein EPAC1, the connection between Piezo2 and this pathway was examined (Eijkelkamp, Linley, et al., 2013). A neuropathic pain model showed that Epac1 mRNA levels increased in the dorsal root ganglia innervating the ipsilateral paw while no change was observed for Epac2. Epac1$^{-/-}$ mouse has significantly attenuated neuropathic pain compared to wild type. Large diameter neurons produce rapidly adapting mechanical currents when distended. The application of 8pCPT, an activator of Epac1, significantly increase MA currents in these neurons and shifted the activation curve leftward.

To further characterize the relationship between Epac1 and Piezo2 both were coexpressed in HEK293 cells and then treated with the activator for Epac1, 8pCPT, resulting a 2.5-fold increase in current and a leftward shift in the activation curve. PKC and PKA activators did not affect the Piezo2 currents indicating that it was a different signaling pathway described for BK activation. Increased intracellular calcium strongly potentiates piezo currents.

3. CONCLUSION

In the brief time since the discovery of piezo channel family by the Patapoutian group, our understanding of how eukaryotic cells respond to mechanical cues from the environment has developed rapidly. A significant amount of work has already emerged describing the structure and properties of the piezo channel. The functional involvement of piezo channels in many physiological responses speaks to the importance of this channel. This is supported by mutations to the channel that alter biophysical properties that result in diseases related to cell mechanics. The world of piezo channels has opened to us new insights to the life of animal cells. What we learn about the properties of piezo can help us to understand the basics of the cell's mechanical response.

ACKNOWLEDGMENT

I thank Fred Sachs, Art Beyder, and Gail Dinter Gottlieb for their helpful comments on this review.

REFERENCES

Akinlaja, J., & Sachs, F. (1998). The breakdown of cell membranes by electrical and mechanical stress. *Biophysical Journal, 75*(1), 247–254. http://dx.doi.org/10.1016/S0006-3495(98)77511-3.

Albuisson, J., Murthy, S. E., Bandell, M., Coste, B., Louis-Dit-Picard, H., Mathur, J., … Patapoutian, A. (2013). Dehydrated hereditary stomatocytosis linked to gain-of-function mutations in mechanically activated PIEZO1 ion channels. *Nature Communications, 4*, 1884. http://dx.doi.org/10.1038/ncomms2899.

Alcaino, C., Farrugia, G., & Beyder, A. (2016). Mechanosensitive piezo channels in the gastrointestinal tract. In P. Gottlieb (Ed.), *Piezo Channels* (vol. 79, pp. 219–244).

Anderson, E. O., Schneider, E. R., & Bagriantsev, S. N. (2016). Piezo2 in cutaneous and proprioceptive mechanotransduction in vertebrates. In P. Gottlieb (Ed.), *Piezo Channels* (vol. 79, pp. 197–217).

Andolfo, I., Alper, S. L., De Franceschi, L., Auriemma, C., Russo, R., De Falco, L., … Iolascon, A. (2013). Multiple clinical forms of dehydrated hereditary stomatocytosis arise from mutations in PIEZO1. *Blood, 121*(19), 3925–3935. http://dx.doi.org/10.1182/blood-2013-02-482489. S3921–S3912.

Anishkin, A., & Kung, C. (2013). Stiffened lipid platforms at molecular force foci. *Proceedings of the National Academy of Sciences of the United States of America, 110*(13), 4886–4892. http://dx.doi.org/10.1073/pnas.1302018110.

Arnadottir, J., & Chalfie, M. (2010). Eukaryotic mechanosensitive channels. *Annual Review of Biophysics, 39*, 111–137. http://dx.doi.org/10.1146/annurev.biophys.37.032807.125836.

Bae, C., Gnanasambandam, R., Nicolai, C., Sachs, F., & Gottlieb, P. A. (2013). Xerocytosis is caused by mutations that alter the kinetics of the mechanosensitive channel PIEZO1. *Proceedings of the National Academy of Sciences of the United States of America, 110*(12), E1162–E1168. http://dx.doi.org/10.1073/pnas.1219777110.

Bae, C., Gottlieb, P. A., & Sachs, F. (2013). Human PIEZO1: Removing inactivation. *Biophysical Journal, 105*(4), 880–886. http://dx.doi.org/10.1016/j.bpj.2013.07.019.

Bae, C., Sachs, F., & Gottlieb, P. A. (2011). The mechanosensitive ion channel Piezo1 is inhibited by the peptide GsMTx4. *Biochemistry, 50*(29), 6295–6300. http://dx.doi.org/10.1021/bi200770q.

Bae, C., Sachs, F., & Gottlieb, P. A. (2015). Protonation of the human PIEZO1 ion channel stabilizes inactivation. *The Journal of Biological Chemistry, 290*(8), 5167–5173. http://dx.doi.org/10.1074/jbc.M114.604033.

Bae, C., Suchyna, T. M., Ziegler, L., Sachs, F., & Gottlieb, P. A. (2016). Human PIEZO1 ion channel functions as a split protein. *PLoS One, 11*(3), e0151289. http://dx.doi.org/10.1371/journal.pone.0151289.

Besch, S. R., Suchyna, T., & Sachs, F. (2002). High-speed pressure clamp. *Pflugers Archiv, 445*(1), 161–166. http://dx.doi.org/10.1007/s00424-002-0903-0.

Beyder, A., Strege, P. R., Reyes, S., Bernard, C. E., Terzic, A., Makielski, J., … Farrugia, G. (2012). Ranolazine decreases mechanosensitivity of the voltage-gated sodium ion channel Na(v)1.5: A novel mechanism of drug action. *Circulation, 125*(22), 2698–2706. http://dx.doi.org/10.1161/CIRCULATIONAHA.112.094714.

Blumenthal, N. R., Hermanson, O., Heimrich, B., & Shastri, V. P. (2014). Stochastic nanoroughness modulates neuron-astrocyte interactions and function via mechanosensing cation channels. *Proceedings of the National Academy of Sciences of the United States of America, 111*(45), 16124–16129. http://dx.doi.org/10.1073/pnas.1412740111.

Borbiro, I., Badheka, D., & Rohacs, T. (2015). Activation of TRPV1 channels inhibits mechanosensitive Piezo channel activity by depleting membrane phosphoinositides. *Science Signaling, 8*(363), ra15. http://dx.doi.org/10.1126/scisignal.2005667.

Borbiro, I., & Rohacs, T. (2016). Regulation of piezo channels by cellular signaling pathways. In P. Gottlieb (Ed.), *Piezo Channels* (vol. 79, pp. 245–261).

Brohawn, S. G., Su, Z., & MacKinnon, R. (2014). Mechanosensitivity is mediated directly by the lipid membrane in TRAAK and TREK-1 K+ channels. *Proceedings of the National Academy of Sciences of the United States of America, 111*(9), 3614–3619. http://dx.doi.org/10.1073/pnas.1320768111.

Cahalan, S. M., Lukacs, V., Ranade, S. S., Chien, S., Bandell, M., & Patapoutian, A. (2015). Piezo1 links mechanical forces to red blood cell volume. *eLife, 4*. http://dx.doi.org/10.7554/eLife.07370.

Christensen, A. P., & Corey, D. P. (2007). TRP channels in mechanosensation: Direct or indirect activation? *Nature Reviews Neuroscience, 8*(7), 510–521. http://dx.doi.org/10.1038/nrn2149.

Cinar, E., Zhou, S., DeCourcey, J., Wang, Y., Waugh, R. E., & Wan, J. (2015). Piezo1 regulates mechanotransductive release of ATP from human RBCs. *Proceedings of the National Academy of Sciences of the United States of America, 112*(38), 11783–11788. http://dx.doi.org/10.1073/pnas.1507309112.

Clapham, D. E., Runnels, L. W., & Strubing, C. (2001). The TRP ion channel family. *Nature Reviews Neuroscience, 2*(6), 387–396. http://dx.doi.org/10.1038/35077544.

Corey, D. P., & Hudspeth, A. J. (1979). Ionic basis of the receptor potential in a vertebrate hair cell. *Nature, 281*(5733), 675–677.

Coste, B., Houge, G., Murray, M. F., Stitziel, N., Bandell, M., Giovanni, M. A., … Patapoutian, A. (2013). Gain-of-function mutations in the mechanically activated ion channel PIEZO2 cause a subtype of distal arthrogryposis. *Proceedings of the National Academy of Sciences of the United States of America, 110*(12), 4667–4672. http://dx.doi.org/10.1073/pnas.1221400110.

Coste, B., Mathur, J., Schmidt, M., Earley, T. J., Ranade, S., Petrus, M. J., … Patapoutian, A. (2010). Piezo1 and Piezo2 are essential components of distinct mechanically activated cation channels. *Science, 330*(6000), 55–60. http://dx.doi.org/10.1126/science.1193270.

Coste, B., Murthy, S. E., Mathur, J., Schmidt, M., Mechioukhi, Y., Delmas, P., … Patapoutian, A. (2015). Piezo1 ion channel pore properties are dictated by C-terminal region. *Nature Communications, 6*, 7223. http://dx.doi.org/10.1038/ncomms8223.

Coste, B., Xiao, B., Santos, J. S., Syeda, R., Grandl, J., Spencer, K. S., … Patapoutian, A. (2012). Piezo proteins are pore-forming subunits of mechanically activated channels. *Nature, 483*(7388), 176–181. http://dx.doi.org/10.1038/nature10812.

Cox, C. D., Bae, C., Ziegler, L., Hartley, S., Nikolova-Krstevski, V., Rohde, P. R., … Martinac, B. (2016). Removal of the mechanoprotective influence of the cytoskeleton reveals

PIEZO1 is gated by bilayer tension. *Nature Communications, 7*, 10366. http://dx.doi.org/10.1038/ncomms10366.
Cox, C. D., Bavi, N., & Martinac, B. (2016). Origin of the force: The force-from-lipids principle applied to piezo channels. In P. Gottlieb (Ed.), *Piezo Channels* (vol. 79, pp. 59–96).
Cui, W., Allen, N. D., Skynner, M., Gusterson, B., & Clark, A. J. (2001). Inducible ablation of astrocytes shows that these cells are required for neuronal survival in the adult brain. *Glia, 34*(4), 272–282.
Dubin, A. E., Schmidt, M., Mathur, J., Petrus, M. J., Xiao, B., Coste, B., … Patapoutian, A. (2012). Inflammatory signals enhance piezo2-mediated mechanosensitive currents. *Cell Reports, 2*(3), 511–517. http://dx.doi.org/10.1016/j.celrep.2012.07.014.
Dupont, S., Morsut, L., Aragona, M., Enzo, E., Giulitti, S., Cordenonsi, M., … Piccolo, S. (2011). Role of YAP/TAZ in mechanotransduction. *Nature, 474*(7350), 179–183. http://dx.doi.org/10.1038/nature10137.
Eijkelkamp, N., Linley, J. E., Torres, J. M., Bee, L., Dickenson, A. H., Gringhuis, M., … Wood, J. N. (2013). A role for Piezo2 in EPAC1-dependent mechanical allodynia. *Nature Communications, 4*, 1682. http://dx.doi.org/10.1038/ncomms2673.
Eijkelkamp, N., Quick, K., & Wood, J. N. (2013). Transient receptor potential channels and mechanosensation. *Annual Review of Neuroscience, 36*, 519–546. http://dx.doi.org/10.1146/annurev-neuro-062012-170412.
Eisenhoffer, G. T., Loftus, P. D., Yoshigi, M., Otsuna, H., Chien, C. B., Morcos, P. A., … Rosenblatt, J. (2012). Crowding induces live cell extrusion to maintain homeostatic cell numbers in epithelia. *Nature, 484*(7395), 546–549. http://dx.doi.org/10.1038/nature10999.
Faucherre, A., Nargeot, J., Mangoni, M. E., & Jopling, C. (2013). piezo2b regulates vertebrate light touch response. *The Journal of Neuroscience, 33*(43), 17089–17094. http://dx.doi.org/10.1523/JNEUROSCI.0522-13.2013.
Ferrari, L. F., Bogen, O., Green, P., & Levine, J. D. (2015). Contribution of Piezo2 to endothelium-dependent pain. *Molecular Pain, 11*. http://dx.doi.org/10.1186/s12990-015-0068-4.
Gailly, P. (2002). New aspects of calcium signaling in skeletal muscle cells: Implications in Duchenne muscular dystrophy. *Biochimica et Biophysica Acta, 1600*(1–2), 38–44.
Ge, J., Li, W., Zhao, Q., Li, N., Chen, M., Zhi, P., … Yang, M. (2015). Architecture of the mammalian mechanosensitive Piezo1 channel. *Nature*. http://dx.doi.org/10.1038/nature15247.
Geng, J., Zhao, Q., Zhang, T., & Xiao, B. (2016). In touch with the mechanosensitive piezo channels: Structure, ion permeation and mechanotransduction. In P. Gottlieb (Ed.), *Piezo Channels* (vol. 79, pp. 159–195).
Gnanasambandam, R., Bae, C., Gottlieb, P. A., & Sachs, F. (2015). Ionic selectivity and permeation properties of human PIEZO1 channels. *PLoS One, 10*(5), e0125503. http://dx.doi.org/10.1371/journal.pone.0125503.
Gnanasambandam, R., Bae, C., Gottlieb, P. A., & Sachs, F. (2016). Functional heteromers of Piezo1. *Biophysics Journal* (unpublished).
Gnanasambandam, R., Gottlieb, P. A., & Sachs, F. (2016). The kinetics and the permeation properties of piezo channels. In P. Gottlieb (Ed.), *Piezo Channels* (vol. 79, pp. 275–307).
Gottlieb, P. A., Bae, C., & Sachs, F. (2012). Gating the mechanical channel Piezo1: A comparison between whole-cell and patch recording. *Channels (Austin), 6*(4), 282–289. http://dx.doi.org/10.4161/chan.21064.
Gottlieb, P., Folgering, J., Maroto, R., Raso, A., Wood, T. G., Kurosky, A., … Honore, E. (2008). Revisiting TRPC1 and TRPC6 mechanosensitivity. *Pflugers Archiv, 455*(6), 1097–1103. http://dx.doi.org/10.1007/s00424-007-0359-3.

Gu, G., Caldwell, G. A., & Chalfie, M. (1996). Genetic interactions affecting touch sensitivity in *Caenorhabditis elegans*. *Proceedings of the National Academy of Sciences of the United States of America, 93*(13), 6577–6582.

Guharay, F., & Sachs, F. (1984). Stretch-activated single ion channel currents in tissue-cultured embryonic chick skeletal muscle. *The Journal of Physiology, 352*, 685–701.

Hamill, O. P., & Martinac, B. (2001). Molecular basis of mechanotransduction in living cells. *Physiological Reviews, 81*(2), 685–740.

Hamill, O. P., & McBride, D. W., Jr. (1997). Mechanogated channels in Xenopus oocytes: Different gating modes enable a channel to switch from a phasic to a tonic mechanotransducer. *The Biological Bulletin, 192*(1), 121–122.

Honore, E. (2007). The neuronal background K2P channels: Focus on TREK-1. *Nature Reviews Neuroscience, 8*(4), 251–261. http://dx.doi.org/10.1038/nrn2117.

Huang, H., Bae, C., Sachs, F., & Suchyna, T. M. (2013). Caveolae regulation of mechanosensitive channel function in myotubes. *PLoS One, 8*(8), e72894. http://dx.doi.org/10.1371/journal.pone.0072894.

Hung, W. C., Yang, J. R., Yankaskas, C. L., Wong, B. S., Wu, P. H., Pardo-Pastor, C., … Konstantopoulos, K. (2016). Confinement sensing and signal optimization via Piezo1/PKA and myosin II Pathways. *Cell Reports, 15*(7), 1430–1441. http://dx.doi.org/10.1016/j.celrep.2016.04.035.

Hyman, A. J., Tumova, S., & Beech, D. J. (2016). Piezo1 channels in vascular development and the sensing of shear stress. In P. Gottlieb (Ed.), *Piezo Channels* (vol. 79, pp. 37–57).

Iggo, A., & Muir, A. R. (1969). The structure and function of a slowly adapting touch corpuscle in hairy skin. *The Journal of Physiology, 200*(3), 763–796.

Ikeda, R., & Gu, J. G. (2014). Piezo2 channel conductance and localization domains in Merkel cells of rat whisker hair follicles. *Neuroscience Letters, 583*, 210–215. http://dx.doi.org/10.1016/j.neulet.2014.05.055.

Ingber, D. E. (2006). Cellular mechanotransduction: Putting all the pieces together again. *The FASEB Journal, 20*(7), 811–827.

Jia, Z., Ikeda, R., Ling, J., Viatchenko-Karpinski, V., & Gu, J. G. (2016). Regulation of Piezo2 mechanotransduction by static plasma membrane tension in primary afferent neurons. *The Journal of Biological Chemistry*. http://dx.doi.org/10.1074/jbc.M115.692384.

Kim, S. E., Coste, B., Chadha, A., Cook, B., & Patapoutian, A. (2012). The role of Drosophila Piezo in mechanical nociception. *Nature, 483*(7388), 209–212. http://dx.doi.org/10.1038/nature10801.

Koser, D. E., Thompson, A. J., Foster, S. K., Dwivedy, A., Pillai, E. K., Sheridan, G. K., … Franze, K. (2016). Mechanosensing is critical for axon growth in the developing brain. *Nature Neuroscience*. http://dx.doi.org/10.1038/nn.4394.

Lansman, J. B. (2007). Mechanosensitive ion channels in dystrophic muscle. *Current Topics in Membranes, 59*, 467–484. http://dx.doi.org/10.1016/S1063-5823(06)59017-9.

Lansman, J. B., & Franco-Obregon, A. (2006). Mechanosensitive ion channels in skeletal muscle: A link in the membrane pathology of muscular dystrophy. *Clinical and Experimental Pharmacology and Physiology, 33*(7), 649–656. http://dx.doi.org/10.1111/j.1440-1681.2006.04393.x.

Lauritzen, I., Chemin, J., Honore, E., Jodar, M., Guy, N., Lazdunski, M., … Jane Patel, A. (2005). Cross-talk between the mechano-gated K2P channel TREK-1 and the actin cytoskeleton. *EMBO Reports, 6*(7), 642–648. http://dx.doi.org/10.1038/sj.embor.7400449.

Lee, W., Guilak, F., & Liedtke, W. (2016). Role of piezo channels in joint health and injury. In P. Gottlieb (Ed.), *Piezo Channels* (vol. 79, pp. 263–273).

Lee, W., Leddy, H. A., Chen, Y., Lee, S. H., Zelenski, N. A., McNulty, A. L., … Liedtke, W. B. (2014). Synergy between Piezo1 and Piezo2 channels confers high-strain mechanosensitivity to articular cartilage. *Proceedings of the National Academy of Sciences of the United States of America, 111*(47), E5114–E5122. http://dx.doi.org/10.1073/pnas.1414298111.
Leipzig, N. D., & Shoichet, M. S. (2009). The effect of substrate stiffness on adult neural stem cell behavior. *Biomaterials, 30*(36), 6867–6878. http://dx.doi.org/10.1016/j.biomaterials.2009.09.002.
Lewis, A. H., & Grandl, J. (2015). Mechanical sensitivity of Piezo1 ion channels can be tuned by cellular membrane tension. *eLife, 4*. http://dx.doi.org/10.7554/eLife.12088.
Li, C., Rezania, S., Kammerer, S., Sokolowski, A., Devaney, T., Gorischek, A., … Schreibmayer, W. (2015). Piezo1 forms mechanosensitive ion channels in the human MCF-7 breast cancer cell line. *Scientific Reports, 5*, 8364. http://dx.doi.org/10.1038/srep08364.
Li, J., Hou, B., Tumova, S., Muraki, K., Bruns, A., Ludlow, M. J., … Beech, D. J. (2014). Piezo1 integration of vascular architecture with physiological force. *Nature, 515*(7526), 279–282. http://dx.doi.org/10.1038/nature13701.
Li, W., Gao, N., & Yang, M. (2016). The structural basis for sensing by the piezo1 protein. In P. Gottlieb (Ed.), *Piezo Channels* (vol. 79, pp. 135–158).
Lolignier, S., Eijkelkamp, N., & Wood, J. N. (2015). Mechanical allodynia. *Pflugers Archiv, 467*(1), 133–139. http://dx.doi.org/10.1007/s00424-014-1532-0.
Maksimovic, S., Nakatani, M., Baba, Y., Nelson, A. M., Marshall, K. L., Wellnitz, S. A., … Lumpkin, E. A. (2014). Epidermal Merkel cells are mechanosensory cells that tune mammalian touch receptors. *Nature, 509*(7502), 617–621. http://dx.doi.org/10.1038/nature13250.
Maneshi, M. M., Gottlieb, P. A., & Hua, S. Z. (2016). A microfluidic approach for studying piezo channels. In P. Gottlieb (Ed.), *Piezo Channels* (vol. 79, pp. 309–334).
Maroto, R., Kurosky, A., & Hamill, O. P. (2012). Mechanosensitive Ca(2+) permeant cation channels in human prostate tumor cells. *Channels (Austin), 6*(4), 290–307. http://dx.doi.org/10.4161/chan.21063.
Martinac, B. (2011). Bacterial mechanosensitive channels as a paradigm for mechanosensory transduction. *Cellular Physiology and Biochemistry, 28*(6), 1051–1060. http://dx.doi.org/10.1159/000335842.
McHugh, B. J., Buttery, R., Lad, Y., Banks, S., Haslett, C., & Sethi, T. (2010). Integrin activation by Fam38A uses a novel mechanism of R-Ras targeting to the endoplasmic reticulum. *Journal of Cell Science, 123*(Pt 1), 51–61. http://dx.doi.org/10.1242/jcs.056424.
McHugh, B. J., Murdoch, A., Haslett, C., & Sethi, T. (2012). Loss of the integrin-activating transmembrane protein Fam38A (Piezo1) promotes a switch to a reduced integrin-dependent mode of cell migration. *PLoS One, 7*(7), e40346. http://dx.doi.org/10.1371/journal.pone.0040346.
Milescu, L. S., Akk, G., & Sachs, F. (2005). Maximum likelihood estimation of ion channel kinetics from macroscopic currents. *Biophysical Journal, 88*(4), 2494–2515.
Miyamoto, T., Mochizuki, T., Nakagomi, H., Kira, S., Watanabe, M., Takayama, Y., … Tominaga, M. (2014). Functional role for Piezo1 in stretch-evoked Ca(2)(+) influx and ATP release in urothelial cell cultures. *The Journal of Biological Chemistry, 289*(23), 16565–16575. http://dx.doi.org/10.1074/jbc.M113.528638.
Morris, C. E. (2011). Voltage-gated channel mechanosensitivity: Fact or friction? *Frontiers in Physiology, 2*, 25. http://dx.doi.org/10.3389/fphys.2011.00025.
Morris, C. E., & Juranka, P. F. (2007). Nav channel mechanosensitivity: Activation and inactivation accelerate reversibly with stretch. *Biophysical Journal, 93*(3), 822–833. http://dx.doi.org/10.1529/biophysj.106.101246.

Nakatani, M., Maksimovic, S., Baba, Y., & Lumpkin, E. A. (2015). Mechanotransduction in epidermal Merkel cells. *Pflugers Archiv, 467*(1), 101–108. http://dx.doi.org/10.1007/s00424-014-1569-0.

Noda, M., Shimizu, S., Tanabe, T., Takai, T., Kayano, T., Ikeda, T., … Numa, S. (1984). Primary structure of *Electrophorus electricus* sodium channel deduced from cDNA sequence. *Nature, 312*(5990), 121–127.

O'Hagan, R., Chalfie, M., & Goodman, M. B. (2005). The MEC-4 DEG/ENaC channel of *Caenorhabditis elegans* touch receptor neurons transduces mechanical signals. *Nature Neuroscience, 8*(1), 43–50. http://dx.doi.org/10.1038/nn1362.

Ostrow, K. L., Mammoser, A., Suchyna, T., Sachs, F., Oswald, R., Kubo, S., … Gottlieb, P. A. (2003). cDNA sequence and in vitro folding of GsMTx4, a specific peptide inhibitor of mechanosensitive channels. *Toxicon, 42*(3), 263–274.

Oswald, R. E., Suchyna, T. M., McFeeters, R., Gottlieb, P., & Sachs, F. (2002). Solution structure of peptide toxins that block mechanosensitive ion channels. *The Journal of Biological Chemistry, 277*(37), 34443–34450. http://dx.doi.org/10.1074/jbc.M202715200.

Patel, A. J., & Honore, E. (2001). Properties and modulation of mammalian 2P domain K+ channels. *Trends in Neurosciences, 24*(6), 339–346.

Patel, A., & Honore, E. (2002). The TREK two P domain K+ channels. *The Journal of Physiology, 539*(Pt 3), 647.

Pathak, M. M., Nourse, J. L., Tran, T., Hwe, J., Arulmoli, J., Le, D. T., … Tombola, F. (2014). Stretch-activated ion channel Piezo1 directs lineage choice in human neural stem cells. *Proceedings of the National Academy of Sciences of the United States of America, 111*(45), 16148–16153. http://dx.doi.org/10.1073/pnas.1409802111.

Peyronnet, R., Martins, J. R., Duprat, F., Demolombe, S., Arhatte, M., Jodar, M., … Patel, A. (2013). Piezo1-dependent stretch-activated channels are inhibited by Polycystin-2 in renal tubular epithelial cells. *EMBO Reports, 14*(12), 1143–1148. http://dx.doi.org/10.1038/embor.2013.170.

Poole, K., Herget, R., Lapatsina, L., Ngo, H. D., & Lewin, G. R. (2014). Tuning Piezo ion channels to detect molecular-scale movements relevant for fine touch. *Nature Communications, 5*, 3520. http://dx.doi.org/10.1038/ncomms4520.

Prole, D. L., & Taylor, C. W. (2013). Identification and analysis of putative homologues of mechanosensitive channels in pathogenic protozoa. *PLoS One, 8*(6), e66068. http://dx.doi.org/10.1371/journal.pone.0066068.

Qi, Y., Andolfi, L., Frattini, F., Mayer, F., Lazzarino, M., & Hu, J. (2015). Membrane stiffening by STOML3 facilitates mechanosensation in sensory neurons. *Nature Communications, 6*, 8512. http://dx.doi.org/10.1038/ncomms9512.

Ranade, S. S., Qiu, Z., Woo, S. H., Hur, S. S., Murthy, S. E., Cahalan, S. M., … Patapoutian, A. (2014). Piezo1, a mechanically activated ion channel, is required for vascular development in mice. *Proceedings of the National Academy of Sciences of the United States of America.* http://dx.doi.org/10.1073/pnas.1409233111.

Ranade, S. S., Syeda, R., & Patapoutian, A. (2015). Mechanically activated ion channels. *Neuron, 87*(6), 1162–1179. http://dx.doi.org/10.1016/j.neuron.2015.08.032.

Ranade, S. S., Woo, S. H., Dubin, A. E., Moshourab, R. A., Wetzel, C., Petrus, M., … Patapoutian, A. (2014). Piezo2 is the major transducer of mechanical forces for touch sensation in mice. *Nature, 516*(7529), 121–125. http://dx.doi.org/10.1038/nature13980.

Retailleau, K., Arhatte, M., Demolombe, S., Jodar, M., Baudrie, V., Offermanns, S., … Duprat, F. (2016). Smooth muscle filamin A is a major determinant of conduit artery structure and function at the adult stage. *Pflugers Archiv.* http://dx.doi.org/10.1007/s00424-016-1813-x.

Retailleau, K., Arhatte, M., Demolombe, S., Peyronnet, R., Baudrie, V., Jodar, M., … Honore, E. (2016). Arterial myogenic activation through smooth muscle filamin A. *Cell Reports, 14*(9), 2050–2058. http://dx.doi.org/10.1016/j.celrep.2016.02.019.

Retailleau, K., Duprat, F., Arhatte, M., Ranade, S. S., Peyronnet, R., Martins, J. R., … Honore, E. (2015). Piezo1 in smooth muscle cells is involved in hypertension-dependent arterial remodeling. *Cell Reports, 13*(6), 1161–1171. http://dx.doi.org/10.1016/j.celrep.2015.09.072.

Saha, K., Keung, A. J., Irwin, E. F., Li, Y., Little, L., Schaffer, D. V., … Healy, K. E. (2008). Substrate modulus directs neural stem cell behavior. *Biophysical Journal, 95*(9), 4426–4438. http://dx.doi.org/10.1529/biophysj.108.132217.

Satoh, K., Hata, M., Takahara, S., Tsuzaki, H., Yokota, H., Akatsu, H., … Yamada, T. (2006). A novel membrane protein, encoded by the gene covering KIAA0233, is transcriptionally induced in senile plaque-associated astrocytes. *Brain Research, 1108*(1), 19–27. http://dx.doi.org/10.1016/j.brainres.2006.06.050.

Schneider, E. R., Mastrotto, M., Laursen, W. J., Schulz, V. P., Goodman, J. B., Funk, O. H., … Bagriantsev, S. N. (2014). Neuronal mechanism for acute mechanosensitivity in tactile-foraging waterfowl. *Proceedings of the National Academy of Sciences of the United States of America, 111*(41), 14941–14946. http://dx.doi.org/10.1073/pnas.1413656111.

Slavchov, R. I., Nomura, T., Martinac, B., Sokabe, M., & Sachs, F. (2014). Gigaseal mechanics: Creep of the gigaseal under the action of pressure, adhesion, and voltage. *The Journal of Physical Chemistry B, 118*(44), 12660–12672. http://dx.doi.org/10.1021/jp506965v.

Suchyna, T. M., Besch, S. R., & Sachs, F. (2004). Dynamic regulation of mechanosensitive channels: Capacitance used to monitor patch tension in real time. *Physical Biology, 1*(1–2), 1–18. http://dx.doi.org/10.1088/1478-3967/1/1/001.

Suchyna, T. M., Johnson, J. H., Hamer, K., Leykam, J. F., Gage, D. A., Clemo, H. F., … Sachs, F. (2000). Identification of a peptide toxin from *Grammostola spatulata* spider venom that blocks cation-selective stretch-activated channels. *The Journal of General Physiology, 115*(5), 583–598.

Suchyna, T. M., Markin, V. S., & Sachs, F. (2009). Biophysics and structure of the patch and the gigaseal. *Biophysical Journal, 97*(3), 738–747. http://dx.doi.org/10.1016/j.bpj.2009.05.018.

Suchyna, T. M., & Sachs, F. (2007). Mechanosensitive channel properties and membrane mechanics in mouse dystrophic myotubes. *The Journal of Physiology, 581*(Pt 1), 369–387. http://dx.doi.org/10.1113/jphysiol.2006.125021.

Sukharev, S. I., Blount, P., Martinac, B., Blattner, F. R., & Kung, C. (1994). A large-conductance mechanosensitive channel in *E. coli* encoded by mscL alone. *Nature, 368*(6468), 265–268. http://dx.doi.org/10.1038/368265a0.

Sukharev, S., & Corey, D. P. (2004). Mechanosensitive channels: Multiplicity of families and gating paradigms. *Science's STKE, 2004*(219), re4. http://dx.doi.org/10.1126/stke.2192004re4.

Sukharev, S., & Sachs, F. (2012). Molecular force transduction by ion channels: Diversity and unifying principles. *Journal of Cell Science, 125*(Pt 13), 3075–3083. http://dx.doi.org/10.1242/jcs.092353.

Sun, Y., Chen, C. S., & Fu, J. (2012). Forcing stem cells to behave: A biophysical perspective of the cellular microenvironment. *Annual Review of Biophysics, 41*, 519–542. http://dx.doi.org/10.1146/annurev-biophys-042910-155306.

Syeda, R., Florendo, M. N., Cox, C. D., Kefauver, J. M., Santos, J. S., Martinac, B., … Patapoutian, A. (2016). Piezo1 channels are inherently mechanosensitive. *Cell Reports, 17*(7), 1739–1746. http://dx.doi.org/10.1016/j.celrep.2016.10.033.

Syeda, R., Xu, J., Dubin, A. E., Coste, B., Mathur, J., Huynh, T., … Patapoutian, A. (2015). Chemical activation of the mechanotransduction channel Piezo1. *eLife, 4*. http://dx.doi.org/10.7554/eLife.07369.

Vasquez, V., Scherrer, G., & Goodman, M. B. (2014). Sensory biology: It takes Piezo2 to tango. *Current Biology, 24*(12), R566–R569. http://dx.doi.org/10.1016/j.cub.2014.05.011.

Woo, S. H., Lukacs, V., de Nooij, J. C., Zaytseva, D., Criddle, C. R., Francisco, A., … Patapoutian, A. (2015). Piezo2 is the principal mechanotransduction channel for proprioception. *Nature Neuroscience, 18*(12), 1756–1762. http://dx.doi.org/10.1038/nn.4162.

Woo, S. H., Lumpkin, E. A., & Patapoutian, A. (2015). Merkel cells and neurons keep in touch. *Trends in Cell Biology, 25*(2), 74–81. http://dx.doi.org/10.1016/j.tcb.2014.10.003.

Woo, S. H., Ranade, S., Weyer, A. D., Dubin, A. E., Baba, Y., Qiu, Z., … Patapoutian, A. (2014). Piezo2 is required for Merkel-cell mechanotransduction. *Nature, 509*(7502), 622–626. http://dx.doi.org/10.1038/nature13251.

Yan, Z., Zhang, W., He, Y., Gorczyca, D., Xiang, Y., Cheng, L. E., … Jan, Y. N. (2013). Drosophila NOMPC is a mechanotransduction channel subunit for gentle-touch sensation. *Nature, 493*(7431), 221–225. http://dx.doi.org/10.1038/nature11685.

Yeung, E. W., & Allen, D. G. (2004). Stretch-activated channels in stretch-induced muscle damage: Role in muscular dystrophy. *Clinical and Experimental Pharmacology and Physiology, 31*(8), 551–556. http://dx.doi.org/10.1111/j.1440-1681.2004.04027.x.

Zarychanski, R., Schulz, V. P., Houston, B. L., Maksimova, Y., Houston, D. S., Smith, B., … Gallagher, P. G. (2012). Mutations in the mechanotransduction protein PIEZO1 are associated with hereditary xerocytosis. *Blood, 120*(9), 1908–1915. http://dx.doi.org/10.1182/blood-2012-04-422253.

Zhang, W., Cheng, L. E., Kittelmann, M., Li, J., Petkovic, M., Cheng, T., … Jan, Y. N. (2015). Ankyrin repeats convey force to gate the NOMPC mechanotransduction channel. *Cell, 162*(6), 1391–1403. http://dx.doi.org/10.1016/j.cell.2015.08.024.

Zhao, Q., Wu, K., Geng, J., Chi, S., Wang, Y., Zhi, P., … Xiao, B. (2016). Ion permeation and mechanotransduction mechanisms of mechanosensitive piezo channels. *Neuron, 89*(6), 1248–1263. http://dx.doi.org/10.1016/j.neuron.2016.01.046.

CHAPTER TWO

Piezo1 Channels in Vascular Development and the Sensing of Shear Stress

A.J. Hyman, S. Tumova and D.J. Beech[1]
University of Leeds, Leeds, United Kingdom
[1]Corresponding author: E-mail: d.j.beech@leeds.ac.uk

Contents

Abstract

A critical point in mammalian development occurs before mid-embryogenesis when the heart starts to beat, pushing blood into the nascent endothelial lattice. This pushing force is a signal, detected by endothelial cells as a frictional force (shear stress) to trigger cellular changes that underlie the essential processes of vascular remodeling and expansion required for embryonic growth. The processes are complex and multifactorial and Piezo1 became a recognized player only 2 years ago, 4 years after Piezo1's initial discovery as a functional membrane protein. Piezo1 is now known to be critical in

Current Topics in Membranes, Volume 79
ISSN 1063-5823
http://dx.doi.org/10.1016/bs.ctm.2016.11.001

murine embryonic development just at the time when the pushing force is first detected by endothelial cells. Murine *Piezo1* gene disruption in endothelial cells is embryonic lethal and mutations in human *PIEZO1* associate with severe disease phenotype due to abnormal lymphatic vascular development. Piezo1 proteins coassemble to form calcium-permeable nonselective cationic channels, most likely as trimers. They are large proteins with little if any resemblance to other proteins or ion channel subunits. The channels appear to sense mechanical force directly, including the force imposed on endothelial cells by physiological shear stress. Here, we review current knowledge of Piezo1 in the vascular setting and discuss hypotheses about how it might serve its vascular functions and integrate with other mechanisms. Piezo1 is a new important player for investigators in this field and promises much as a basis for better understanding of vascular physiology and pathophysiology and perhaps also discovery of new therapies.

1. INTRODUCTION

Vascular systems are crucial for the development and survival of mammals and other animals. Early in development the embryo switches from an avascular state reliant on diffusion to a vascular state with complex systems of blood vessels. Orchestrated events enable the switch, controlled by chemical signaling but also, importantly, the physiological force of shear stress. Initially, there is vasculogenesis when endothelial cells migrate, differentiate, and form tube-like structures which remodel into mature blood vessels. This is complemented by angiogenesis as vascular networks arise from preexisting vessels and the distinct vessel types emerge, forming arteries, arterioles, veins, venules, capillaries, and lymphatics depending on recruitment of additional cell types: fibroblasts, neurones, pericytes, and smooth muscle cells.

As indicated, an important physiological parameter governing vascular development is shear stress, a frictional force arising from blood flow along the endothelial cells lining the nascent tubes and developing vessels. The endothelial cells must sense this force to trigger molecular events, which enable vascular remodeling and thus vascular development and embryonic development itself. There has been much interest in shear stress-sensing and the downstream mechanisms not only because of their importance in development but also apparently related mechanisms exist in the adult vasculature and cardiovascular disease (Chiu & Chien, 2011; Cunningham & Gotlieb, 2005). Recent advances in understanding the shear stress-sensing mechanisms of endothelial cells have revealed a critical mechanosensitive Ca^{2+}-permeable nonselective cationic channel, the Piezo1 channel, which is important for shear stress sensing and critical in murine vascular development (Li et al., 2014; Ranade et al., 2014). Genetic studies suggest that disruption of Piezo1 in humans has particular impact on lymphatic vessel development leading to severe phenotypes (Fotiou et al., 2015; Lukacs et al., 2015). This chapter focuses on the Piezo1 discovery and its contextualization.

2. SHEAR STRESS IN VASCULAR DEVELOPMENT

Early in embryogenesis, endothelial and hematopoietic precursor cells emerge in the embryonic mesoderm (Caprioli et al., 2001). They migrate to form a vascular plexus with focal clusters—blood islands; in the mouse this occurs at embryonic day 6.5–7.0 (E6.5–7.0). The peripheral cells of the islands are angioblasts, which differentiate into endothelial cells whereas the inner cells are hematopoietic, differentiating into blood cells. Marker proteins for endothelial and hemotopoietic precursors such as cluster of differentiation (CD) 31, CD34, and vascular endothelial growth factor receptor 2 (VEGFR2) are identifiable at this stage (Yamaguchi, Dumont, Conlon, Breitman, & Rossant, 1993). The angioblasts then migrate and form focal adhesions with each other, resulting in a honeycomb-like structure—a lattice known as the vascular plexus. Angioblasts migrate and coalesce to form cords, which later form the dorsal aorta and cardinal veins. The formation of these networks is largely driven by growth factors (Coultas, Chawengsaksophak, & Rossant, 2005). Angiogenesis then occurs through sprouting and intussusception, where transluminal tissue columns split preexisting vessels into new vessels. Angiogenesis starts at E9.5 and is responsible for the majority of embryonic blood vessels and vascularization of the major organs derived from the ectoderm–mesoderm.

The hemodynamic force of shear stress is important and perhaps even essential for the remodeling of the nascent network into a vascular tree, which can then efficiently perfuse the yolk sac and embryo to allow growth and development. The importance of shear stress has been highlighted by studies of genes required for the heart beat. Mice with disrupted myosin light chain 2 or Na^{+}–Ca^{2+} exchanger 1 do not show vascular remodeling in the yolk sac and display embryonic lethality (Huang et al., 2003; Wakimoto et al., 2000). Such effects could be attributed to other factors such as lack of circulating oxygen, nutrients, and growth factors but further evidence of the importance of shear stress has been obtained by manipulating blood viscosity in the embryo. Sequestering erythroblasts in the blood islands using polymerized acrylamide has resulted in reduced hematocrit and therefore blood viscosity. This approach ablated vascular remodeling in the yolk sac (Lucitti et al., 2007). Moreover, mice deficient in hematopoiesis, which have low blood hematocrit and viscosity, also show ablated yolk sac remodeling (He et al., 2008). The findings suggest importance of shear stress sensing by endothelial cells in triggering vascular remodeling after the heart starts to beat, without which the embryo is growth-retarded or fails to develop.

3. ION CHANNELS IN SHEAR STRESS SENSING

For endothelial cells to respond to shear stress, they must be able to detect it and transduce it into an appropriate response. A challenging question to answer has been: How do the cells sense shear stress? In many ways the question is still far from completely answered. Before addressing Piezo1 in this context we briefly review other ion channel mechanisms, which have been suggested as direct sensors of the force, components of an integrated force-sensing system, or elements of the overall cellular response. Other candidate mechanisms including ligand-dependent and independent receptors and adhesion molecules have been reviewed elsewhere (Ando & Yamamoto, 2013; Conway & Schwartz, 2012; Johnson, Mather, & Wallace, 2011; Li, Haga, & Chien, 2005).

Ion channels located to the plasma membrane have long been considered as candidates for the shear stress sensor or sensors, preceding the Piezo1 discovery. As we discuss in Section 1.3, ion channels seem particularly well suited to sensing physical forces and are strongly suggested as key players in the sensing of shear stress. There are several ideas about how ion channels might be responsive to shear stress. There could be direct interactions between the physical drag of shear stress and the ion channel (although the energy required might be beyond what physiological shear stress can deliver). Ion channels could be activated indirectly because they are linked to the cytoskeleton, which becomes strained as the shear stress impacts on endothelial cell shape and the pull of the cytoskeleton then opens a gate in the ion channel. Ion channels also could be activated because their open conformation is coupled to the fluidity, structure, or composition of the cell membrane, which apparently changes in response to shear stress (Butler & Norwich, 2001; Yamamoto & Ando, 2013).

3.1 K^+ channels and Cl^- channels

Channels with selectivity for K^+ were among the first reported responders to shear stress (Hoger, Ilyin, Forsyth, & Hoger, 2002; Jacobs et al., 1995; Olesen, Clapham, & Davies, 1988) but the idea that K^+ channels are the shear stress sensor of endothelial cells is not dominant in current models. The early studies reported inward rectifier K^+ current to be stimulated by shear stress (Hoger et al., 2002; Jacobs et al., 1995; Olesen et al., 1988). The current was activated by 0.1 dyn/cm^2, so the underlying channels appear to have been very sensitive to shear stress. Furthermore, inwardly rectifying K^+ channels contribute to flow-induced vasodilation in mouse mesenteric arteries (Ahn et al., 2016). Other K^+ channels have been found to be indirectly affected by shear stress, as demonstrated by slow effects on the expression of both ATP-sensitive and Ca^{2+}-activated K^+ channels (Brakemeier, 2003; Chatterjee, Al-Mehdi, Levitan, Stevens, & Fisher, 2003); these are presumably downstream consequences of the activation of some other protein sensor. The net consequence of the K^+ channel effects is hyperpolarization of the endothelial cell membrane potential, which drives Ca^{2+} entry and promotes Ca^{2+}-dependent processes such as the generation of nitric oxide. This is consistent with observations of increased intracellular Ca^{2+} in response to shear stress as discussed in the following section. It has been suggested that Cl^- channel activation follows inward rectifier K^+ channel activation and this would be expected to reset the membrane potential (Barakat, Leaver, Pappone, & Davies, 1999).

3.2 Ca^{2+}-permeable channels

There is substantial evidence from early and recent studies that elevation of the intracellular Ca^{2+} concentration is one of the earliest events in the response to shear stress; it might even be the first event. Shear stress evokes both Ca^{2+} entry and Ca^{2+} release from intracellular stores (Ando, Komatsuda, & Kamiya, 1988; Ando, Ohtsuka, & Korenaga, 1993; Li et al., 2014; Schwarz, Droogmans, & Nilius, 1992). There are apparently direct effects of shear stress on a Ca^{2+} entry mechanism or mechanisms, a theme to which we return with Piezo1 channels. Reports by Lansman, Hallam, and Rink (1987) and Brakemeier, Eichler, Hopp, Kohler, and Hoyer (2002) are early examples describing mechanical activation of Ca^{2+}-permeable channels in endothelial cells; both studies showed the channels to be nonselective cationic channels, which means that they are not only Ca^{2+} permeable but also Na^+ permeable. Several studies since have sought to identify the molecular identity of the channels as a key step for understanding how shear stress is sensed and transduced.

P2X4 proteins assemble in trimers to form ATP-activated Ca^{2+}-permeable nonselective cationic channels (Browne, Jiang, & North, 2010). Mice with disrupted P2X4 showed striking disruption of endothelial cell response to shear stress, suggesting that P2X4 channels have a major role (Yamamoto et al., 2006). However, although shear stress has modulatory effects on channel gating (Yamamoto, Korenaga, Kamiya, & Ando, 2000) the mechanism of activation of P2X4 channels by shear stress is likely to be indirect, depending on ATP release from the cells (Yamamoto et al., 2007, 2003). The mechanism by which shear stress causes ATP release appears to be dependent on Piezo1 and mediated by pannexin channels (Wang et al., 2016).

TRPV4 proteins assemble in what are thought to be tetramers to form Ca^{2+}-permeable nonselective cationic channels, which are activated by arachidonic acid and other factors (Everaerts, Nilius, & Owsianik, 2010). As with P2X4, disruption of the *Trpv4* gene strongly inhibited responses to shear stress in mice (Hartmannsgruber et al., 2007). The TRPV4 channels apparently lack direct sensitivity to shear stress and are instead activated indirectly (Köhler & Hoyer, 2007). The factor activating the channels would seem to be arachidonic acid but, again, the mechanism by which shear stress causes arachidonic acid release is unclear. The molecular composition of these channels is also more complex than initially thought, apparently involving also TRPC1 and TRPP2 proteins within the same complex and perhaps within the tetramer itself (Du et al., 2014; Heckel et al., 2015; Ma et al., 2010).

TRPP2, also known as polycystin-2, has been suggested to have a special role in primary cilia, which occur in many cell types including apparently endothelial cells and is implicated in sensing fluid flow (AbouAlaiwi et al., 2009; Egorova, van der Heiden, Poelmann, & Hierck, 2012). A key role has been suggested for this TRPP2 in sensing shear stress and mediating Ca^{2+} and downstream nitric oxide elevation (AbouAlaiwi et al., 2009). How this TRPP2 relates to the TRPP2 involved in TRPV4 responses is unclear.

4. Piezo1 CHANNELS

4.1 Piezo1 channel properties

Functional Piezo proteins were discovered in a targeted short interfering RNA screen against mechanically activated ionic currents in the N2A neuroblastoma cell line (Coste et al., 2010); this work initially did not seem relevant to vascular biology, yet it turns out to be important. The Piezo family consists only of two members, Piezo1 and Piezo2. Both proteins confer mechanically activated nonselective cationic currents with fast activation kinetics when cells overexpressing the proteins are struck by a glass stylus (Coste et al., 2010). Channel inactivation then follows rapidly. Here, we focus on Piezo1 because of its importance in vascular development. Roles of Piezo2 are not however excluded in the vascular context as Piezo2 was recently suggested to be important in endothelium-dependent pain and tumor angiogenesis (Ferrari, Bogen, Green, & Levine, 2015; Yang et al., 2016).

Although structural modeling comparisons have suggested distant relationships to acid-sensing ion channels and P2X receptors (channels), in detail Piezo proteins bear little resemblance to other known ion channel subunits (Ge et al., 2015; Volkers, Mechioukhi, & Coste, 2014). The predicted molecular mass of Piezo1 is 286 kDa and the ion channel itself is an assembly of multiple Piezo1 proteins, so Piezo1 channels are among the largest of ion channels (Ge et al., 2015). Initially, it seemed that the channel was a tetramer (Coste et al., 2012) but cryo-electron microscopy with a structural resolution at 4.8 Å indicated a trimer, with the overall channel looking like a triblade propeller with a small hole in the middle (Ge et al., 2015). The C-termini are together at the central pore. Over the top is a large functionally important extracellular cap. Peripheral N-terminal regions extend away from the central region like blades and might be critical for sensing membrane tension (Zhao et al., 2016).

Specific inhibitors of Piezo1 are currently lacking but a frequently used inhibitor is the nonspecific agent ruthenium red, which blocks mouse Piezo1 channels with an IC_{50} of about 5 μM (Coste et al., 2012). It has an extracellular effect at negative voltages, consistent with action as a pore blocker (Coste et al., 2012). The commonly used toxin inhibitor of mechanosensitive channels, GsMTx4, also inhibits the channels (Bae, Sachs, & Gottlieb, 2011) but it might not bind Piezo1 directly, rather modulating it via the lipid bilayer (Suchyna et al., 2000, 2004). A large-scale screen of synthetic compounds revealed an apparently specific agonist for Piezo1 channels, namely Yoda1 (Syeda et al., 2015). Yoda1 did not activate Piezo2 and had no effect on Piezo1-depleted erythrocytes, consistent with it being specific for Piezo1 (Cahalan et al., 2015).

This chapter focuses on vascular functions of Piezo1 but it is important to recognize that Piezo1 is not specific to the vasculature. Roles of Piezo1 in erythrocyte integrity, cell migration, bladder urothelial pressure sensing, and collecting duct osmoregulation also exist (Cahalan et al., 2015; Li et al., 2015; Martins et al., 2016; McHugh et al., 2010; Miyamoto et al., 2014).

4.2 Discovery of Piezo1's role in shear stress sensing

Because of our interest in Ca^{2+}-permeable and nonselective cationic channels in vascular biology we knew even before the discovery of Piezo1 in N2A cells that Ca^{2+} entry was one of the earliest events after the initiation of shear stress, suggesting that a yet unidentified Ca^{2+}-entry channel might be the shear stress sensor itself. Brakemeier et al. had convincingly described a shear stress-sensitive endothelial nonselective cationic channel in 2002, but the molecular identity remained unknown (Brakemeier et al., 2002). Since we noticed that the unitary conductance of this channel was similar to that of the channel in N2A cells at about 25 pS (Coste et al., 2010), quite distinct from that of other candidate channels (P2X4 9 pS, TRPV4 60 pS, TRPC1 16 pS, TRPP2-TRPC1 40 pS), immediately we focused on determining the relevance of Piezo1 to endothelial cells and the Brakemeier channel. We had persuasive evidence within a year, submitted a comprehensive article for publication in November 2013 and it was published the following year (Li et al., 2014). A competing shorter article was submitted while ours was in late-stage review and then published within a month via the member's route of *PNAS*. It reached similar conclusions to our article although without vascular-specific knockout and other mechanistic insight (Ranade et al., 2014).

4.3 Necessity for endothelial Piezo1 after the embryonic heart beats

Many proteins are involved in shear stress responses studied in vitro where the shear stress can be controlled precisely. Some of these proteins are considered briefly in the preceding sections. Key questions are whether these proteins are able to act as shear stress sensors and if so, whether they serve such sensing roles in physiology—in the living animal. It is a challenge to answer these questions. Although we should be careful not to be too rigid in our expectations, as a starting point it seems reasonable to expect that a sensor, or component thereof, should be critical for vascular maturation in development, which depends on shear stress sensing. We can test this expectation by examining the details of embryonic vasculature in knockout mice just after the embryonic heart starts to beat and drive blood flow into the nascent endothelial lattice shortly after E8.5. Disruption of the gene encoding a shear stress sensor protein should not impair endothelial cell formation or the initial endothelial lattice, which is independent of shear stress, but should prevent the subsequent remodeling and maturation of these endothelial cells into large vessels because these processes are driven by shear stress. Such maturation is necessary to meet the demands of organ development and expansion from a specific time point, E9.5. Therefore, knockout of a key sensor protein should be embryonic lethal around E9.5. Few candidate shear stress sensors meet this expectation. Despite the data supporting

P2X4 and TRPV4 Ca^{2+}-permeable channels as key players, knockouts for these proteins do not disrupt embryonic vasculature and are not lethal (Hartmannsgruber et al., 2007; Yamamoto et al., 2006). Knockout of TRPP2 is lethal but not until E13.5 and its endothelial-specific disruption failed to yield a striking phenotype (Garcia-Gonzalez et al., 2010; Wu et al., 2000).

Strikingly, global homozygous and endothelial-specific disruption of Piezo1 both led to similar phenotypes: disrupted embryonic vasculature at E9.5 and pronounced whole embryo growth retardation by E10.5 and lethality (Li et al., 2014). Endothelial cell formation and the initial endothelial lattice were not prevented but the subsequent remodeling and maturation of the endothelial cells into large vessels were profoundly disturbed (Li et al., 2014). These features point to Piezo1 as a critical component of the endothelial shear stress sensor. Because a similar phenotype can result from embryonic heart failure we checked by ultrasound and were able to show that the heart beat was normal in Piezo1 knockouts (Li et al., 2014). Therefore the evidence favors Piezo1 as the endothelial sensor of blood flow as the embryonic heart beat drives blood flow into the nascent endothelial tubes, a sensor which then drives remodeling and vessel maturation (Fig. 1).

4.4 Requirement for Piezo1 in endothelial cell Ca^{2+} entry and nonselective cationic current

To test if Piezo1 is necessary for shear stress-evoked Ca^{2+} entry, its activity would ideally be measured in vivo in the embryo as the heart starts to beat, but this is currently beyond technical capability. Instead we isolated endothelial cells from mouse embryos, loaded them with Ca^{2+} indicator dye, and applied shear stress in vitro. In this way we could indeed show shear stress-evoked Ca^{2+} entry and, importantly, its dependence on Piezo1 (Li et al., 2014). Similar observations of Piezo1-dependent shear stress-evoked Ca^{2+} entry were made in human umbilical vein endothelial cells and human liver endothelial cells from patients, suggesting common principles in humans and adult endothelial cells (Li et al., 2014).

An advantage of in vitro studies is that patch-clamp techniques can be applied for measuring transmembrane ionic currents directly. These studies revealed that the native endothelial cell Piezo1-dependent mechanism was an ion channel with unitary conductance of 25 pS, linear current–voltage relationship, and Cl^--independent reversal potential close to 0 mV (Li et al., 2014). These are expected characteristics of Piezo1 channels.

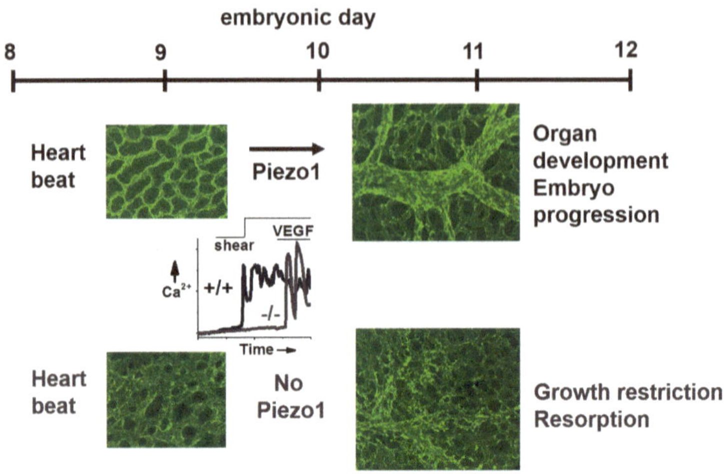

Figure 1 Timeline of murine yolk sac vascular development with and without Piezo1. Images of yolk sacs at ~E9.5 on the left and ~E10.5 on the right stained for CD31, a marker of endothelial cells. Inset is a plot of intracellular Ca^{2+} in embryonic endothelial cells isolated from wild-type (+/+) and Piezo1 knockout (−/−) mice in response to shear stress. Piezo1−null embryonic endothelial cells did not respond to shear stress but were viable and responsive to VEGF. After initiation of the heartbeat the yolk sac vasculature showed remodeling in Piezo1-containing vessels allowing for the progression of organ development and survival. Without Piezo1 the vasculature failed to remodel and there was growth restriction and resorption. *Adapted from Li, J., Hou, B., Tumova, S., Muraki, K., Bruns, A., Ludlow, M. J, … Beech, D. J. (2014). Piezo1 integration of vascular architecture with physiological force.* Nature, 515.

Unexpected was the relatively slow activation rate in response to mechanical strain applied to membrane patches and the apparent absence of inactivation (Li et al., 2014). Our hypothesis is that such slow kinetics reflect modulation of endogenous endothelial Piezo1 channels by another protein, the nature of which remains to be identified.

4.5 The potential simplicity of Piezo1 as a stand-alone sensor

The study of endothelial shear stress has been plagued by mystique shrouding the nature of the force sensor itself. Does sensing akin to how a receptor senses a ligand exist for sensing of the shear stress force? Does a specific protein sense the force or is force-sensing the property of a system, such as an architecture of cooperating proteins and cofactors? A multitude of mechanisms has been proposed, entailing a long list of candidate sensors—proteins, lipids, and complexes thereof. One approach to this multitude is to consider that all or at least many of the candidate sensors are indeed sensors or components of sensors—that multiple hypotheses are true and that various mechanisms integrate together; confer backup systems; fine-tune and adapt; or enable detection of a spectrum of different shear stresses in terms of amplitude, directionality, and degrees of chaos. The old philosophy of Ockham's razor nevertheless encourages simplicity, in which case only one of the mechanisms or a small number would be considered correct ("entities must not be multiplied beyond necessity"). Does the discovery of Piezo1 in endothelial shear stress sensing enable us to simplify—not in terms of what can be involved in the overall response to shear stress or what is possible in vitro but in terms of the nature of the physiological shear stress sensor in vivo?

Ion channels have emerged as proteins, which are particularly well suited to sensing physical parameters such as transmembrane voltage, body temperature, and innocuous and noxious environmental temperatures. This suitability appears to extend to mechanical force parameters, as elegantly reviewed recently (Ranade, Syeda, & Patapoutian, 2015). Therefore, our starting point is the simplest of hypotheses that Piezo1 directly senses shear stress and is the endothelium's only shear stress sensor.

If Piezo1 does indeed directly sense shear stress, it might be possible to reconstitute shear stress sensing by Piezo1 alone. We do not yet know if this is possible but overexpression of Piezo1 in HEK 293 cells was sufficient to confer shear stress-evoked Ca^{2+} entry and nonselective cationic current on these otherwise shear stress-resistant cells (Li et al., 2014; Ranade et al., 2014). These cells of course contained endogenous proteins and other factors, so they were not null backgrounds. The endogenous factors may have supported Piezo1 and acted concertedly with Piezo1 to confer shear stress sensitivity. It is notable that high shear stress was needed to evoke Piezo1-dependent nonselective cationic currents in HEK 293 cells (Ranade et al., 2014), more than is needed to evoke responses in endothelial cells (Li et al., 2014). This difference could be due to an important cofactor being in limited supply in HEK 293 cells. Consistent with this thinking we have since found that Ca^{2+} entry responses are not reliably reconstituted by Piezo1 in all batches of HEK 293 cells, which could be explained by different batches of HEK 293 cells containing different amounts of the putative cofactor (Hyman, unpublished observations). Moreover, reconstitution of Piezo1 into lipid bilayers revealed constitutively active channels, which were apparently not mechanically sensitive (Coste et al., 2012). These observations suggest that Piezo1 is probably not sufficient alone as a mechanical force sensor but they do not rule out Piezo1 alone as the direct sensor and transducer because Piezo1 might simply need a certain environment, a particular context, to be tuned for mechanical force-sensing. The environment might be a collection of local factors, which set the initial opening probability of Piezo1 channels. These factors might be lipids, or cytoskeletal or other proteins. If the opening probability of the channels is initially too low or too high, the channels might not respond to the mechanical force or they might respond but show no functional readout such as a change in ionic current or Ca^{2+} entry.

4.6 Challenges for the Piezo1 sensor hypothesis

One challenge, and not only for the Piezo1 hypothesis, is the presence of glycocalyx (Reitsma, Slaaf, Vink, van Zandvoort, & oude Egbrink, 2007; Tarbell, Simon, & Curry, 2014). This glycoprotein–polysaccharide structure on the luminal surface of endothelium reaches at least 100 nm into the extracellular space, coating the cells with an apparent barrier between flowing fluid and the membrane proteins and lipid bilayer (Reitsma et al., 2007; Tarbell et al., 2014). In this situation, how can Piezo1 detect the fluid flow? Of course one explanation might be that our Piezo1 hypothesis is wrong and that the shear stress is actually sensed by a protein, which is a component of the glycocalyx, such as a syndecan. The strengths and weaknesses of the glycocalyx hypothesis have been reviewed (Reitsma et al., 2007; Tarbell et al., 2014) but it is notable that syndecan-1 and syndecan-4 knockout mice are viable and, to the best of our knowledge, no genetic manipulation of the glycocalyx is lethal. Syndecan-4 is nevertheless important for alignment of endothelial cells to the direction of shear stress (Baeyens et al., 2014). Mechanical effects of shear stress on the glycocalyx might be transmitted to underlying membrane proteins such as Piezo1, itself a glycoprotein.

Another challenge is the emerging evidence that Piezo1 channels are sensors of membrane tension (Cox et al., 2016; Lewis & Grandl, 2015). If this is the only mechanical-sensing capability of Piezo1, does it mean that shear stress is detected indirectly via changes in membrane tension? Indeed, rapid changes in the order of membrane lipids have been suggested in endothelial cells responding to shear stress (Yamamoto & Ando, 2013). Alternatively, shear stress might increase tension in the bilayer via leverage or compression mediated through the glycocalyx. These different models should be tested experimentally.

4.7 How might activation of Piezo1 channels lead to vascular remodeling?

We are quite far from piecing together how Piezo1 channel activation ultimately results in the complex and orchestrated process of vascular remodeling in embryonic development. But it is striking that even in vitro Piezo1 activation is necessary for alignment of endothelial cells to the direction of shear stress. How this directional repositioning can result from activation of plasma membrane Ca^{2+} entry channels is currently unknown.

As a route to obtaining insight into the downstream systems, we used titanium dioxide trapping coupled with mass spectrometry (phosphoproteomics) to identify regulated proteins affected by Piezo1 depletion in vitro under static and shear stress conditions and in vivo in the embryo (Li et al., 2014). In silico pathway analysis of the arising data highlighted clusters of proteins from actin cytoskeleton and focal adhesions. Relevance of these proteins was also indicated by denser actin structures in Piezo1-depleted cells and accumulation of Piezo1–GFP at the leading apical lamellipodia where focal adhesion turnover becomes important as the endothelial cell adjusts to achieve alignment (Li et al., 2002). We were particularly struck by significant effects of Piezo1 depletion on calpain-2 and many of its known substrates, which are important in the structure of the actin cytoskeleton and focal adhesions. We therefore speculated that a calpain-2 system might be downstream of Piezo1, consistent with calpain-2 as a Ca^{2+}-activated proteolytic enzyme, calpain's known association with Piezo1 in HeLa cells, and calpain's suggested roles in focal adhesion turnover and endothelial cell alignment to shear stress (McHugh et al., 2010; Miyazaki, Honda, & Ohata, 2007). Intriguingly, knockout of a critical regulator of calpain-2, calpain small subunit 1, is known to disturb vascular development at E10.5 (Arthur, Elce, Hegadorn, Williams, & Greer, 2000). Consistent with this calpain hypothesis, we found that calpain activity was significantly lower in Piezo1-depleted embryos, shear stress-evoked increases in calpain activity were abolished by the GsMTx4 inhibitor of Piezo1 activity, and Piezo1–GFP localized to dissolving focal adhesions at the trailing edge of the cell as shear stress was applied (Li et al., 2014). The importance of Piezo1-mediated Ca^{2+} entry and downstream Ca^{2+} activation of calpain-2 was further indicated by the sensitivity of endothelial cell alignment to the absence of extracellular Ca^{2+} or the presence of calpain inhibitors (Li et al., 2014). These observations led us to hypothesize that shear stress-activated Piezo1 drives vascular remodeling partly by enabling localized Ca^{2+} entry, which activates calpain-2 to cleave actin cytoskeletal and focal adhesion proteins, loosening up the cell to move at points where the Piezo1 channels are most active (Fig. 2).

4.8 Mutations in human *PIEZO1* gene

A small number of homozygous and compound heterozygous mutations in the *PIEZO1* gene have been associated with generalized lymphatic dysplasia with a high incidence of nonimmune hydrops fetalis and childhood onset of facial and four limb lymphoedema (Fotiou et al., 2015). The effects of the *PIEZO1* mutations were most severe prenatally. Another group identified a splicing variant leading to early truncation and a nonsynonymous missense variant associated with persistent lymphoedema in congenital lymphatic dysplasia (Lukacs et al., 2015). Lukacs et al. noted that the phenotype was

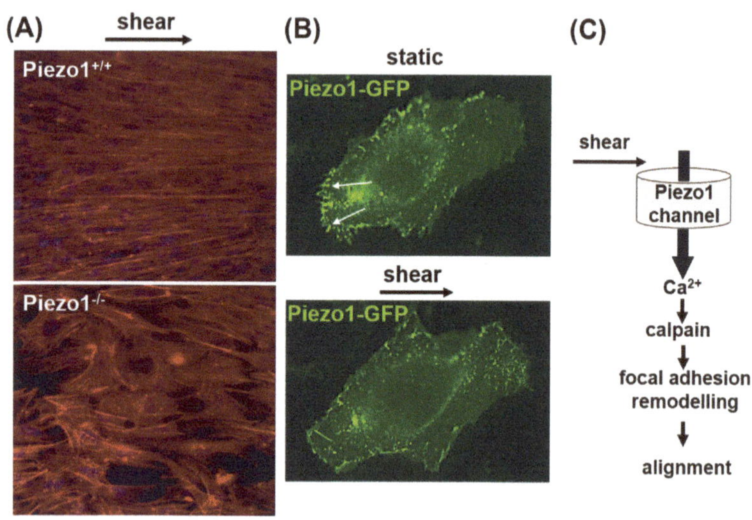

Figure 2 Calpain hypothesis for Piezo1-dependent endothelial cell remodeling in response to shear stress. (A) Embryonic endothelial cells isolated from Piezo1-null ($Piezo1^{-/-}$) and wild-type ($Piezo1^{+/+}$) mice. The actin cytoskeleton was stained using rhodamine–phalloidin [red] and nuclei were stained blue. $Piezo1^{+/+}$ endothelial cells displayed alignment to shear stress whereas $Piezo1^{-/-}$ cells did not. (B) Human umbilical vein endothelial cells transfected with Piezo1–GFP [green]. Exposure to shear stress caused dissipation of Piezo1–GFP-positive focal adhesions at the rear of the cell (*white arrows* in the static condition). (C) Diagrammatic representation of the pathway suggested to exist between shear stress and focal adhesions. *Adapted from Li, J., Hou, B., Tumova, S., Muraki, K., Bruns, A., Ludlow, M. J., … Beech, D. J. (2014). Piezo1 integration of vascular architecture with physiological force.* Nature, 515.

milder than they expected based on the Piezo1 knockout mouse studies, suggesting either that residual protein function or compensatory effects could be sufficient to maintain vascular development in humans. In the Fotiou et al. study, however, death in utero was observed, which is, at least superficially, similar to the murine phenotype. Nevertheless, both studies broadly support the findings from mice and cultured human cells, which suggest a major role of Piezo1 in vascular development. In humans this role seems to be most prominent in lymphatic vascular development, which like other types of vascular development is thought to depend strongly on the detection of shear stress.

4.9 Relationships of Piezo1 to other candidate shear stress sensors

A prominent concept in the field is that force impacting on CD31 (PECAM-1) in homo- or heterophilic interactions activates a Src kinase, which recruits VEGFR2 in the presence of vascular endothelial-cadherin to deliver downstream effects such as activation of integrin, phosphoinositide-3-kinase, and endothelial nitric oxide synthase (Conway & Schwartz, 2012; Tzima et al., 2005). The pivot in this model is CD31, yet CD31 knockout mice are alive and do not exhibit disrupted embryonic vasculature (Conway & Schwartz, 2012; Duncan et al., 1999). The phenotype of these mice is relatively mild, although there are indications of changes in shear stress sensing in adults (Bagi et al., 2005). These findings show that CD31 is not a critical shear stress sensor in embryonic vasculature and that shear stress sensing can occur efficiently without CD31. A role for CD31 might of course be compensated in the embryo and CD31 might have roles, which are different in the adult. Our hypotheses are that Piezo1 is the shear stress sensor, which was previously unrecognized in the earlier CD31 studies and that CD31 is not a direct sensor but an important component of a signaling network, which is somehow intimately linked to Piezo1. Because Piezo1 has only recently been discovered, there is still much research to do to understand how it might relate to CD31 and other players in the shear stress response.

As described earlier, TRPP2 is an another protein implicated in shear stress sensing. Intriguingly, it has also been described as an inhibitor of Piezo1 channels in renal epithelial cells (Peyronnet et al., 2013). Whether it has such a role in endothelial cells and shear stress sensing is unclear. TRPP2 was a hit in our phosphoproteomic studies of Piezo1-depleted human umbilical vein endothelial cells (Li et al., 2014), showing about

25% downregulation in confirmation studies; we could not however find localization of Piezo1–GFP to the primary cilium (Tumova, unpublished data). Ca^{2+} signaling in primary cilium mechanosensation was recently disputed along with the capability of primary cilia to sense mechanical stimuli (Delling et al., 2016).

5. CONCLUSIONS AND FUTURE DIRECTIONS

Piezo1 was first identified in activated astrocytes associated with senile plaques (Satoh et al., 2006), a Chinese hamster ovary cell screen for modulators of integrin signaling (McHugh et al., 2010) and a neuroblastoma cell screen for mediators of ionic current evoked by mechanical hitting (Coste et al., 2010). These do not seem fertile grounds for transforming our understanding of vascular biology. Yet Piezo1 is undoubtedly now recognized as a major new player in shear stress sensing and vascular development: independent reports describe its critical role in murine vascular development and others strongly suggest its importance in human vascular development—most obviously in lymphatic vasculature. These roles of Piezo1 exist because of its endothelial cell expression. Piezo1 channel is arguably the strongest candidate for the long-sought Ca^{2+}-permeable ion channel, which detects shear stress and triggers a cascade of downstream effects. Importantly, there is a case for Piezo1 channels being directly mechanically sensitive, albeit almost certainly requiring and being modulated by cellular context. Functionality in isolation would be too much to expect, so interactions with other proteins and factors will likely be increasingly revealed through further study. There is also much to be learned about the relevance of Piezo1 across all vascular physiology and pathophysiology, not just in the restricted world of vascular development. Because of lethality in the global and endothelial-specific Piezo1 knockouts there is a need to induce Piezo1 disruption in the adult mouse to circumvent developmental complications. Such studies hold much promise and we can look forward to the outcomes here too. Last but not least, further Piezo1 studies may pave the way for therapies in rare but debilitating conditions such as nonimmune hydrops fetalis. Piezo1 has brought important new perspectives to the fields of shear stress sensing and vascular development and it promises much more in the future.

ACKNOWLEDGMENTS

We are grateful to the Medical Research Council, Biotechnology and Biological Sciences Research Council, and the Wellcome Trust for research support.

REFERENCES

AbouAlaiwi, W., Takahashi, M., Mell, B., Jones, T., Ratnam, S., Kolb, R., & Nauli, S. (2009). Ciliary polycystin-2 is a mechanosensitive calcium channel involved in nitric oxide signaling cascades. *Circulation Research, 104*(7), 860–869.

Ahn, S. J., Fancher, I. S., Bian, J.-T., Zhang, C. X., Schwab, S., Gaffin, R., ... Levitan, I. (2016). Inwardly-rectifying K^+ channels are major contributors to flow-induced vasodilation in resistance arteries. *Journal of Physiology*. http://dx.doi.org/10.1113/JP273255 (accepted author manuscript).

Ando, J., Komatsuda, T., & Kamiya, A. (1988). Cytoplasmic calcium response to fluid shear stress in cultured vascular endothelial cells. *In Vitro Cellular & Developmental Biology, 24*(9), 871–877.

Ando, J., Ohtsuka, A., & Korenaga, R. (1993). Wall shear stress rather than shear rate regulates cytoplasmic Ca^{++} responses to flow in vascular endothelial cells. *Biochemical and Biophysical Research Communications, 190*(3), 716–723.

Ando, J., & Yamamoto, K. (2013). Flow detection and calcium signalling in vascular endothelial cells. *Cardiovascular Research, 99*(2), 260–268.

Arthur, J. S. C., Elce, J. S., Hegadorn, C., Williams, K., & Greer, P. A. (2000). Disruption of the murine calpain small subunit gene, Capn4: Calpain is essential for embryonic development but not for cell growth and division. *Molecular and Cellular Biology, 20*(12), 4474–4481.

Bae, C., Sachs, F., & Gottlieb, P. (2011). The mechanosensitive ion channel Piezo1 is inhibited by the peptide GsMTx4. *Biochemistry, 50*(29), 6295–6300.

Baeyens, N., Mulligan-Kehoe, M. J., Corti, F., Simon, D. D., Ross, T. D., Rhodes, J. M., ... Schwartz, M. A. (2014). Syndecan 4 is required for endothelial alignment in flow and atheroprotective signaling. *Proceedings of the National Academy of Sciences of the United States of America, 111*(48), 17308–17313.

Bagi, Z., Frangos, J. A., Yeh, J.-C., White, C. R., Kaley, G., & Koller, A. (2005). PECAM-1 mediates NO-dependent dilation of arterioles to high temporal gradients of shear stress. *Arteriosclerosis, Thrombosis, and Vascular Biology, 25*(8), 1590–1595.

Barakat, A., Leaver, E., Pappone, P., & Davies, P. (1999). A flow-activated chloride-selective membrane current in vascular endothelial cells. *Circulation Research, 85*(9), 820–828.

Brakemeier, S. (2003). Shear stress-induced up-regulation of the intermediate-conductance Ca^{2+}-activated K^+ channel in human endothelium. *Cardiovascular Research, 60*(3), 488–496.

Brakemeier, S., Eichler, I., Hopp, H., Kohler, R., & Hoyer, J. (2002). Up-regulation of endothelial stretch-activated cation channels by fluid shear stress. *Cardiovascular Research, 53*(1), 209–218.

Browne, L. E., Jiang, L.-H., & North, R. A. (2010). New structure enlivens interest in P2X receptors. *Trends in Pharmacological Sciences, 31*(5), 229–237.

Butler, P., & Norwich, G. (2001). Shear stress induces a time-and position-dependent increase in endothelial cell membrane fluidity. *American Journal of Physiology. Cell Physiology, 280*, C962–C969.

Cahalan, S. M., Lukacs, V., Ranade, S. S., Chien, S., Bandell, M., & Patapoutian, A. (2015). Piezo1 links mechanical forces to red blood cell volume. *eLife, 4*, e07370.

Caprioli, A., Minko, K., Drevon, C., Eichmann, A., Dieterlen-Lièvre, F., & Jaffredo, T. (2001). Hemangioblast commitment in the avian allantois: Cellular and molecular aspects. *Developmental Biology, 238*(1), 64–78.

Chatterjee, S., Al-Mehdi, A.-B., Levitan, I., Stevens, T., & Fisher, A. B. (2003). Shear stress increases expression of a KATP channel in rat and bovine pulmonary vascular endothelial cells. *American Journal of Physiology. Cell Physiology, 285*(4), C959–C967.

Chiu, J., & Chien, S. (2011). Effects of disturbed flow on vascular endothelium: Pathophysiological basis and clinical perspectives. *Physiological Reviews, 91*(1), 327–387.

Conway, D., & Schwartz, M. A. (2012). Lessons from the endothelial junctional mechanosensory complex. *F1000 Biology Reports, 4*, 1.

Coste, B., Mathur, J., Schmidt, M., Earley, T. J., Ranade, S., Petrus, M. J., … Patapoutian, A. (2010). Piezo1 and Piezo2 are essential components of distinct mechanically activated cation channels. *Science, 330*(6000), 55–60.

Coste, B., Xiao, B., Santos, J. S., Syeda, R., Grandl, J., Spencer, K. S., … Patapoutian, A. (2012). Piezo proteins are pore-forming subunits of mechanically activated channels. *Nature, 483*(7388), 176–181.

Coultas, L., Chawengsaksophak, K., & Rossant, J. (2005). Endothelial cells and VEGF in vascular development. *Nature, 438*(7070), 937–945.

Cox, C. D., Bae, C., Ziegler, L., Hartley, S., Nikolova-Krstevski, V., Rohde, P. R., … Martinac, B. (2016). Removal of the mechanoprotective influence of the cytoskeleton reveals PIEZO1 is gated by bilayer tension. *Nature Communications, 7*, 10366.

Cunningham, K. S., & Gotlieb, A. I. (2005). The role of shear stress in the pathogenesis of atherosclerosis. *Laboratory Investigation, 85*(1), 9–23.

Delling, M., Indzhykulian, A. A., Liu, X., Li, Y., Xie, T., Corey, D. P., & Clapham, D. E. (2016). Primary cilia are not calcium-responsive mechanosensors. *Nature, 531*(7596), 656–660.

Du, J., Ma, X., Shen, B., Huang, Y., Birnbaumer, L., & Yao, X. (2014). TRPV4, TRPC1, and TRPP2 assemble to form a flow-sensitive heteromeric channel. *FASEB Journal*, 1–9.

Duncan, G. S., Andrew, D. P., Takimoto, H., Kaufman, S. A., Yoshida, H., Spellberg, J., … Mak, T. W. (1999). Genetic evidence for functional redundancy of platelet/endothelial cell adhesion molecule-1 (PECAM-1): CD31-deficient mice reveal PECAM-1-dependent and PECAM-1-independent functions. *Journal of Immunology, 162*(5), 3022–3030.

Egorova, A. D., van der Heiden, K., Poelmann, R. E., & Hierck, B. P. (2012). Primary cilia as biomechanical sensors in regulating endothelial function. *Differentiation, 83*(2), S56–S61.

Everaerts, W., Nilius, B., & Owsianik, G. (2010). The vanilloid transient receptor potential channel TRPV4: From structure to disease. *Progress in Biophysics and Molecular Biology, 103*(1), 2–17.

Ferrari, L. F., Bogen, O., Green, P., & Levine, J. D. (2015). Contribution of Piezo2 to endothelium-dependent pain. *Molecular Pain, 11*, 65.

Fotiou, E., Martin-Almedina, S., Simpson, M. A., Lin, S., Gordon, K., Brice, G., … Ostergaard, P. (2015). Novel mutations in PIEZO1 cause an autosomal recessive generalized lymphatic dysplasia with non-immune hydrops fetalis. *Nature Communications, 6*, 8085.

Garcia-Gonzalez, M. A., Outeda, P., Zhou, Q., Zhou, F., Menezes, L. F., Qian, F., … Watnick, T. (2010). Pkd1 and Pkd2 are required for normal placental development. *PLoS One, 5*(9).

Ge, J., Li, W., Zhao, Q., Li, N., Chen, M., Zhi, P., … Yang, M. (2015). Architecture of the mammalian mechanosensitive Piezo1 channel. *Nature, 527*.

Hartmannsgruber, V., Heyken, W.-T. T., Kacik, M., Kaistha, A., Grgic, I., Harteneck, C., … Köhler, R. (2007). Arterial response to shear stress critically depends on endothelial TRPV4 expression. *PLoS One, 2*(9), e827.

He, C., Hu, H., Braren, R., Fong, S.-Y., Trumpp, A., Carlson, T. R., & Wang, R. A. (2008). c-myc in the hematopoietic lineage is crucial for its angiogenic function in the mouse embryo. *Development, 135*(14), 2467–2477.

Heckel, E., Boselli, F., Roth, S., Krudewig, A., Belting, H.-G., Charvin, G., & Vermot, J. (2015). Oscillatory flow modulates mechanosensitive klf2a expression through trpv4 and trpp2 during heart valve development. *Current Biology, 25*(10), 1354–1361.

Hoger, J. H., Ilyin, V. I., Forsyth, S., & Hoger, A. (2002). Shear stress regulates the endothelial Kir2.1 ion channel. *Proceedings of the National Academy of Sciences of the United States of America, 99*(11), 7780–7785.

Huang, C., Sheikh, F., Hollander, M., Cai, C., Becker, D., Chu, P.-H., … Chen, J. (2003). Embryonic atrial function is essential for mouse embryogenesis, cardiac morphogenesis and angiogenesis. *Development, 130*(24), 6111–6119.

Jacobs, E. R., Cheliakine, C., Gebremedhin, D., Birks, E. K., Davies, P. F., & Harder, D. R. (1995). Shear activated channels in cell-attached patches of cultured bovine aortic endothelial cells. *Pflügers Archiv, 431*(1), 129–131.

Johnson, B. D., Mather, K. J., & Wallace, J. P. (2011). Mechanotransduction of shear in the endothelium: Basic studies and clinical implications. *Vascular Medicine, 16*(5), 365–377.

Köhler, R., & Hoyer, J. (2007). *Role of TRPV4 in the mechanotransduction of shear stress in endothelial cells*. CRC Press/Taylor & Francis.

Lansman, J., Hallam, T., & Rink, T. (1987). Single stretch-activated ion channels in vascular endothelial cells as mechanotransducers? *Nature, 325*(6107), 811–813.

Lewis, A. H., & Grandl, J. (2015). Mechanical sensitivity of Piezo1 ion channels can be tuned by cellular membrane tension. *eLife, 4*.

Li, C., Rezania, S., Kammerer, S., Sokolowski, A., Devaney, T., Gorischek, A., … Schreibmayer, W. (2015). Piezo1 forms mechanosensitive ion channels in the human MCF-7 breast cancer cell line. *Scientific Reports, 5*, 8364.

Li, J., Hou, B., Tumova, S., Muraki, K., Bruns, A., Ludlow, M. J., … Beech, D. J. (2014). Piezo1 integration of vascular architecture with physiological force. *Nature, 515*.

Li, S., Butler, P., Wang, Y., Hu, Y., Han, D. C., Usami, S., … Chien, S. (2002). The role of the dynamics of focal adhesion kinase in the mechanotaxis of endothelial cells. *Proceedings of the National Academy of Sciences of the United States of America, 99*(6), 3546–3551.

Li, Y.-S. J., Haga, J. H., & Chien, S. (2005). Molecular basis of the effects of shear stress on vascular endothelial cells. *Journal of Biomechanics, 38*(10), 1949–1971.

Lucitti, J. L., Jones, E. A. V., Huang, C., Chen, J., Fraser, S. E., & Dickinson, M. E. (2007). Vascular remodeling of the mouse yolk sac requires hemodynamic force. *Development, 134*(18), 3317–3326.

Lukacs, V., Mathur, J., Mao, R., Bayrak-Toydemir, P., Procter, M., Cahalan, S. M., … Krock, B. L. (2015). Impaired PIEZO1 function in patients with a novel autosomal recessive congenital lymphatic dysplasia. *Nature Communications, 6*, 8329.

Ma, X., Qiu, S., Luo, J., Ma, Y., Ngai, C.-Y., Shen, B., … Yao, X. (2010). Functional role of vanilloid transient receptor potential 4-canonical transient receptor potential 1 complex in flow-induced Ca^{2+} influx. *Arteriosclerosis, Thrombosis, and Vascular Biology, 30*(4), 851–858.

Martins, J. R., Penton, D., Peyronnet, R., Arhatte, M., Moro, C., Picard, N., … Demolombe, S. (2016). Piezo1-dependent regulation of urinary osmolarity. *Pflugers Archiv, 468*.

McHugh, B. J., Buttery, R., Lad, Y., Banks, S., Haslett, C., & Sethi, T. (2010). Integrin activation by Fam38A uses a novel mechanism of R-Ras targeting to the endoplasmic reticulum. *Journal of Cell Science, 123*(Pt 1), 51–61.

Miyamoto, T., Mochizuki, T., Nakagomi, H., Kira, S., Watanabe, M., Takayama, Y., … Tominaga, M. (2014). Functional role for Piezo1 in stretch-evoked Ca^{2+} influx and ATP release in urothelial cell cultures. *The Journal of Biological Chemistry, 289*(23), 16565–16575.

Miyazaki, T., Honda, K., & Ohata, H. (2007). Requirement of Ca^{2+} influx- and phosphatidylinositol 3-kinase-mediated m-calpain activity for shear stress-induced endothelial cell polarity. *American Journal of Physiology. Cell Physiology, 293*(4), C1216–C1225.

Olesen, S., Clapham, D., & Davies, P. (1988). Haemodynamic shear stress activates a K^+ current in vascular endothelial cells. *Nature, 331*(6152), 168–170.

Peyronnet, R., Martins, J. R., Duprat, F., Demolombe, S., Arhatte, M., Jodar, M., ... Patel, A. (2013). Piezo1-dependent stretch-activated channels are inhibited by polycystin-2 in renal tubular epithelial cells. *EMBO Reports, 14*(12), 1143–1148.

Ranade, S. S., Syeda, R., & Patapoutian, A. (2015). Mechanically activated ion channels. *Neuron, 87*(6), 1162–1179.

Ranade, S. S., Qiu, Z., Woo, S.-H., Hur, S. S., Murthy, S. E., Cahalan, S. M., ... Patapoutian, A. (2014). Piezo1, a mechanically activated ion channel, is required for vascular development in mice. *Proceedings of the National Academy of Sciences of the United States of America, 111*(28), 10347–10352.

Reitsma, S., Slaaf, D. W., Vink, H., van Zandvoort, M. A. M. J., & oude Egbrink, M. G. A. (2007). The endothelial glycocalyx: Composition, functions, and visualization. *Pflügers Archiv, 454*(3), 345–359.

Satoh, K., Hata, M., Takahara, S., Tsuzaki, H., Yokota, H., Akatsu, H., ... Yamada, T. (2006). A novel membrane protein, encoded by the gene covering KIAA0233, is transcriptionally induced in senile plaque-associated astrocytes. *Brain Research, 1108*(1), 19–27.

Schwarz, G., Droogmans, G., & Nilius, B. (1992). Shear stress induced membrane currents and calcium transients in human vascular endothelial cells. *Pflügers Archiv, 421*(4), 394–396.

Suchyna, T., Johnson, J., Hamer, K., Leykam, J., Gage, D., Clemoc, H., ... Sachs, F. (2000). Identification of a peptide toxin from *Grammostola spatulata* spider venom that blocks cation-selective stretch-activated channels. *The Journal of General Physiology, 115*(5), 583–598.

Suchyna, T. M., Tape, S. E., Koeppe, R. E., Andersen, O. S., Sachs, F., & Gottlieb, P. A. (2004). Bilayer-dependent inhibition of mechanosensitive channels by neuroactive peptide enantiomers. *Nature, 430*(6996), 235–240.

Syeda, R., Xu, J., Dubin, A. E., Coste, B., Mathur, J., Huynh, T., ... Patapoutian, A. (2015). Chemical activation of the mechanotransduction channel Piezo1. *eLife, 4*.

Tarbell, J. M., Simon, S. I., & Curry, F.-R. E. (2014). Mechanosensing at the vascular interface. *Annual Review of Biomedical Engineering, 16*(1), 505–532.

Tzima, E., Irani-Tehrani, M., Kiosses, W. B., Dejana, E., Schultz, D. A., Engelhardt, B., ... Schwartz, M. A. (2005). A mechanosensory complex that mediates the endothelial cell response to fluid shear stress. *Nature, 437*(7057), 426–431.

Volkers, L., Mechioukhi, Y., & Coste, B. (2014). Piezo channels: From structure to function. *Pflugers Archiv, 467*.

Wakimoto, K., Kobayashi, K., Kuro-O, M., Yao, A., Iwamoto, T., Yanaka, N., ... Komuro, I. (2000). Targeted disruption of Na^+/Ca^{2+} exchanger gene leads to cardiomyocyte apoptosis and defects in heartbeat. *The Journal of Biological Chemistry, 275*(47), 36991–36998.

Wang, S., Chennupati, R., Kaur, H., Iring, A., Wettschureck, N., & Offermanns, S. (2016). Endothelial cation channel PIEZO1 controls blood pressure by mediating flow-induced ATP release. *Journal of Clinical Investigation, 126*(12), 4527–4536. http://dx.doi.org/10.1172/JCI87343.

Wu, G., Markowitz, G. S., Li, L., D'Agati, V. D., Factor, S. M., Geng, L., ... Somlo, S. (2000). Cardiac defects and renal failure in mice with targeted mutations in Pkd2. *Nature Genetics, 24*(1), 75–78.

Yamaguchi, T. P., Dumont, D. J., Conlon, R. A., Breitman, M. L., & Rossant, J. (1993). flk-1, an flt-related receptor tyrosine kinase is an early marker for endothelial cell precursors. *Development, 118*(2), 489–498.

Yamamoto, K., & Ando, J. (2013). Endothelial cell and model membranes respond to shear stress by rapidly decreasing the order of their lipid phases. *Journal of Cell Science, 126*(Pt 5), 1227–1234.

Yamamoto, K., Korenaga, R., Kamiya, A., & Ando, J. (2000). Fluid shear stress activates Ca^{2+} influx into human endothelial cells via P2X4 purinoceptors. *Circulation Research, 87*(5), 385–391.

Yamamoto, K., Shimizu, N., Obi, S., Kumagaya, S., Taketani, Y., Kamiya, A., & Ando, J. (2007). Involvement of cell surface ATP synthase in flow-induced ATP release by vascular endothelial cells. *American Journal of Physiology. Heart and Circulatory Physiology, 293*(3), 1646–1653.

Yamamoto, K., Sokabe, T., Matsumoto, T., Yoshimura, K., Shibata, M., Ohura, N., … Ando, J. (2006). Impaired flow-dependent control of vascular tone and remodeling in P2X4-deficient mice. *Nature Medicine, 12*(1), 133–137.

Yamamoto, K., Sokabe, T., Ohura, N., Nakatsuka, H., Kamiya, A., & Ando, J. (2003). Endogenously released ATP mediates shear stress-induced Ca^{2+} influx into pulmonary artery endothelial cells. *American Journal of Physiology. Heart and Circulatory Physiology, 285*(2), H793–H803.

Yang, H., Liu, C., Zhou, R.-M., Yao, J., Li, X.-M., Shen, Y., … Jiang, Q. (2016). Piezo2 protein: A novel regulator of tumor angiogenesis and hyperpermeability. *Oncotarget, 7.*

Zhao, Q., Wu, K., Geng, J., Chi, S., Wang, Y., Zhi, P., … Xiao, B. (2016). Ion permeation and mechanotransduction mechanisms of mechanosensitive Piezo channels. *Neuron, 89*(6), 1248–1263.

CHAPTER THREE

Origin of the Force: The Force-From-Lipids Principle Applied to Piezo Channels

C.D. Cox*[,§], N. Bavi*[,§] and B. Martinac*[,§,1]

*Victor Chang Cardiac Research Institute, Darlinghurst, NSW, Australia
[§]University of New South Wales, Darlinghurst, NSW, Australia
[1]Corresponding Author: E-mail: b.martinac@victorchang.edu.au

Contents

Abstract

Piezo channels are a ubiquitously expressed, principal type of molecular force sensor in eukaryotes. They enable cells to decode a myriad of physical stimuli and are essential components of numerous mechanosensory processes. Central to their physiological role is the ability to change conformation in response to mechanical force. Here we

Current Topics in Membranes, Volume 79
ISSN 1063-5823
http://dx.doi.org/10.1016/bs.ctm.2016.09.001

discuss the evolutionary origin of Piezo in relation to other MS channels in addition to the force that gates Piezo channels. In particular, we discuss whether Piezo channels are inherently mechanosensitive in accordance with the *force-from-lipid* paradigm which has been firmly established for bacterial MS channels and two-pore domain K^+ (K_{2P}) channels. We also discuss the evidence supporting a reliance on or direct interaction with structural scaffold proteins of the cytoskeleton and extracellular matrix according to the *force-from-filament* principle. In doing so, we explain the false dichotomy that these distinctions represent. We also discuss the possible unifying models that shed light on channel mechanosensitivity at the molecular level.

1. INTRODUCTION

Mechanotransduction is an ancient sensing mechanism involving mechanosensitive (MS) ion channels, whose function as highly efficient biological force-sensing macromolecules, is tightly coupled to the mechanics of their surrounding cell membrane. Together with the cytoskeleton, integrins, G-protein-coupled receptors and enzymes they represent the major molecular mechanosensors tasked with transducing the mechanical stimuli exerted on membranes of living cells into electrical and/or chemical intracellular signals (Hamill & Martinac, 2001; Ingber, 2006). In eukaryotes, the speed with which these channels can transduce electrical signals allows them to play a central role in sensory physiology, including being integral in the higher processes of touch, hearing, fluid balance and blood pressure regulation (Martinac, 2004). Force induced conformational changes in these channels result in electrical signal transduction mediated via ion fluxes determined by both channel properties and electrochemical gradients. The studies of MS ion channels that have been conducted over the last 30 or more years, from their conception (Corey & Hudspeth, 1979; Katz, 1950) to proof of their existence (Brehm, Kullberg, & Moody-Corbett, 1984; Guharay & Sachs, 1984; Hamill, 1983), have greatly contributed to our understanding of the molecular workings of the entities underlying the physiology of mechanosensory transduction (Martinac & Cox, 2016). However, a complete understanding of the molecular force sensing mechanisms at play remains elusive.

A recent exciting discovery is that of Piezo channels, a new class of a long sought after bona fide eukaryotic MS channels that underlie mechanically evoked activity in numerous human cell types (Coste et al., 2010, 2012). Despite their relatively young age, in scientific terms, Piezo channels have already been shown to play essential roles in numerous physiological processes ranging from touch sensation to vascular development (Li et al., 2014; Ranade et al., 2014; Woo et al., 2014). At the heart of their molecular activity is the ability to convert the ubiquitous physical forces bombarding all cells into meaningful cellular responses. Thus, a central question relevant to all MS ion channels, including Piezos, is how is the force transmitted to the channel in order to instigate the necessary conformational change for gating? These stimuli may be transmitted directly to the channel via bilayer deformation ("force-from-lipids principle") (Kung, 2005; Martinac, Adler, & Kung, 1990; Teng, Loukin, Anishkin, & Kung, 2015) or via tethering of the channel to cytoskeletal elements or the extracellular matrix ("force-from-filament principle") (Arnadottir & Chalfie, 2010; Katta, Krieg, & Goodman, 2015).

We believe that all mechanotransduction processes that involve channels can be reduced to a *force-from-lipids* paradigm (Kung, 2005; Martinac et al., 1990). As discussed further in the text this reductionist view may also apply to ion channels that at first glance seem to be gated by force at attached tethers according to a *force-from-filament paradigm* (Katta et al., 2015; Zhang et al., 2015). In these cases structural and scaffold proteins (ECM and/or cytoskeleton) merely extend the dynamic range of these molecular force transducers and focus the mechanical stimuli to cellular sites where these mechanosensors are located. This may apply to specialized mechanoreceptor cells such as hair cells and sensory neurons in the skin. Given that all ion channels are required to change conformation to gate (Morris, 2011; Sachs, 2010) we can create a "mechanical continuum" on which all channels sit from those that are highly sensitive to mechanical stress such as Piezo channels to those that are almost insensitive, such as the bacterial K^+ channel KcsA. The position of the channels on this continuum is labile and can be modified by their cellular localization and the presence or absence of structural scaffold proteins that can share or focalize the applied force (Fig. 1). This endows cells and tissues with an exquisitely large repertoire to fine-tune their spatio-temporal sensitivity to applied mechanical stimuli.

This chapter focuses on the molecular force transduction mechanism in Piezo channels and more specifically whether the *force-from-lipids* paradigm is applicable to this class of ion channels. A large portion of the work discussed here relates to Piezo1 but given the high degree of sequence similarity with Piezo2, we believe it relatively easy to draw significant parallels.

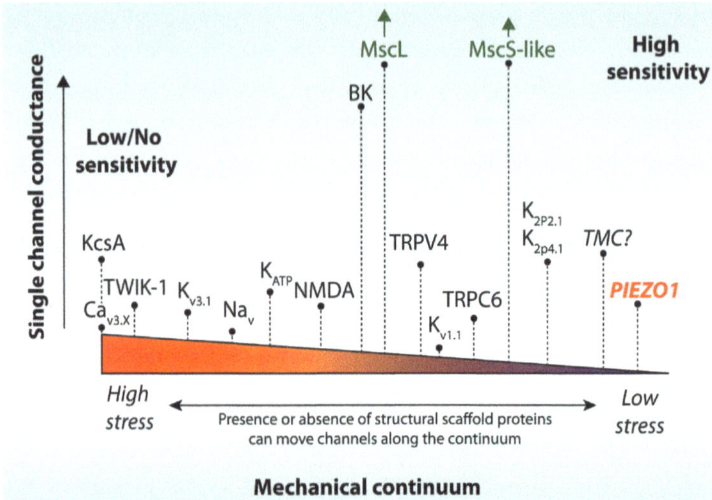

Figure 1 ***Ion channel mechanical continuum.*** All ion channels can be placed on a mechanical continuum from highly sensitive (e.g., Piezo1/2) to those that are almost insensitive to applied force (e.g., KcsA). The position of most channels on this continuum is labile and can be modified by the presence or absence of specific structural scaffold proteins providing each individual cell with an ability to tightly and spatially regulate its sensitivity to mechanical force. *Unknown whether transmembrane-like channels (TMCs) are bona fide ion conducting pores.

2. ORIGIN OF THE "FORCE-FROM-LIPIDS PRINCIPLE"

The force-from-lipids principle was first proposed some 30 years ago in conjunction with activation of bacterial MS channels by amphipaths in giant spheroplasts of *Escherichia. coli* (Martinac et al., 1990). Amphipaths are chemically unrelated substances soluble in both water and organic solvents. With regards to their solubility properties phospholipids, sphingolipids and sterols, for example, which represent major constituents of biological membranes, belong to the same amphipathic class of substances. The amphipathic nature of lipid molecules is essential for the existence of all life forms existing on Earth. This is because the lipid bilayer as the core structure of cell membranes, is a product of the interaction of lipid molecules with water resulting in the "hydrophobic effect" (Chandler, 2005). Since the very first experiments examining the effect of amphipaths on bacterial MS channels, an array of amphipathic compounds have been shown to also have similar effects on eukaryotic MS channels (e.g., TREK-1, TRAAK, Piezo1) (Maingret, Patel, Lesage, Lazdunski, & Honore, 2000;

Syeda et al., 2015). The final proof for the *force-from-lipids* being sufficient for activation and modulation of MS channels by mechanical force came from experiments showing that MscL reconstituted into lipid bilayers retained its mechanosensitivity similar to its activity in the native bacterial membrane (Häse, Le Dain, & Martinac, 1995; Sukharev, Blount, Martinac, Blattner, & Kung, 1994). With this, liposomal reconstitution of purified MS channel proteins and examination of their function by the patch-clamp technique became a "gold standard" for establishing inherent mechanosensitivity of ion channels. This is because this experimental paradigm excludes any extraneous cellular components, thus leaving the lipid bilayer as the only medium transmitting the mechanical force necessary for gating. In fact, the MscL protein was isolated and identified at the molecular level by tracking the MS activity of reconstituted chromatographic column fractions starting from *E. coli* lysate (Sukharev et al., 1994; Sukharev, Martinac, Arshavsky, & Kung, 1993). Since then liposome reconstitution has helped to demonstrate that a number of eukaryotic MS ion channels are also inherently mechanosensitive, most notably TREK and TRAAK (Berrier et al., 2013; Brohawn, Campbell, & MacKinnon, 2014; Brohawn, Su, & MacKinnon, 2014; Dong et al., 2015) and possibly Piezo1 (Cox et al., 2016; Syeda et al., 2015) indicating evolutionary conservation of this gating modality from prokaryotes to eukaryotes.

2.1 Evolutionary origins of mechanosensitive channels

Given the vastly different structures of the bona fide MS channels currently identified (MscL, MscS, ENaC, Piezo, TREK, TRAAK, TRPN) (Martinac & Cox, 2016) it is very likely that mechanosensitivity has been produced at numerous points during evolution via convergent means.

So, what was the first family of MS channels? Given the centrality of water to life and the osmotic forces that primordial proto-cells would have encountered at the birth of life on our planet, it is likely that membrane proteins similar to the bacterial channels MscL and MscS were the first to develop (Martinac, 2011, 2012). MscL seems to be confined almost completely to bacterial and archaeal species whereas MscS-like channels are more widely distributed and can be found in plants, some fungi and numerous eukaryotic human parasites most notably *Trypanosoma*, *Toxoplasma* and *Plasmodium* (Cox, Nakayama, Nomura, & Martinac, 2015; Pivetti et al., 2003).

The likely evolutionary paradigm for both MscL-like and MscS-like channels is that their primary function was as "emergency release valves" enabling primordial cells to survive sudden changes in environmental osmolarity (Booth, Miller, Muller, & Lehtovirta-Morley, 2015; Naismith & Booth, 2012). MscL-like channels remained "faithful" to this basic function throughout bacterial and archaeal evolution, which is consistent with their absence from almost all bacteria that are native to marine environments. In contrast MscS-like channels have developed more complex physiological functions with the emergence of higher cell-walled organisms, including fungi and plants (Cox et al., 2015; Hamilton, Schlegel, & Haswell, 2015; Malcolm & Maurer, 2012). Working as an "evolutionary tinkerer" Nature (Jacob, Science, 1977) has allowed MscS-like channels to evolve significantly in order to fulfill more specialized cellular functions dictated by natural selection. This occurred through alterations of the pre-existing MscS-like channel scaffold resulting in proteins with vastly modified structures. This is clearly illustrated by the fact that members of the MscS-like channel superfamily contain distinct functional motifs such as EF-hand calcium-binding motifs (Nakayama, Yoshimura, & Iida, 2012).

2.2 Evolutionary timeline of MS channels

So in the evolution of different life forms on Earth, MS ion channels such as MscL and MscS are likely descendants of the primordial sensors of mechanical force that evolved as primary signaling molecules supporting mechanosensory physiology of living organisms. Although MscS-like proteins have served well as templates for MS channels in eukaryotic cell-walled organisms such as fungi and plants (Hamilton et al., 2015; Nakayama et al., 2012), these MS channels have apparently not provided sufficiently good templates for further evolution of molecular force sensors in multicellular animal-like organisms. Instead, more specialized forms of mechanotransduction in higher organisms have employed more complex molecular entities to transduce physical stimuli including TRP-type channels and Piezo channels (Table 1). However, as we will see the evolutionary origins of many of these channel types including Piezo channels is likely to stretch back to the most ancient of metazoans and perhaps even begins at the evolutionary boundary between uni- and multicellularity.

2.3 Piezo evolution

Given the presence of Piezo channels in some of the earliest known metazoans such as sponges, it is interesting to envisage the importance of these ion channels in the generation of multicellularity. Complex multicellular organisms only evolved in a handful of eukaryotic groups, namely: animals, fungi,

Table 1 Distribution of MscS-like, TRP-like and Piezo-like channels throughout eukaryotic organisms

Taxonomic group	Species name	TRP	MscS-like	Piezo
Animalia - Mammalia	*Homo sapiens*	27	0	2
	Mus musculus	28	0	2
Animalia- Actinopterygii	*Danio rerio*	27	0	3
Animalia - Insecta	*Drosophila melanogaster*	17	0	1
Animalia - Hydrozoa	*Hydra vulgaris*	5	0	1
Animalia - Anthozoa	*Nematostella vectensis*	1	1	1
Animalia - Ascidiacea	*Ciona intestinalis*	27	0	2
Animalia - Chromoadorea	*Caenorhabditis elegans*	17	0	1
Fungi	*Schizosaccharomyces pombe*	1	2	0
	Saccharomyces cerevisiae	1	0	0
	Rozella allomycis	0	1	1
Amoebozoa - Dictyostelia	*Dictyostelium discoideum*	0*	2	1
Choanozoa - Choanoflagellate	*Salpingoeca rosetta*	0*	3	2
Plantae - Chlorophyceae	*Chlamydomonas reinharditii*	19	7	0
Plantae - Prasinophyceae	*Ostreococcus lucimarinus*	0	3	1
Plantae - Rhodophyta (Red algae)	*Cyanidioschyzon merolae*	0	2	0
Plantae - Embryophyte	*Arabidopsis thaliana*	0	10	1
	Physcomitrella patens	8	16	1
Alveolata - Oligohymenophorea	*Ichthyophthirius multifiliis*	1	0	2
Chromoalveolata - Coscinodiscophycea (Diatoms)	*Thalassiosira pseudonana*	5	6	0*
Chromoalveolata - Phaeophyceae (Brown algae)	*Ectocarpus siliculosus*	2	3	0*
Chromoalveolata - Oomycete	*Phytophtora sojae*	9	3	1

*, represents that a low sequence similar homologue may exist.

land plants and brown, red and green algae (Niklas, 2014). Of these organisms only fungi and brown algae seem to lack Piezo homologues (Table 1). Interestingly, Piezo homologues can also be found in unicellular organisms widely thought to be the immediate evolutionary step before true complex multicellularity. These include *Capsaspora*, the choanoflagellate *Monosiga* and the polyphyletic group of slime molds, all of which are unicelullar free-living organisms that can congregate into basic multicellular societies (Niklas,

2014). This leads us to the proposition that Piezo proteins may well have been a necessary cellular component allowing the development of multicellularity and cell—cell communication, playing a role in sensing neighboring cells (Hung et al., 2016). The potential for Piezo being a part of the Hippo signaling pathway acting as a mechanical signal transducer further fans the flames of such a speculation, given that Piezo1 activation seems to activate YAP/TAZ transcription factors (Pathak et al., 2014).

3. APPLICABILITY OF FORCE-FROM-LIPIDS TO EUKARYOTIC CHANNELS

Both the *force-from-lipid* and *force-from-filament* principles act in mechanotransduction processes in animal and human cells. In this section, we look at the evidence for *force-from-lipid* gating in eukaryotic channels by first looking at the two-pore domain potassium channels TREK-1 and TRAAK and then at both TRPV4 and Piezo1 channels.

3.1 Two-pore domain K^+ (K_{2P}) channels

TRAAK ($K_{2P4.1}$) and TREK-1 ($K_{2P2.1}$), have been unequivocally shown to be solely gated by membrane tension (Berrier et al., 2013; Brohawn, Su, et al., 2014) similar to the bacterial channels MscL and MscS (Moe & Blount, 2005; Sukharev, 2002; Sukharev et al., 1994) and, as we will later discuss, also potentially Piezo1 (Cox et al., 2016). Thus, in addition to bacterial MS channels, the "force-from-lipid" principle also applies to the predominantly neuronally expressed TRAAK and TREK-1 channels. These eukaryotic channels are polymodal sensors of diverse physicochemical cues including pH, voltage, temperature, lipids and polyunsaturated fatty acids, as well as membrane tension (Honore, 2007).

It is of note that TRAAK activity in cells apposes Piezo1 activity (Brohawn, Su, et al., 2014), i.e., in cells in which both Piezo1 and TRAAK are expressed, Piezo1 depolarizes and TRAAK repolarizes or hyperpolarizes the cell. Moreover, although the initial activation threshold of both Piezo1 and TREK-1 is in the range of the lipid—glass adhesion tension, 0.5—4 mN/m (Opsahl & Webb, 1994), the open probability of TREK-1 increases less than Piezo1 in response to a specific increase of the applied force i.e., the slope of the Boltzmann fit to the activation curve is shallower (Bae, Gnanasambandam, Nicolai, Sachs, & Gottlieb, 2013; Maksaev, Milac, Anishkin, Guy, & Sukharev, 2011). The difference between their sensitivity to mechanical force is likely due to their specific structure and the way they interact with membrane lipids. Although the role of lipids for gating K_{2P} channels is clear, the exact gating mechanism is the subject of intense research.

Crystal structures of several MS channels including K_{2P}s in different conformations, have provided invaluable insight into the gating mechanisms of this family of channels (Brohawn, Campbell, et al., 2014; Brohawn, del Marmol, & MacKinnon, 2012; Dong et al., 2015; Lolicato, Riegelhaupt, Arrigoni, Clark, & Minor, 2014). From the X-ray crystal structure of the non-conductive conformation of TRAAK, it has been concluded that lipids play a significant role in ion conduction since lipid acyl chains seem to access two fenestrations in the C-terminal part of the channel with the initial suggestion that this physically inhibits ion passage acting as the gate. In this model in the open state, the lipid acyl chain becomes replaced by the end part of the TM4 helix. However, this is likely to represent an unlikely extreme but the idea that lipid molecules sterically hinder an essential gating movement, in this case movement of the TM4 helix, has found more support (Brohawn, Campbell, et al., 2014). The question becomes; is the lipid a passive bystander merely moving out of the way to allow gating or is it driving gating by pulling structural elements of the channel? (We will further elaborate on this issue in Section 4.1.1). This may be particularly relevant if the end of TM4 tightly interacts with the lipid acyl chains. The importance of TM4 helix is widely appreciated given that a separate study on TRAAK has shown that TM4 movement is accompanied by a buckling of TM2 at a conserved GXG sequence (Lolicato et al., 2014).

In TREK1 however, the adsorption (coupling) of the C-terminus, which is physically connected to the TM4 helix, to the membrane results in gating of TREK-1 (Honoré, 2007; Treptow & Klein, 2010). This mechanism is widely supported by electrophysiology experiments showing that progressive deletion of the C-terminal domain of TREK-1 gradually reduces the force sensitivity of this channel (Bagriantsev, Clark, & Minor, 2012; Honoré, 2007; Patel et al., 1998).

Amphipaths are often used to assess the activation of MS channels by the bilayer mechanism (Martinac et al., 1990; Perozo, Cortes, Sompornpisut, Kloda, & Martinac, 2002; Perozo, Kloda, Cortes, & Martinac, 2002). It is thus noteworthy that even at low doses, arachidonic acid and extracellular lysophospholipids, including lysophosphatidylcholine, facilitate the activation of both TREK-1 and TRAAK (Maingret et al., 2000). It will be therefore interesting to see whether these molecules have the same or similar effects on Piezo channel gating. We will discuss this in more detail in Section 3.3.

3.2 TRPV4

The mechanosensitivity of TRP channels in general is controversial, in particular when considering members of the TRPC channel family (Christensen & Corey, 2007; Gottlieb et al., 2008). However, while TRP channels may not represent the molecular entity underlying auditory transduction (previously suggested to be TRPA1), they almost certainly play central roles in many mechanosensory processes, the question often is whether they represent the primary mechanotransducer. TRPV4 ion channels are reported to function as mechanosensors in many different cell types including neurosensory cells responsive to systemic osmotic pressure, astrocytes, chondrocytes, sensory Merkel cells as well as renal and hepatic cell types (Martinac & Cox, 2016; Zheng, 2013). In addition, TRPV4 channels play a central role in mechanotransduction in bone and cartilage where they participate in extracellular matrix production by chondrocytes in response to mechanical stimulation. This is supported by the fact that TRPV4 antagonists such as GSK205 inhibit and TRPV4 agonists such as GSK1016790A stimulate matrix production in the absence of mechanical stimulation (O'Conor, Leddy, Benefield, Liedtke, & Guilak, 2014). As polymodally regulated ion channels TRPV4 channels are regulated by heat as well as amphipaths such as metabolites of arachidonic acid (Cohen, 2005; Watanabe et al., 2003). However, deletion of ankyrin repeats reduced TRPV4 channel sensitivity to lowering of extracellular osmolarity (Liedtke et al., 2000), which suggested that channel activation occurred largely according to the *force-from-filament* principle via the ankyrin tether similar to TRPN channels (Zhang et al., 2015) and possibly TRPV1 (Prager-Khoutorsky, Khoutorsky, & Bourque, 2014). In contrast, other studies indicate that TRPV4 is inherently mechanosensitive (Loukin, Zhou, Su, Saimi, & Kung, 2010; Teng, Loukin, Anishkin, & Kung, 2015b) and that channel activation is mediated through a specific interaction between L596 (in the S4-S5 linker) and lipids of the inner-leaflet. While this mechanism somewhat resembles the interaction between E306 residue and the inner-leaflet of the lipid bilayer in TREK-1 gating we lack direct evidence that TRPV4 can respond to membrane tension in other heterologous systems aside from *Xenopus* oocytes such as HEK293 or CHO cells.

While it seems clear that TRPV4 is involved in a number of mechanosensory processes, it is still unclear how mechanical stimuli is transmitted to the channel. While work from the Kung laboratory indicates a direct link with the lipid bilayer other work suggests more reliance on tethers. Unlike the K_{2P} channel family, no definitive evidence exists for TRPV4, or any other TRP channel, supporting a *force-from-lipid* type gating model. So what is the evidence that Piezo can directly sense force from the lipid bilayer? Is it unequivocal like data presented on K_{2P} channels or less robust and widely agreed upon like members of the TRP channel family such as TRPV4?

3.3 Piezo1

Piezo1 channels have recently been reported to be gated in HEK293 cell membrane blebs (Fig. 2) (Cox et al., 2016). Membrane blebs are rounded membrane protrusions that lack cortical cytoskeleton and play important roles in apoptosis, cytokinesis and certain forms of cell motility

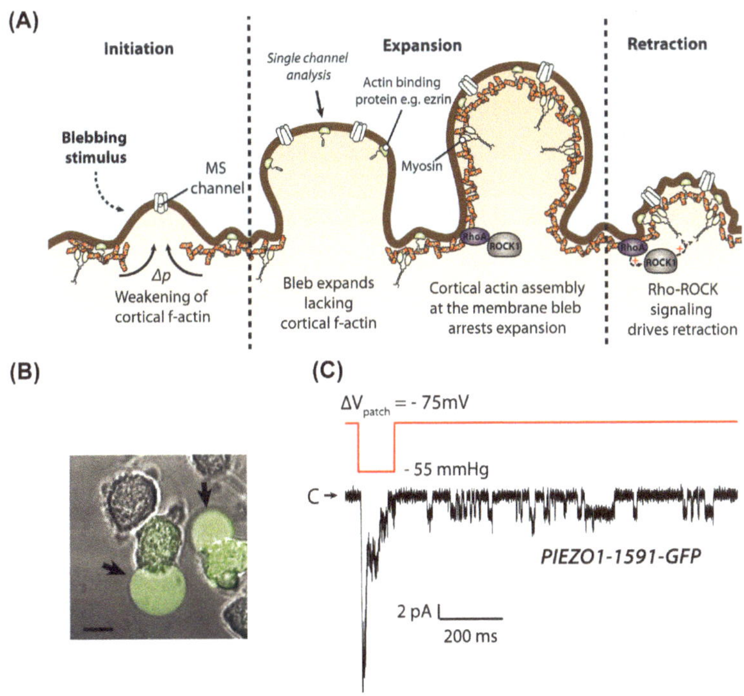

Figure 2 ***Human Piezo1 activation in HEK293 cell membrane blebs.*** (A) Generation scheme of a membrane bleb following a three stage process. Initiation can be induced by numerous stimuli including changes in external osmolarity and then expansion occurs largely devoid of cytoskeletal elements. (B) Image of a HEK293 cell membrane bleb. It is at this point that the bleb is amenable to patch-clamp electrophysiology. (C) Activity of human Piezo1-1591-GFP fusion protein in a HEK293 bleb-attached patch in response to a square wave pressure pulse applied by a high speed pressure clamp.

(Charras, Coughlin, Mitchison, & Mahadevan, 2008). Membrane blebs represent an interesting experimental paradigm that exists between studying an MS channel in its native cellular environment with an intact cytoskeleton and the ultimate reductionist approach of reconstituting purified channel protein in an "artificial" bilayer in the absence of any extraneous cellular components. It is important to note that TREK-1 was shown to be gated in membrane blebs (Honoré, Patel, Chemin, Suchyna, & Sachs, 2006; Maksaev et al., 2011) prior to its reconstitution in lipid bilayers. This provides strong support that Piezo1 is also gated by the bilayer mechanism similar to bacterial MS channels. This activity in blebs persisted in colchicine (prevents microtubule assembly) and cytochalasin D (actin depolymerization) treated cells further affirming a lack of reliance on structural scaffold proteins. Furthermore, human Piezo1 was also shown to exhibit a dimension change of around $\sim 8\ nm^2$ during opening comparable to dimensional changes of the inherent mechanosensitive channels MscS (Shaikh, Cox, Nomura, & Martinac, 2014) and MscL (Bae et al., 2013; Cox et al., 2016), supporting the finding that Piezo1 is sensing bilayer tension similar to bacterial MS channels as a typical "TΔA" channel (for further information see Hamill & Martinac, 2001).

Piezo1 seems to be gated by tensions in the range of 1–3 mN/m and like other tension sensing channels (e.g., MscS (Nomura, Cox, Bavi, Sokabe, & Martinac, 2015) responds to both positive and negative pressure applied to a patch pipette (Cox et al., 2016; Lewis & Grandl, 2015). The studies here use Laplace's law to calculate tensions but we should note that this is a gross over simplification of the mechanical properties of cells and provides an upper limit of the tension required to gate the channel. Inhibition of Piezo1 channel activity by the spider toxin GsMTx-4 (discussed further in Chapter 7) also points towards a membrane tension based gating mechanism (Bae, Sachs, & Gottlieb, 2011). Although similar experiments have not been done with Piezo2 channels, given the high degree of sequence similarity with Piezo1, it is likely they are also gated by membrane tension (Jia, Ikeda, Ling, Viatchenko-Karpinski, & Gu, 2016).

One way to unequivocally determine the inherent mechanosensitivity of Piezo channels and test the effect of lipids, is to purify and reconstitute Piezo into "artificial" bilayers akin to experiments undertaken with MscL, MscS, TREK1/2 and TRAAK (Berrier et al., 2013; Nomura et al., 2012). The methods, which can be used include patch-clamp recording from liposome membrane patches (N. Bavi et al., 2016; Nomura et al., 2015; Sukharev, 2002), planar lipid bilayers and droplet bilayer systems (Coste et al., 2012;

Najem et al., 2015; Syeda et al., 2015). The function of purified mouse Piezo1 has been shown in reconstituted bilayers of diphytanoylphosphocholine (DPhPC) with one leaflet doped with the conically shaped lipid phosphatidic acid (DOPA) (Coste et al., 2012; Syeda et al., 2015). Initially this was carried out in order to prove that they were indeed the pore-forming subunit of a native MS channel. However, thus far only spontaneous activity has been observed in droplet interface bilayers which interestingly requires an asymmetric bilayer composition as no activity is seen in symmetrical bilayers. Multiple labs have failed to elicit mechanically gated currents in standard reconstitution experiments, which remains the final hurdle to proving the inherent mechanosensitivity of Piezo1 channels.

4. WHICH FORCE IS THE RIGHT FORCE?

So if Piezo channels are gated by force transmitted directly from the bilayer (in many circumstances focused by the cytoskeleton and extracellular matrix) what is the most relevant force?

The answer to this question for each individual MS channel is of course intimately associated with their structure and how this structure interacts with their surrounding (i.e., lipids, extracellular matrix and cytoskeleton). The literature about what is the right force for activation of MS channels is sometimes misleading. There are three distinct types of stress in Nature, which are stretch/compression, pure bending and pure shear. Our intent here is to first make a distinction between the applied stimuli and the generated force/stress. This is because regardless of the overt mechanical stimulus applied (e.g., fluid flow, indentation), the external stimuli may turn into alternate force/stress types depending on the shape and material (mechanical) properties of the cells and their membranes. For example, there are an ever growing battery of external stimuli that are commonly used for MS channel activation expertly reviewed by Patapoutian and colleagues (Ranade, Syeda, & Patapoutian, 2015). However, it is likely that many of the described methods produce the same type of stress potentially over different length scales. For example, negative pressure (suction) applied to a patch pipette and indentation or cell poking both induce an increase in membrane tension while one is compressive whereas the other is a tensile force. As another example, the flow of a fluid over a cell surface may at first seem to create shear stress but it has been shown that the stimulating shear stress is negligible compared to the produced membrane tension. For example, if the in vivo blood shear is 1 Pa (N/m^2), then the tension in the cell membrane can be 10^4 Pa or larger (Liu, Yen, & Fung, 1994). Given the response of Piezo1 to shear stress and its role in vascular physiology this is a particularly pertinent point (the Piezo channel response to shear stress is further discussed in Chapter 2) (Li et al., 2014).

4.1 Membrane tension-induced gating

Stress/pressure distributions within the bilayer are difficult to measure experimentally, though they have been estimated for different lipid compositions using a variety of computational approaches such as molecular dynamics (MD) simulations (Cantor, 1999; Gullingsrud and Schulten, 2004). Even in "pure" bilayers, the transbilayer pressure profile is largely inhomogeneous (Fig. 3A). This stress heterogeneity in bilayers originates from the amphipathic nature of the lipid molecules and the presence of water. Thus, different lipid bilayers may have different pressure profiles depending on their compositions and their surrounding environments (e.g., ionic concentration and pH). The hydrophilic lipid headgroups are squeezed together to prevent water exposure of the hydrophobic tails and repel each other, both electrostatically and sterically. Thus, an acute lateral tension develops at the water-lipid interior interface. A typical surface

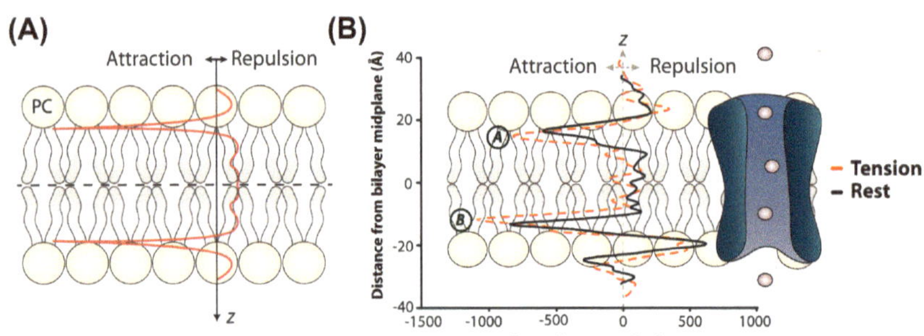

Figure 3 ***The transbilayer pressure profile.*** (A) An idealized symmetrical transbilayer pressure profile showing characteristic negative peaks at the water-lipid interface and repulsive positive peaks in the headgroup and tail region. (B) Transbilayer pressure profiles from molecular dynamics simulation of a POPE bilayer with *Escherichia coli* MscL inserted in it (*Black line* is the lateral pressure profile at rest, and red is in the presence of applied tension). Note how in the presence of the protein the pressure profile in a symmetrical lipid bilayer has become distinctly asymmetric. Peak A and peak B represent the rise in the pressure profile at lipid solvent interface (for more details see N. Bavi et al., 2016).

tension in each monolayer is estimated to be 50 mN/m (Gullingsrud & Schulten, 2004; N. Bavi et al., 2016). The lipid tails also tend to repel each other to maximize their entropy and occupy a greater bilayer volume. This way, they balance the tension generated at the water-lipid interface. A typical value for the peak pressure inside the lipid bilayer is about 300 atm whereas around the head groups it can be up to 1000 atm (Cantor, 1997; Gullingsrud & Schulten, 2004). So, considering such huge pressure values across the bilayer thickness, it is highly unsurprising that membrane proteins in general can be sensitive to the intrinsic transbilayer pressure profile (Jensen & Mouritsen, 2004; Lee, 2004). It should be noted however, that proteins can reciprocally redistribute the transbilayer pressure profile depending on their interaction with the lipid (N. Bavi et al., 2016; Cantor, 1997, 1999; Lundbaek, Collingwood, Ingolfsson, Kapoor, & Andersen, 2010). In the case of MS channels, which are gated by membrane tension mediated by the surrounding lipids, the bilayer tension is distributed heterogeneously, at both microscopic (Bavi et al., 2014) and atomistic levels (Gullingsrud & Schulten, 2004). However, neither the scales at which relevant variation takes place nor the effect of varying lipid composition or tension has yet been investigated experimentally in molecular/atomic details. Some information has however been gleaned using MD simulations as an atomistic microscope. For example, Fig. 3B illustrates the changes in the pressure profile of a POPE lipid bilayer in the presence of *E. coli* MscL, with and without applied tension (N. Bavi et al., 2016). Firstly of note is the asymmetry in the pressure profile and the reduction of the pressure peaks in the hydrophobic core after MscL has been inserted into the bilayer. As shown, when the bilayer has been stretched, the pressure profile of the lipid in the presence of MscL changes particularly at the lipid—water interface (Peak A and Peak B). In this scenario, compared to a non-stressed bilayer and based on the area under the pressure profile, one may estimate the equivalent bilayer tension required for the activation of MscL (See N. Bavi et al., 2016 for more details).

We should also consider that bilayer pressure profile asymmetry is an inherent characteristic of biological membranes, where lipid composition of the monolayers is vastly different and tightly regulated, compounded by the fact that the membrane is studded with connections to the extracellular matrix and cytoskeleton. Interestingly as mentioned, purified Piezo1 seems to require an asymmetric pressure profile in order to function in droplet interface bilayers, hinting at the fact that the transbilayer pressure profile is a driver of Piezo1 gating.

4.1.1 Tension-induced gating: pulling or entropy driven?

If generation of membrane tension drives MS channel gating it still remains to be seen how this force is transmitted to the channel? Here two paradigms have begun to develop (discussed in more detail in Martinac and Cox, 2016). The first involves the "pulling" of structural elements of the channel via direct lipid–protein interactions. A prime example of such a model is MscL. Numerous residues including the amphipathic N-terminus have been shown to interact with the bilayer driving the closed to open transition. For example, lipids protrude into an inter-subunit cavity in the closed state and as tension is applied they drive gating by dragging the N-terminal helix to open the pore (Fig. 4A,B) (N. Bavi et al., 2016; Yoshimura, Nomura, & Sokabe, 2004).

The annular lipids that drive gating of course are removed from the cavities during opening but their strong interaction with the N-terminus is maintained (Fig. 5). It is this continued close interaction that provides the necessary momentum for gating.

Another model which seems to be gaining further traction is that as tension is applied, lipids are removed from protein pockets or cavities that were previously preventing structural rearrangement. The rearrangement can then be driven by an inherent strain in the protein (possibly applied by the transbilayer pressure profile). Thus we have an entropy driven gating paradigm where tension enables removal of lipids from these pockets permitting gating.

This has been suggested for the gating of the bacterial channel MscS (Pliotas et al., 2015) (Fig. 4C,D). A similar mechanism was put forward for MscS by Malcolm, Blount, and Maurer (2015) and termed the "jack-in-the-box" gating paradigm referring to the fact that removal or reduction in certain regions of the transbilayer pressure profile allows the channel to "spring" into the open state (We should note however that the gating of MscS is arguably better described by a gating paradigm presented by the Sukharev lab (Akitake, Anishkin, Liu, & Sukharev, 2007; Belyy, Anishkin, Kamaraju, Liu, & Sukharev, 2010)). This entropy driven model also has support in the K_{2P} fraternity, as discussed in Section 3.1, where lipids prevent the movement of TM4. Interestingly, such a paradigm may also be relevant in the gating cycle of TRP channels. A recent seminal article from the Julius and Cheng labs has revealed a cryo-EM structure of TRPV1 in a native bilayer using lipid nanodiscs (Gao, Cao, Julius, & Cheng, 2016). Using a toxin activator from the Chinese bird spider, they show that the toxin dislodges a lipid from a protein pocket enabling a structural transition to the

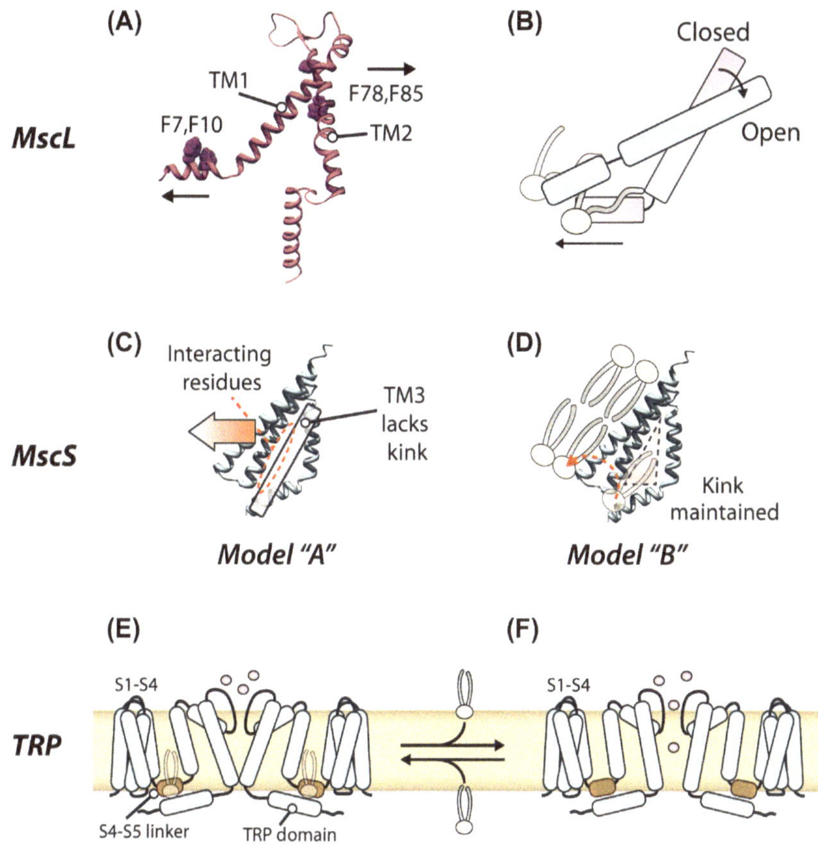

Figure 4 ***Membrane tension-induced gating, pulling structural elements or removal of lipids from protein pockets?*** (A) *Escherichia coli* MscL monomer structure with important lipid interacting residues has been shown. (B) Cartoon of the movement of the N-terminus being "dragged" by lipids as a result of tight lipid–protein interaction (interaction energy shown in Fig. 5). There are two different models for MscS channel gating where (C) represents a model suggested by the Sukharev lab in which TM3 is upright and gating is driven by force transmission from the bilayer to the gate via tight TM2-TM3 interactions, more akin with the ideas presented for MscL (Model A) (Belyy et al., 2010). The other model shown in (D) relies on movement of lipids from protein cavities that allows gating (Model B) (Pliotas et al., 2015). This movement of lipids from protein cavities has been suggested also as a gating model for TRPV1 in response to toxin binding and suggested for gating induced by heat, where the stimulus displaces lipids in a 'pocket' (E) allowing transition to the open state shown in (F) (Gao et al., 2016). It is a small leap to suggest that tension-induced removal from these pockets, as suggested for MscS, may underlie mechanical gating of TRP channels (although we should note this is a controversial area).

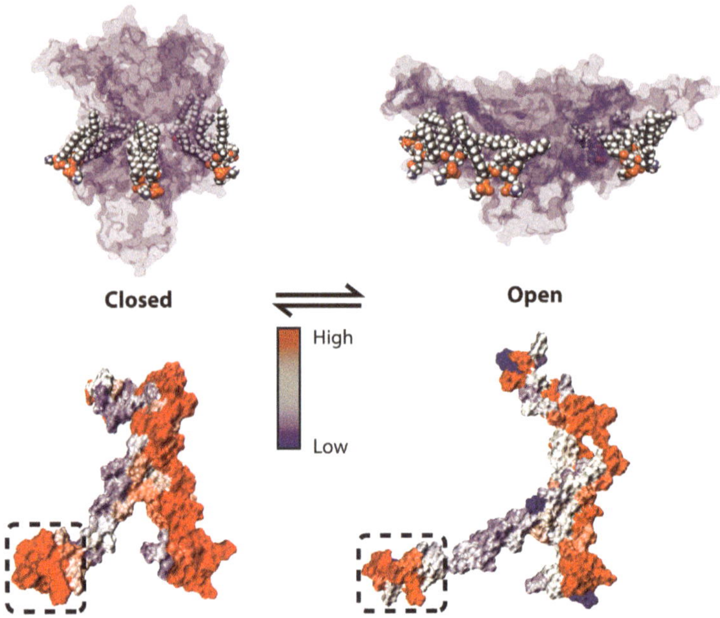

Figure 5 ***Lipids pull the N-terminus of MscL and have a continual strong interaction.*** Upper panel shows open (left) and closed (right) structure of MscL from an MD simulation with annular POPE lipids protruding into the structure. Lower panel shows the degree of interaction between the protein and the lipid (Blue – low, Red, high). The N-terminus has a strong interaction with the lipid throughout the gating cycle despite moving out of the inter-subunit cavity.

open state. They speculate entropy driven removal of lipids from this region by heat may also underlie TRPV1 heat induced gating. It is thus a small leap to suggest that much like the proposal for *E. coli* MscS that tension-induced removal of lipids from this cavity may underpin the gating of some mechanically-gated TRPs (Fig. 4E,F). It has long been recognized that the transbilayer pressure profile can influence the shape of membrane proteins inserted into a bilayer inducing a degree of resting strain (Cantor, 1999; Lee, 2004). However, it is difficult to see in these models how removal of lipids from hydrophobic lipid filled cavities would impart *no* strain on protein structure. For example, the TRPV1 model from the Julius lab (Fig. 4E,F) would also fit with data presented by the Kung laboratory discussed above on TRPV4, where residues in the S4-S5 linker interact with lipids driving mechanically induced gating.

While the presence of lipids in these cavities is likely to hinder the necessary structural arrangements for gating in a number of channel classes it seems likely that some strain must be imparted on the protein structure by their movement. Thus both the movement of the lipid and its interaction with channel residues are likely to be essential for tension-induced gating.

So what do we know about lipid interactions with Piezo channels? The answer is very little. Firstly, we know that Piezo1 is sensitive to cholesterol content and the cholesterol binding protein STOML3 (Qi et al., 2015) and may even directly bind cholesterol (Hulce, Cognetta, Niphakis, Tully, & Cravatt, 2013) (discussed further in Section 4.3). We also know that channel activity is sensitive to the presence of PIP_2 and possibly DAG (Borbiro, Badheka, & Rohacs, 2015).

Probing important mechanosensing residues in Piezo channels and in particular potential lipid–protein interaction sites is difficult. One way to characterize the mechanosensitivity of eukaryotic MS channels such as Piezo is to use MscL or one of its gain-of-function (GOF) mutants (more sensitive to force) as a reference. For example from the MscL response, the membrane tension required for the activation of eukaryotic MS channels can be estimated, also partly negating the dependence of results on pipette geometry. This has already been successfully done in two studies of Piezo1 channels (Bae et al., 2013; Cox et al., 2016). This could easily be applied to point mutations in Piezo channels allowing easy quantitative comparison of their sensitivity to force. This kind of methodology using a channel as a reference point has been widely used in the study of MscL and MscS point mutations (Akitake et al., 2007; Belyy et al., 2010; Edwards et al., 2005; Iscla, Wray, Eaton, & Blount, 2015; Zhong & Blount, 2014; Zhong, Yang, & Blount, 2014).

4.2 Membrane local curvature

Curvature is pervasive in biological systems (Zimmerberg & Kozlov, 2006). Its formation and sensing are essential for the normal functioning of all cells. Now the question is given the large size and the bowed structure of Piezo1 (which will be described in Section 5), are Piezos molecular curvature sensors? In this respect, it is important to understand the scale of curvature being discussed (Fig. 6). The level of curvature in a standard patch-clamp experiment has two scales global (visible under the microscope $R > 0.1\ \mu m$) and local ($R < 0.1\ \mu m$) curvature (Fig. 6). The global curvature has little energetic consequence for the channel whereas the local curvature can cause substantial stress. For example, let's consider a scenario where an MS channel such as Piezo1, was located in a membrane invagination of very large local curvature characteristic of caveolae (this is a possibility

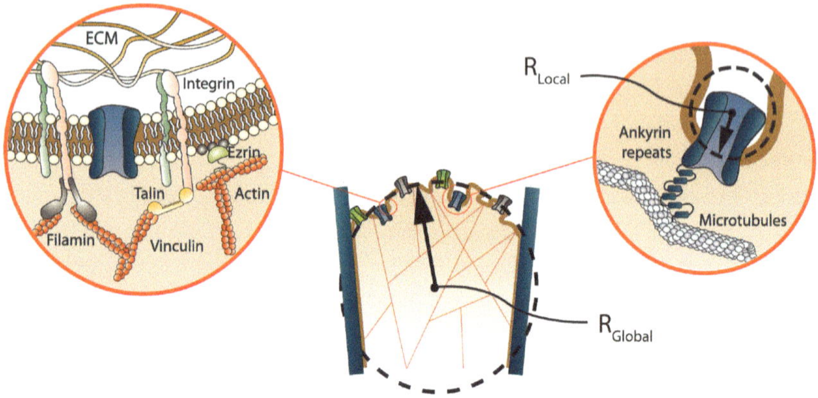

Figure 6 ***Different scales of curvature in a patch-clamp experiment.*** This figure draws attention to the distinction between the curvature of a patch membrane (global patch curvature - R_{Global}) and the curvature relevant to the size of the channel (local curvature — R_{Local}), such as microvilli and caveolae. It has been shown that only local curvatures, with the radius "R" in the order of ~25 nm, can modulate the function of an MS channel like MscL. This figure also describes different extracellular and cytoskeletal elements that may be present around a plasma membrane patch. These elements may cause some highly curved membrane folds by directly pulling on a specific part of a membrane protein or on the lipid raft adjacent to the protein.

given the links with cholesterol and STOML3 discussed in Section 4.3). If the cell membrane is stretched the invagination may flatten. This change in membrane conformation from a highly curved state (caveola) to a flat state can result in k_BT energy sufficient to activate an MS channel (Bavi, Cox, et al., 2016). Despite its cup-like shape robust localization in caveolae of Piezo, or any other channel for that matter, has not been demonstrated. However, the presence of caveolae almost certainly modifies the response of MS channels (Huang, Bae, Sachs, & Suchyna, 2013). Further work is required to determine whether local curvature is a relevant stimulus for Piezo channels.

4.2.1 Amphipaths

The transbilayer pressure profile and tension can also be modulated by amphipaths which introduce local frustration. Asymmetric incorporation of amphipaths or co-surfactants into one leaflet of lipid bilayers also potentially results in positive (outward) or negative curvature (inward) and can thin the bilayer (Bavi, Cox, et al., 2016; Bavi, Vossoughi, Naghdabadi, & Jamali, 2016; Yoo & Cui, 2009). Almost all of the studies in this area suggest that MS channel modulation by asymmetric insertion of amphipaths is due to their effect on local curvature of the bilayer. However, this has not been conclusively experimentally shown and we should note that other mechanisms have been proposed, such as effects on line tension (Mukherjee et al., 2014).

Whether amphipaths cause relevant local curvature is not widely established but what is certain is that due to uneven partitioning of amphipaths into a lipid monolayer there is a redistribution of mechanical tension between the two monolayers, resulting in a change in open probability of many MS channels (MscL, MscS and TREK-1). This is potentially the case for Piezo channels too given purified Piezo1 can only be activated in asymmetric liposomal membranes. It is important to note that amphipaths are effective only when added asymmetrically. So if Piezo channels are gated by the *force-from-lipids* paradigm are they affected by amphipaths? Experimentally chlorpromazine applied externally does not seem to modulate Piezo channels (Lee et al., 2014). However, application of conically shaped lipids (DOPA) in droplet interface bilayers is sufficient to gate reconstituted purified mouse Piezo1 protein. Piezo1 activity is also modulated by the conversion of PIP_2 to diacylglycerol, lipids with the opposite conical shapes. It is unknown whether this is due to a mechanical effect on the bilayer or other signaling cascades or even the loss of structural scaffolding links (discussed further in Chapter 9). Thus far there is no information on whether other conically shaped molecules can gate Piezo channels such as arachidonic acid, its metabolites or lysolipids. Answers to these questions are central to providing a molecular understanding of Piezo channel function.

4.3 Role of the cytoskeleton and extracellular matrix in membrane tension-induced gating

Unlike in spheroplasts or lipid bilayers, measuring the induced tension in cell membranes is often complex because of the linkages between the plasma membrane and the cortical cytoskeleton. It includes both membrane protein-cytoskeleton and lipid—cytoskeleton interactions and all these interactions contribute to the total surface energy. It is of note that for some cell types like keratinocytes, the cytoskeleton bears almost all of the applied stress such that membrane tension is far lower than would be estimated by Laplace's law (Dai & Sheetz, 1999). The problem for mechanobiologists is that these contributions are generally inseparable (Sachs, 2015). If the cell membrane is perturbed, the beneath cytoskeletal network also deforms and therefore contributes to the surface energy. At the same time if, e.g., a filament pulls on the membrane or a part of a protein again the surface energy incorporates all these factors. This is why measurement of the membrane tension or the cytoskeletal share of the generated surface energy due to an external stimulus is difficult to determine precisely. The only current handle we have on such processes is from using genetically encoded stress sensors, in the form of cytoskeletal elements (For more information see (Guo, Sachs, & Meng, 2014)). More relevant to our discussion in this chapter, it is also difficult to determine under an applied external stimulus, whether the relevant force that activates the MS channels is bilayer tension or the force transmitted from the cytoskeleton or extracellular matrix.

The origin of forces that gate MS channels in vitro and in vivo has been speculated upon and probed since the initial discovery of MS channels in chick skeletal muscle when the force gating the channels was proposed to originate from the cytochalasin-resistant cytoskeletal network (Guharay & Sachs, 1984). It continued with the "gating spring model" describing auditory mechanotransduction (Howard, Roberts, & Hudspeth, 1988; Hudspeth, 1989; Markin & Hudspeth, 1995). Alternatively, the bilayer model originates from the discovery of MS channels in bacteria (Martinac, Buechner, Delcour, Adler, & Kung, 1987) and was proposed for mechanotransduction of MS channels in prokaryotic cells (Martinac et al., 1990). While the reductionist approaches based on the *bilayer or tether dichotomy* are useful in deciphering the minimal structural elements that confer mechanosensitivity to membrane proteins, we should be mindful that MS channels are hardly, if ever, found in isolation from scaffold proteins that underlie the cytoskeleton and extracellular matrix in mammalian cells. Thus, while the *force-from-lipid* and *force-from-filament* paradigms provide basic physical principles that drive MS channel gating, it is essential that we understand how these scaffold proteins interact and modulate MS channels if we are to really understand mechanosensory transduction at a whole organ and indeed whole organism level.

Eukaryotic cells of higher organisms possess an excess membrane area in the form of ruffles, folds, and microvilli, which protect the membrane bilayer from excessive strain (Fig. 6). The excess membrane area is supported by the extracellular skeleton (extracellular matrix), which serves as external scaffolding, and the cortical cytoskeleton acting as a cross-linked gel (Spagnoli, Beyder, Besch, & Sachs, 2008), whose contraction and expansion locally regulate the mechanosensitivity of MS channels (Hamill & McBride, 1997). To add to the structural complexity of the cell membrane the lipid bilayer consists not only of phospholipids, including phosphatidylcholine, phosphatidylethanolamine, phosphatidylserine and phospho-inositides as its major components, but also of cholesterol- and sphingolipid-enriched lipid rafts (Anishkin & Kung, 2013; Rajendran & Simons, 2005), which may represent a platform for focalization of mechanical force. Cholesterol, in combination with sphingomyelin and specialized proteins, results in a more ordered phase/domain in the bilayer. Such a stiffened platform may suppress mechanical noise and confine force transmission to MS channels without the need for direct connection to integrins or other protein tethers (Anishkin & Kung, 2013). For instance stomatin-like protein-3 (STOML3), which is localized in cholesterol-rich lipid rafts, binds to cholesterol and thus sensitizes Piezo channels (Poole, Herget, Lapatsina, Ngo, & Lewin, 2014; Qi et al., 2015). Qi et al. demonstrated that in mouse sensory neurons, depletion of cholesterol and deficiency of STOML3 interdependently desensitizes Piezo1 and Piezo2 by softening the membrane elasticity modulus. In a separate study, it has been shown that STOML3 sensitizes Piezo1 and Piezo2 currents to ~10 nm indention, a relevant stimulus to gentle touch sensation in sensory neurons. Thus not only do we have a transverse heterogeneity (pressure profile) but also a longitudinal heterogeneity (lipid rafts/domains). This cholesterol dependent regulation has also been shown for MS channels in other cellular systems (Chubinskiy-Nadezhdin, Negulyaev, & Morachevskaya, 2011). Although this data does not unequivocally prove that MS channel inhibition and lipid raft disruption are causally linked.

Tethering to the extracellular matrix and/or cytoskeleton directly or indirectly via membrane cholesterol-rich domains may contribute to channel mechanosensitivity not only in specialized mechanoreceptors but also in non-specialized cells associated with mechanosensation (Anishkin & Kung, 2013; Chalfie, 2009). Piezo1 has been reported to regulate integrin activation by recruiting the small GTPase R-Ras to the endoplasmic reticulum, which activates the calcium-activated protease calpain by increasing Ca^{2+} release from cytoplasmic stores (McHugh, Murdoch, Haslett, & Sethi, 2012) with a similar cascade suggested in vascular endothelial cells initiated by shear stress (Li, Hou, & Beech, 2015; Li et al., 2014). This is an important finding given the fundamental role that integrins play in physically coupling extracellular matrix molecules (e.g., fibronectin) and the cytoskeleton in cell-extracellular matrix adhesion. They also confine and transfer mechanical stimuli throughout tissues. This coupling occurs through focal adhesions, which mediate directly or indirectly transmission of forces to membrane proteins, including MS channels (Wang, Butler, & Ingber, 1993).

Using MscL as a reference in a heterologous expression system (HEK293), Cox et al. have demonstrated that for both MscL and Piezo1, the cytoskeleton of cells had a mechanoprotective function (Morris, 2001). In other words MscL and Piezo1 were less sensitive to membrane tension in cell-attached patches compared to bleb patches (Cox et al., 2016). Cytochalasin treatment also made Piezo1 more sensitive in cell-attached patches in a separate study (Gnanasambandam, Bae, Gottlieb, & Sachs, 2015). Commensurate with such observations increased expression of the actin binding protein filamin A reduced the sensitivity of Piezo1 in primary vascular smooth muscle cells in response to a physiological mechanical load (Retailleau et al., 2015). Seemingly in contrast to this, in the whole-cell configuration cytochalasin D reduced Piezo1 currents elicited by cell indentation (Gottlieb, Bae, & Sachs, 2012). However, as discussed in Section 4, we believe this is because the cytoskeleton is important in distributing the force applied by these external stimuli. This is supported by the fact that osmotically stressing cytochalasin D treated cells subsequently boosted Piezo1 currents. Further in depth analysis of cell mechanics is necessary to clarify the reason for this apparent discrepancy.

4.4 Force-from-lipids or force-from-filament: a false dichotomy?

Mechanosensory transduction is based on a highly dynamic interplay between the lipid bilayer and a large variety of membrane and scaffold proteins with Piezo channels being a case in point. Structural scaffold proteins not only share and redistribute applied forces (Guo et al., 2014) but themselves can be redistributed to counter act the applied forces (Luo, Mohan, Iglesias, & Robinson, 2013; Schiffhauer et al., 2016).

Is MS channel tethering to extracellular matrix and/or cytoskeleton as the "*force-from-filament*" pathway separable from the MS channel to bilayer pathway ("*force-from-lipids*")? In other words, does the distinction between *bilayer or tether* become a false dichotomy only acting as a useful construct to help us understand how mechanical force may distribute between different components required for MS channel activation? In all likelihood both are essential in all mechanosensory transduction processes and teasing out which is most important at the physiological level is an arduous task that requires the use of force sensing reporters such as those already documented (Guo et al., 2014; Martinac & Cox, 2016) and the generation of new and novel modalities.

5. ARCHITECTURAL BLUEPRINT OF PIEZO CHANNELS

Piezos are very large transmembrane proteins, whose primary amino acid sequence comprises over 2500 amino acids. The initial hydropathy plot analysis predicted 30–40 transmembrane segments for the secondary structure of each Piezo monomer (Coste et al., 2010, 2012). Electrophysiological and pharmacological studies of purified mouse Piezo1 (mPiezo1) protein reconstituted into lipid bilayers indicated that Piezo1 functions as the pore-forming subunit of a largely non-selective cation channel, initially thought to be a homotetramer (Coste et al., 2012). Recently, the structure of full-length (2547 residues) mPiezo1 has been determined using cryo-electron microscopy (cryo-EM) at a resolution of 4.8 Å (Ge et al., 2015). The structure reveals a homotrimeric assembly resembling that of the trimeric ASIC1 structure (Jasti, Furukawa, Gonzales, & Gouaux, 2007).This oligomeric discrepancy is likely to have arisen from the use of a GST tag in the initial studies, which are known to dimerize thus indicating a larger oligomeric assembly. The cryo-EM structure of mPiezo1 further revealed that the transmembrane region consisted of around 14 rather than 30–40 segments per monomer. However, this is not conclusive as only the pore region (C-terminal ~360 amino acids) is of sufficient resolution to unequivocally assign TM helices. The structure also possesses three extracellular "blade" regions and takes on a propeller-like arrangement, with a central pore module. The pore seems to be formed by the last ~360 residues (residues 2189–2547) including the last two of the 14 resolved TM helices (inner helix – IH; Outer helix - OH), the C-terminal extracellular domain (CED) of ~240 residues (*C. elegans* homologous region crystallized by the Rees group (Kamajaya, Kaiser, Lee, Reid, & Rees, 2014)) and the intracellular C-terminal domain (CTD) of ~70 residues of each monomer. In addition, specific residues within these regions have been identified which determine the unitary conductance, ion selectivity for divalent and monovalent cations as well as pore blockage by ruthenium red (Zhao et al., 2016). This includes some residues that seem not to directly line the pore such as mouse E2133 (Coste et al., 2015). Work from the Xiao group suggests the pore is lined predominantly by the IH based on cysteine accessibility however, this data cannot rule out, much like tetrameric ion channels, that the pore is lined partially by both of the final TM helices i.e., both IH and OH (Ge et al., 2015; Zhao et al., 2016). A separate ion selectivity study demonstrated that human

Piezo1 (hPiezo1) expressed in HEK293 cells was permeable to both monovalent and divalent ions with the conductance for K^+ being the highest, intermediate and similar for Cs^+ and Na^+ and lowest for Li^+. Divalent ions, except for Mn^{2+}, as well as large ions such as TMA and TEA permeated the channel suggesting the narrowest region of the pore to be ~8 Å in diameter (Gnanasambandam et al., 2015).

5.1 Putative molecular force-sensing domains

Concerning Piezo1 mechanosensitivity the peripheral domains of the channel apparently play a major role. These domains are composed of the extracellular "blades", 12 or more peripheral helices (PHs) in each monomer, the intracellular beam-like structures and highly conserved "anchor" domains shown as a triangle in Fig. 7 (Ge et al., 2015; Zhao et al., 2016). The blades may represent specific molecular sensors for shear stress but as discussed shear may be largely converted into membrane tension. Evidence for the peripheral domains forming intrinsic mechanosensitive modules of Piezo1 was provided by showing that a hybrid channel, in which the N-terminal ~2000 amino acids were attached to the mechano-insensitive mASIC1 pore module generated a mechanically gated channel responsive to the ASIC blocker amiloride (Zhao et al., 2016). This demonstrates that Piezo1 channels possess intrinsic mechanosensitive structural elements separate from their pore modules. Interestingly this N-terminal portion also contains the highly conserved anchor domain with an individual helix that runs parallel to the bilayer (Ge et al., 2015; Prole & Taylor, 2013). The molecular mechanism of how *force-from-lipids* is translated into Piezo channel gating remains to be resolved. However, despite its complex structure perhaps lessons can be learnt from other more widely studied MS channels.

5.2 A Unifying structural mechanism for mechanosensitivity

Here using examples particularly of the K_{2P} channel family we can see that the *force-from-lipids* concept can apply to eukaryotic channels. Thus it should not come as a great shock to find that Piezos may also sense forces directly transmitted from the bilayer. However, to understand how, despite their very different molecular structures, these MS channels can transduce these bilayer forces into structural rearrangements it is important to look for a possible structurally lowest "common denominator" that confers mechanosensitivity to these ion channels. These similar gating elements may represent general structural mechanisms that underlie the optimal transformation of

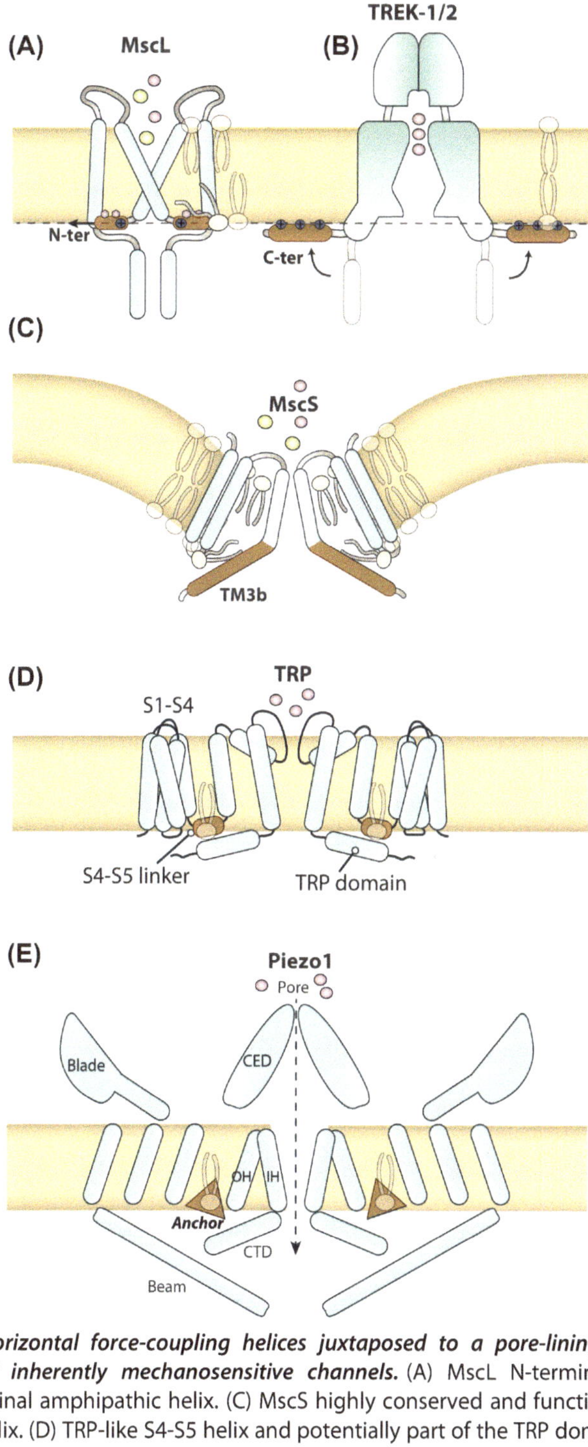

Figure 7 ***Horizontal force-coupling helices juxtaposed to a pore-lining helix as a hallmark of inherently mechanosensitive channels.*** (A) MscL N-terminal helix. (B) TREK C-terminal amphipathic helix. (C) MscS highly conserved and functionally essential TM3b helix. (D) TRP-like S4-S5 helix and potentially part of the TRP domain. (E) Here the horizontal force coupling helix may be part of the anchor domain the most highly conserved of Piezo regions. It is juxtaposed to a potential pore-lining helix. In addition it runs parallel to helices in the CTD similar to the S4-S5 linker and the TRP domain in TRP channels as shown above.

mechanical force into conformational rearrangement without the necessity for sequence conservation.

One commonality that we see between the structures is a horizontal bilayer interacting helix. These horizontal membrane-coupling helices must be juxtaposed to a pore-lining helix, like the N-terminal domain of MscL, and appear to be a hallmark of MS channels gated according to the *force-from-lipid* paradigm (Fig. 7) (N. Bavi et al., 2016). These amphipathic helices can be partially buried in the bilayer as shown for the N-terminus of MscL. This may also be the case for TM3b of MscS and the S4-S5 linker of TRPV4 (J. Teng, S. H. Loukinet al., 2015). As discussed recent work by the Julius and Cheng labs suggests a functionally important lipid binding pocket in this region of TRPV1 (Gao et al., 2016). Alternatively, these force-coupling helices may be adsorbed onto the membrane surface through electrostatic interactions. Such regions are seen in TREK-1/2 and TRAAK where the lower half of these channels seem to have a close association with the lipid bilayer (Brohawn et al., 2012; Brohawn, Su, et al., 2014; Dong et al., 2015; Lolicato et al., 2014). In addition, the C-terminal domain in these force sensitive K_{2P} channels, which is not seen in any of the crystal structures, is predicted to be an extended amphipathic helix functionally necessary for mechanical activation (Honoré, 2007; Honoré, Maingret, Lazdunski, & Patel, 2002). Regardless of the complexity of the gating cycles of these individual classes of MS channels, these horizontal force-transmitting helices have repeatedly been shown to play integral roles in the transmission of mechanical stimuli according to the *force-from-lipid* concept. Due to the innumerable lipid types and the divergent ways that these coupling helices can interact (buried, adsorbed) there is little requirement for sequence conservation despite the fact they play an almost indistinguishable mechanistic role in gating (N. Bavi et al., 2016). Given that recent evidence strongly suggests Piezo1 is gated according to the *force-from-lipids* paradigm (Cox et al., 2016; Syeda et al., 2015), we propose taking into account the functional data from the Xiao group (Zhao et al., 2016) and the conservation in this region (Prole & Taylor, 2013), that the $\alpha 4^{anchor}$, which runs almost parallel to the inner bilayer leaflet (Ge et al., 2015), may also serve the same purpose in the gating of Piezo1 channels as the N-terminus does for MscL. This is further supported by the fact that this region may also bind cholesterol (Hulce et al., 2013). Mechanistically speaking we speculate that the anchor domain and the CTD play similar roles to the S4-S5 linker and TRP domain in TRP channels (Fig. 7).

The most important part here is that the horizontal force coupling helix is juxtaposed to the pore-lining helix. With regards to MS channels activated by a tether/filament, it is proposed that the force is conveyed via an intracellular or extracellular filament to the pore and in many scenarios, this may be a more efficient way to transfer force compared with membrane tension. But we should note that this does not mean that these channels are not activated by the *force-from-lipids* also (As stated clearly in the elegant paper on TRPN1 by Zhang et al., 2015). This is because mechanical force causes both lipids and MS channels to move relative to each other, which results in a change of the transbilayer pressure profile. As discussed in Section 4.2 above, any change in transbilayer pressure profile acts on an MS channel by affecting its structural dynamics and consequently its conformational equilibria. Thus the *force-from-lipids* principle can be understood as a change in the transbilayer pressure profile, which can activate an MS channel by either originating from direct action of mechanical force on the membrane bilayer or indirect action of the force on an intracellular or extracellular filament pulling on the MS channel. This is why we believe that all mechanotransduction processes involving channels can be reduced to a *force-from-lipids* paradigm (Kung, 2005; Martinac et al., 1990).

6. CONCLUSIONS AND FUTURE PERSPECTIVES

It is clear that the *force-from-lipids* concept can apply to eukaryotic channels and that it may well apply to Piezo channels. The discovery of Piezo1 and Piezo2 ion channels has opened a new and exciting avenue towards the understanding of both physiological and pathophysiological mechanotransduction processes in higher eukaryotes. However, unequivocal proof for their inherent mechanosensitivity in the form of gating in response to mechanical stimuli in reconstituted bilayers or gating in response to amphipaths is still lacking. Nevertheless, gating in membrane blebs and asymmetric droplet interface bilayers strongly negates a complete reliance on structural scaffold proteins of the cytoskeleton or extracellular matrix. So we have a scenario where the lipid bilayer alone seems sufficient to mechanically gate Piezo channels (Cox et al., 2016; Syeda et al., 2015), whereas structural scaffold proteins are necessary to modulate and fine-tune its gating (Poole et al., 2014; Qi et al., 2015; Retailleau et al., 2016) depending on the tissue specific sensitivity that is required. The very recent elucidation of the 3D structure of the mouse Piezo1 channel by cryo-EM provides a major tool for deciphering the molecular workings of this unusually large membrane protein. There is still much work to be done to uncover the molecular mechanobiological secrets that these channels hold. Among these questions are; Given the presence of Piezo channels in some of the earliest known metazoans such as sponges, are they involved in the generation of multicellularity? What is the evolutionary necessity for the large size and complex architecture? Does this suggest further functions, maybe as a ligand-gated ion channel given the work on the small molecule agonist Yoda-1? Is it because Piezo1 and Piezo2 may interact with a large number of membrane and structural components other than bilayer lipids? This seems almost certain and we await the discovery of further binding partners. Does the cup-like shape in the cryo-EM structure represent a native conformation and suggest localization to a region of high local curvature? May they also sense membrane curvature as a mechanical stimulus? The interactions with cholesterol and STOML3 would at least support such a notion. As a corollary, what lipids directly interact with Piezo? Is the anchor essential for the sensing of physical stimuli? Does activation by shear stress and membrane tension follow distinct pathways or is there a unifying gating paradigm? These and many other questions about the function of these channels in mechanophysiology and mechanopathology await to be answered. Given the intensity and extensiveness of research currently being undertaken to answer many of these questions, however, the answers may be revealed in the not too distant future.

REFERENCES

Akitake, B., Anishkin, A., Liu, N., & Sukharev, S. (2007). Straightening and sequential buckling of the pore-lining helices define the gating cycle of MscS. *Nature Structural & Molecular Biology, 14*(12), 1141–1149. http://dx.doi.org/10.1038/nsmb1341.

Anishkin, A., & Kung, C. (2013). Stiffened lipid platforms at molecular force foci. *Proceedings of the National Academy of Sciences of the United States of America, 110*(13), 4886–4892. http://dx.doi.org/10.1073/pnas.1302018110.

Arnadottir, J., & Chalfie, M. (2010). Eukaryotic mechanosensitive channels. *Annual Review of Biophysics, 39*, 111–137. http://dx.doi.org/10.1146/annurev.biophys.37.032807.125836.

Bae, C., Gnanasambandam, R., Nicolai, C., Sachs, F., & Gottlieb, P. A. (2013). Xerocytosis is caused by mutations that alter the kinetics of the mechanosensitive channel PIEZO1. *Proceedings of the National Academy of Sciences of the United States of America, 110*(12), E1162–E1168. http://dx.doi.org/10.1073/pnas.1219777110.

Bae, C., Sachs, F., & Gottlieb, P. A. (2011). The mechanosensitive ion channel Piezo1 is inhibited by the peptide GsMTx4. *Biochemistry, 50*(29), 6295–6300. http://dx.doi.org/10.1021/bi200770q.

Bagriantsev, S. N., Clark, K. A., & Minor, D. L., Jr. (2012). Metabolic and thermal stimuli control K(2P)2.1 (TREK-1) through modular sensory and gating domains. *The EMBO Journal, 31*(15), 3297–3308. http://dx.doi.org/10.1038/emboj.2012.171.

Bavi, N., Cortes, M., Cox, C. D., Rohde, P. R., Liu, W., Deitmer, J. W., … Martinac, B. (2016). The role of MscL amphipathic N terminus indicates a blueprint for bilayer-mediated gating of mechanosensitive channels. *Nature Communications* (accepted).

Bavi, N., Nakayama, Y., Bavi, O., Cox, C. D., Qin, Q. H., & Martinac, B. (2014). Biophysical implications of lipid bilayer rheometry for mechanosensitive channels. *Proceedings of the National Academy of Sciences of the United States of America, 111*(38), 13864–13869. http://dx.doi.org/10.1073/pnas.1409011111.

Bavi, O., Cox, C. D., Vossoughi, M., Naghdabadi, R., Jamali, Y., & Martinac, B. (2016). Influence of global and local membrane curvature on mechanosensitive ion channels: A finite element approach. *Membranes (Basel), 6*(1). http://dx.doi.org/10.3390/membranes6010014.

Bavi, O., Vossoughi, M., Naghdabadi, R., & Jamali, Y. (2016). The combined effect of hydrophobic mismatch and bilayer local bending on the regulation of mechanosensitive ion channels. *PLoS One, 11*(3), e0150578. http://dx.doi.org/10.1371/journal.pone.0150578.

Belyy, V., Anishkin, A., Kamaraju, K., Liu, N., & Sukharev, S. (2010). The tension-transmitting 'clutch' in the mechanosensitive channel MscS. *Nature Structural & Molecular Biology, 17*(4), 451–458. http://dx.doi.org/10.1038/nsmb.1775.

Berrier, C., Pozza, A., de Lacroix de Lavalette, A., Chardonnet, S., Mesneau, A., Jaxel, C., … Ghazi, A. (2013). The purified mechanosensitive channel TREK-1 is directly sensitive to membrane tension. *The Journal of Biological Chemistry, 288*(38), 27307–27314. http://dx.doi.org/10.1074/jbc.M113.478321.

Booth, I. R., Miller, S., Muller, A., & Lehtovirta-Morley, L. (2015). The evolution of bacterial mechanosensitive channels. *Cell Calcium, 57*(3), 140–150. http://dx.doi.org/10.1016/j.ceca.2014.12.011.

Borbiro, I., Badheka, D., & Rohacs, T. (2015). Activation of TRPV1 channels inhibits mechanosensitive Piezo channel activity by depleting membrane phosphoinositides. *Science Signaling, 8*(363), ra15. http://dx.doi.org/10.1126/scisignal.2005667.

Brehm, P., Kullberg, R., & Moody-Corbett, F. (1984). Properties of non-junctional acetylcholine receptor channels on innervated muscle of *Xenopus laevis*. *The Journal of Physiology, 350*, 631–648.

Brohawn, S. G., Campbell, E. B., & MacKinnon, R. (2014). Physical mechanism for gating and mechanosensitivity of the human TRAAK K+ channel. *Nature, 516*(7529), 126–130. http://dx.doi.org/10.1038/nature14013.

Brohawn, S. G., del Marmol, J., & MacKinnon, R. (2012). Crystal structure of the human K2P TRAAK, a lipid- and mechano-sensitive K+ ion channel. *Science, 335*(6067), 436–441. http://dx.doi.org/10.1126/science.1213808.

Brohawn, S. G., Su, Z., & MacKinnon, R. (2014). Mechanosensitivity is mediated directly by the lipid membrane in TRAAK and TREK1 K+ channels. *Proceedings of the National Academy of Sciences of the United States of America, 111*(9), 3614–3619. http://dx.doi.org/10.1073/pnas.1320768111.

Cantor, R. S. (1997). The lateral pressure profile in membranes: A physical mechanism of general anesthesia. *Biochemistry, 36*(9), 2339–2344. http://dx.doi.org/10.1021/bi9627323.

Cantor, R. S. (1999). Lipid composition and the lateral pressure profile in bilayers. *Biophysical Journal, 76*(5), 2625–2639. http://dx.doi.org/10.1016/S0006-3495(99)77415-1.

Chalfie, M. (2009). Neurosensory mechanotransduction. *Nature Reviews Molecular Cell Biology, 10*(1), 44–52. http://dx.doi.org/10.1038/nrm2595. pii:nrm2595.

Chandler, D. (2005). Interfaces and the driving force of hydrophobic assembly. *Nature, 437*(7059), 640–647. http://dx.doi.org/10.1038/nature04162.

Charras, G. T., Coughlin, M., Mitchison, T. J., & Mahadevan, L. (2008). Life and times of a cellular bleb. *Biophysical Journal, 94*(5), 1836–1853. http://dx.doi.org/10.1529/biophysj.107.113605.

Christensen, A. P., & Corey, D. P. (2007). TRP channels in mechanosensation: Direct or indirect activation? *Nature Reviews Neuroscience, 8*(7), 510–521. http://dx.doi.org/10.1038/nrn2149.

Chubinskiy-Nadezhdin, V. I., Negulyaev, Y. A., & Morachevskaya, E. A. (2011). Cholesterol depletion-induced inhibition of stretch-activated channels is mediated via actin rearrangement. *Biochemical and Biophysical Research Communications, 412*(1), 80–85. http://dx.doi.org/10.1016/j.bbrc.2011.07.046.

Cohen, D. M. (2005). TRPV4 and the mammalian kidney. *Pflugers Archiv, 451*(1), 168–175. http://dx.doi.org/10.1007/s00424-005-1456-9.

Corey, D. P., & Hudspeth, A. J. (1979). Response latency of vertebrate hair cells. *Biophysical Journal, 26*(3), 499–506. http://dx.doi.org/10.1016/S0006-3495(79)85267-4.

Coste, B., Mathur, J., Schmidt, M., Earley, T. J., Ranade, S., Petrus, M. J., … Patapoutian, A. (2010). Piezo1 and Piezo2 are essential components of distinct mechanically activated cation channels. *Science, 330*(6000), 55–60. http://dx.doi.org/10.1126/science.1193270.

Coste, B., Murthy, S. E., Mathur, J., Schmidt, M., Mechioukhi, Y., Delmas, P., & Patapoutian, A. (2015). Piezo1 ion channel pore properties are dictated by C-terminal region. *Nature Communications, 6*, 7223. http://dx.doi.org/10.1038/ncomms8223.

Coste, B., Xiao, B., Santos, J. S., Syeda, R., Grandl, J., Spencer, K. S., … Patapoutian, A. (2012). Piezo proteins are pore-forming subunits of mechanically activated channels. *Nature, 483*(7388), 176–181. http://dx.doi.org/10.1038/nature10812.

Cox, C. D., Bae, C., Ziegler, L., Hartley, S., Nikolova-Krstevski, V., Rohde, P. R., … Martinac, B. (2016). Removal of the mechanoprotective influence of the cytoskeleton reveals PIEZO1 is gated by bilayer tension. *Nature Communications, 7*, 10366. http://dx.doi.org/10.1038/ncomms10366.

Cox, C. D., Nakayama, Y., Nomura, T., & Martinac, B. (2015). The evolutionary 'tinkering' of MscS-like channels: Generation of structural and functional diversity. *Pflugers Archiv.* http://dx.doi.org/10.1007/s00424-014-1522-2.

Dai, J., & Sheetz, M. P. (1999). Membrane tether formation from blebbing cells. *Biophysical Journal, 77*(6), 3363–3370. http://dx.doi.org/10.1016/S0006-3495(99)77168-7.

Dong, Y. Y., Pike, A. C., Mackenzie, A., McClenaghan, C., Aryal, P., Dong, L., … Carpenter, E. P. (2015). K2P channel gating mechanisms revealed by structures of TREK-2 and a complex with Prozac. *Science, 347*(6227), 1256–1259. http://dx.doi.org/10.1126/science.1261512.

Edwards, M. D., Li, Y., Kim, S., Miller, S., Bartlett, W., Black, S., … Booth, I. R. (2005). Pivotal role of the glycine-rich TM3 helix in gating the MscS mechanosensitive channel. *Nature Structural & Molecular Biology, 12*(2), 113–119. http://dx.doi.org/10.1038/nsmb895.

Gao, Y., Cao, E., Julius, D., & Cheng, Y. (2016). TRPV1 structures in nanodiscs reveal mechanisms of ligand and lipid action. *Nature, 534*, 347–351.

Ge, J., Li, W., Zhao, Q., Li, N., Chen, M., Zhi, P., … Yang, M. (2015). Architecture of the mammalian mechanosensitive Piezo1 channel. *Nature, 527*(7576), 64–69. http://dx.doi.org/10.1038/nature15247.

Gnanasambandam, R., Bae, C., Gottlieb, P. A., & Sachs, F. (2015). Ionic selectivity and permeation properties of human PIEZO1 channels. *PLoS One, 10*(5), e0125503. http://dx.doi.org/10.1371/journal.pone.0125503.

Gottlieb, P., Folgering, J., Maroto, R., Raso, A., Wood, T. G., Kurosky, A., … Honoré, E. (2008). Revisiting TRPC1 and TRPC6 mechanosensitivity. *Pflugers Archiv, 455*(6), 1097–1103. http://dx.doi.org/10.1007/s00424-007-0359-3.

Gottlieb, P. A., Bae, C., & Sachs, F. (2012). Gating the mechanical channel Piezo1: A comparison between whole-cell and patch recording. *Channels (Austin), 6*(4), 282–289. http://dx.doi.org/10.4161/chan.21064.

Guharay, F., & Sachs, F. (1984). Stretch-activated single ion channel currents in tissue-cultured embryonic chick skeletal muscle. *The Journal of Physiology, 352*, 685–701.

Gullingsrud, J., & Schulten, K. (2004). Lipid bilayer pressure profiles and mechanosensitive channel gating. *Biophysical Journal, 86*(6), 3496–3509. http://dx.doi.org/10.1529/biophysj.103.034322.

Guo, J., Sachs, F., & Meng, F. (2014). Fluorescence-based force/tension sensors: A novel tool to visualize mechanical forces in structural proteins in live cells. *Antioxidants & Redox Signaling, 20*(6), 986–999. http://dx.doi.org/10.1089/ars.2013.5708.

Hamill, O. P. (1983). Potassium and chloride current in red blood cells. In B. Sakmann, & E. Neher (Eds.), *Single-channel Recording* (pp. 451–471). New York.

Hamill, O. P., & Martinac, B. (2001). Molecular basis of mechanotransduction in living cells. *Physiological Reviews, 81*(2), 685–740.

Hamill, O. P., & McBride, D. W., Jr. (1997). Induced membrane hypo/hyper-mechanosensitivity: A limitation of patch-clamp recording. *Annual Review of Physiology, 59*, 621–631.

Hamilton, E. S., Schlegel, A. M., & Haswell, E. S. (2015). United in diversity: Mechanosensitive ion channels in plants. *Annual Review of Plant Physiology, 66*, 113–137. http://dx.doi.org/10.1146/annurev-arplant-043014-114700.

Häse, C. C., Le Dain, A. C., & Martinac, B. (1995). Purification and functional reconstitution of the recombinant large mechanosensitive ion channel (MscL) of *Escherichia coli*. *The Journal of Biological Chemistry, 270*(31), 18329–18334.

Honoré, E. (2007). The neuronal background K2P channels: Focus on TREK1. *Nature Reviews Neuroscience, 8*(4), 251–261. http://dx.doi.org/10.1038/nrn2117. pii:nrn2117.

Honoré, E., Maingret, F., Lazdunski, M., & Patel, A. J. (2002). An intracellular proton sensor commands lipid- and mechano-gating of the K(+) channel TREK-1. *The EMBO Journal, 21*(12), 2968–2976. http://dx.doi.org/10.1093/emboj/cdf288.

Honoré, E., Patel, A. J., Chemin, J., Suchyna, T., & Sachs, F. (2006). Desensitization of mechano-gated K2P channels. *Proceedings of the National Academy of Sciences of the United States of America, 103*(18), 6859–6864. http://dx.doi.org/10.1073/pnas.0600463103.

Howard, J., Roberts, W. M., & Hudspeth, A. J. (1988). Mechanoelectrical transduction by hair cells. *Annual Review of Biophysics and Biophysical Chemistry, 17*, 99–124. http://dx.doi.org/10.1146/annurev.bb.17.060188.000531.

Huang, H., Bae, C., Sachs, F., & Suchyna, T. M. (2013). Caveolae regulation of mechanosensitive channel function in myotubes. *PLoS One, 8*(8), e72894. http://dx.doi.org/10.1371/journal.pone.0072894.

Hudspeth, A. J. (1989). How the ear's works work. *Nature, 341*(6241), 397–404. http://dx.doi.org/10.1038/341397a0.

Hulce, J. J., Cognetta, A. B., Niphakis, M. J., Tully, S. E., & Cravatt, B. F. (2013). Proteome-wide mapping of cholesterol-interacting proteins in mammalian cells. *Nature Methods, 10*(3), 259–264. http://dx.doi.org/10.1038/nmeth.2368.

Hung, W. C., Yang, J. R., Yankaskas, C. L., Wong, B. S., Wu, P. H., Pardo-Pastor, C., … Konstantopoulos, K. (2016). Confinement sensing and signal optimization via Piezo1/PKA and myosin II pathways. *Cell Reports, 15*(7), 1430–1441. http://dx.doi.org/10.1016/j.celrep.2016.04.035.

Ingber, D. E. (2006). Cellular mechanotransduction: Putting all the pieces together again. *TheFASEB Journal, 20*(7), 811–827. http://dx.doi.org/10.1096/fj.05–5424rev. pii: 20/7/811.

Iscla, I., Wray, R., Eaton, C., & Blount, P. (2015). Scanning MscL channels with targeted post-translational modifications for functional alterations. *PLoS One, 10*(9), e0137994. http://dx.doi.org/10.1371/journal.pone.0137994.

Jasti, J., Furukawa, H., Gonzales, E. B., & Gouaux, E. (2007). Structure of acid-sensing ion channel 1 at 1.9 A resolution and low pH. *Nature, 449*(7160), 316–323. http://dx.doi.org/10.1038/nature06163.

Jensen, M. O., & Mouritsen, O. G. (2004). Lipids do influence protein function-the hydrophobic matching hypothesis revisited. *Biochimica et Biophysica Acta, 1666*(1–2), 205–226. http://dx.doi.org/10.1016/j.bbamem.2004.06.009.

Jia, Z., Ikeda, R., Ling, J., Viatchenko-Karpinski, V., & Gu, J. G. (2016). Regulation of Piezo2 mechanotransduction by static plasma membrane tension in primary afferent neurons. *The Journal of Biological Chemistry, 291*(17), 9087–9104. http://dx.doi.org/10.1074/jbc.M115.692384.

Kamajaya, A., Kaiser, J. T., Lee, J., Reid, M., & Rees, D. C. (2014). The structure of a conserved piezo channel domain reveals a topologically distinct beta sandwich fold. *Structure, 22*(10), 1520–1527. http://dx.doi.org/10.1016/j.str.2014.08.009.

Katta, S., Krieg, M., & Goodman, M. B. (2015). Feeling force: Physical and physiological principles enabling sensory mechanotransduction. *The Annual Review of Cell and Developmental Biology, 31*, 347–371. http://dx.doi.org/10.1146/annurev-cellbio-100913-013426.

Katz, B. (1950). Depolarization of sensory terminals and the initiation of impulses in the muscle spindle. *The Journal of Physiology, 111*(3–4), 261–282.

Kung, C. (2005). A possible unifying principle for mechanosensation. *Nature, 436*(7051), 647–654. http://dx.doi.org/10.1038/nature03896. pii:nature03896.

Lee, A. G. (2004). How lipids affect the activities of integral membrane proteins. *Biochimica et Biophysica Acta, 1666*(1–2), 62–87. http://dx.doi.org/10.1016/j.bbamem.2004.05.012.

Lee, W., Leddy, H. A., Chen, Y., Lee, S. H., Zelenski, N. A., McNulty, A. L., … Liedtke, W. B. (2014). Synergy between Piezo1 and Piezo2 channels confers high-strain mechanosensitivity to articular cartilage. *Proceedings of the National Academy of Sciences of the United States of America, 111*(47), E5114–E5122. http://dx.doi.org/10.1073/pnas.1414298111.

Lewis, A. H., & Grandl, J. (2015). Mechanical sensitivity of Piezo1 ion channels can be tuned by cellular membrane tension. *Elife, 4*. http://dx.doi.org/10.7554/eLife.12088.

Li, J., Hou, B., & Beech, D. J. (2015). Endothelial Piezo1: Life depends on it. *Channels (Austin), 9*(1), 1–2. http://dx.doi.org/10.4161/19336950.2014.986623.

Li, J., Hou, B., Tumova, S., Muraki, K., Bruns, A., Ludlow, M. J., … Beech, D. J. (2014). Piezo1 integration of vascular architecture with physiological force. *Nature, 515*(7526), 279–282. http://dx.doi.org/10.1038/nature13701.

Liedtke, W., Choe, Y., Marti-Renom, M. A., Bell, A. M., Denis, C. S., Sali, A., … Heller, S. (2000). Vanilloid receptor-related osmotically activated channel (VR-OAC), a candidate vertebrate osmoreceptor. *Cell, 103*(3), 525–535. pii:S0092-8674(00)00143-4.

Liu, S. Q., Yen, M., & Fung, Y. C. (1994). On measuring the third dimension of cultured endothelial cells in shear flow. *Proceedings of the National Academy of Sciences of the United States of America, 91*(19), 8782–8786.

Lolicato, M., Riegelhaupt, P. M., Arrigoni, C., Clark, K. A., & Minor, D. L., Jr. (2014). Transmembrane helix straightening and buckling underlies activation of mechanosensitive and thermosensitive K(2P) channels. *Neuron, 84*(6), 1198–1212. http://dx.doi.org/10.1016/j.neuron.2014.11.017.

Loukin, S., Zhou, X., Su, Z., Saimi, Y., & Kung, C. (2010). Wild-type and brachyolmia-causing mutant TRPV4 channels respond directly to stretch force. *The Journal of Biological Chemistry, 285*(35), 27176–27181. http://dx.doi.org/10.1074/jbc.M110.143370.

Lundbaek, J. A., Collingwood, S. A., Ingolfsson, H. I., Kapoor, R., & Andersen, O. S. (2010). Lipid bilayer regulation of membrane protein function: Gramicidin channels as molecular force probes. *Journal of the Royal Society*, 7(44), 373–395. http://dx.doi.org/10.1098/rsif.2009.0443.

Luo, T., Mohan, K., Iglesias, P. A., & Robinson, D. N. (2013). Molecular mechanisms of cellular mechanosensing. *Nature Materials, 12*(11), 1064–1071. http://dx.doi.org/10.1038/nmat3772.

Maingret, F., Patel, A. J., Lesage, F., Lazdunski, M., & Honoré, E. (2000). Lysophospholipids open the two-pore domain mechano-gated K(+) channels TREK-1 and TRAAK. *The Journal of Biological Chemistry, 275*(14), 10128–10133.
Maksaev, G., Milac, A., Anishkin, A., Guy, H. R., & Sukharev, S. (2011). Analyses of gating thermodynamics and effects of deletions in the mechanosensitive channel TREK-1: Comparisons with structural models. *Channels (Austin), 5*(1), 34–42.
Malcolm, H. R., Blount, P., & Maurer, J. A. (2015). The mechanosensitive channel of small conductance (MscS) functions as a Jack-in-the box. *Biochimica et Biophysica Acta, 1848*(1 Pt A), 159–166. http://dx.doi.org/10.1016/j.bbamem.2014.10.022.
Malcolm, H. R., & Maurer, J. A. (2012). The mechanosensitive channel of small conductance (MscS) superfamily: Not just mechanosensitive channels anymore. *Chembiochem, 13*(14), 2037–2043. http://dx.doi.org/10.1002/cbic.201200410.
Markin, V. S., & Hudspeth, A. J. (1995). Gating-spring models of mechanoelectrical transduction by hair cells of the internal ear. *Annual Review of Biophysics and Biomolecular Structure, 24*, 59–83. http://dx.doi.org/10.1146/annurev.bb.24.060195.000423.
Martinac, B. (2004). Mechanosensitive ion channels: Molecules of mechanotransduction. *Journal of Cell Science, 117*(Pt 12), 2449–2460. http://dx.doi.org/10.1242/jcs.01232.
Martinac, B. (2011). Bacterial mechanosensitive channels as a paradigm for mechanosensory transduction. *Cell Physiology and Biochemistry, 28*(6), 1051–1060.
Martinac, B. (2012). Mechanosensitive ion channels: An evolutionary and scientific tour de force in mechanobiology. *Channels (Austin), 6*(4), 211–213. http://dx.doi.org/10.4161/chan.22047.
Martinac, B., Adler, J., & Kung, C. (1990). Mechanosensitive ion channels of *E. coli* activated by amphipaths. *Nature, 348*(6298), 261–263. http://dx.doi.org/10.1038/348261a0.
Martinac, B., Buechner, M., Delcour, A. H., Adler, J., & Kung, C. (1987). Pressure-sensitive ion channel in *Escherichia coli*. *Proceedings of the National Academy of Sciences of the United States of America, 84*(8), 2297–2301.
Martinac, B., & Cox, C. D. (2016). Mechanosensory transduction: Focus on ion channels. *Comprehensive Biophysics*. Elsevier.
McHugh, B. J., Murdoch, A., Haslett, C., & Sethi, T. (2012). Loss of the integrin-activating transmembrane protein Fam38A (Piezo1) promotes a switch to a reduced integrin-dependent mode of cell migration. *PLoS One, 7*(7), e40346. http://dx.doi.org/10.1371/journal.pone.0040346.
Moe, P., & Blount, P. (2005). Assessment of potential stimuli for mechano-dependent gating of MscL: Effects of pressure, tension, and lipid headgroups. *Biochemistry, 44*(36), 12239–12244. http://dx.doi.org/10.1021/bi0509649.
Morris, C. E. (2001). Mechanoprotection of the plasma membrane in neurons and other non-erythroid cells by the spectrin-based membrane skeleton. *Cellular & Molecular Biology Letters, 6*(3), 703–720.
Morris, C. E. (2011). Voltage-gated channel mechanosensitivity: Fact or friction? *Frontiers in Physiology, 2*, 25. http://dx.doi.org/10.3389/fphys.2011.00025.
Mukherjee, N., Jose, M. D., Birkner, J. P., Walko, M., Ingolfsson, H. I., Dimitrova, A., … Kocer, A. (2014). The activation mode of the mechanosensitive ion channel, MscL, by lysophosphatidylcholine differs from tension-induced gating. *The FASEB Journal, 28*(10), 4292–4302. http://dx.doi.org/10.1096/fj.14-251579.
Naismith, J. H., & Booth, I. R. (2012). Bacterial mechanosensitive channels—MscS: Evolution's solution to creating sensitivity in function. *Annual Review of Biophysics, 41*, 157–177. http://dx.doi.org/10.1146/annurev-biophys-101211-113227.
Najem, J. S., Dunlap, M. D., Rowe, I. D., Freeman, E. C., Grant, J. W., Sukharev, S., & Leo, D. J. (2015). Activation of bacterial channel MscL in mechanically stimulated droplet interface bilayers. *Scientific Reports, 5*, 13726. http://dx.doi.org/10.1038/srep13726.

Nakayama, Y., Yoshimura, K., & Iida, H. (2012). Organellar mechanosensitive channels in fission yeast regulate the hypo-osmotic shock response. *Nature Communications, 3*, 1020. http://dx.doi.org/10.1038/ncomms2014.

Niklas, K. J. (2014). The evolutionary-developmental origins of multicellularity. *American Journal of Botany, 101*(1), 6–25. http://dx.doi.org/10.3732/ajb.1300314.

Nomura, T., Cox, C. D., Bavi, N., Sokabe, M., & Martinac, B. (2015). Unidirectional incorporation of a bacterial mechanosensitive channel into liposomal membranes. *The FASEB Journal, 29*(10), 4334–4345. http://dx.doi.org/10.1096/fj.15-275198.

Nomura, T., Cranfield, C. G., Deplazes, E., Owen, D. M., Macmillan, A., Battle, A. R., … Martinac, B. (2012). Differential effects of lipids and lyso-lipids on the mechanosensitivity of the mechanosensitive channels MscL and MscS. *Proceedings of the National Academy of Sciences of the United States of America, 109*(22), 8770–8775. http://dx.doi.org/10.1073/pnas.1200051109.

O'Conor, C. J., Leddy, H. A., Benefield, H. C., Liedtke, W. B., & Guilak, F. (2014). TRPV4-mediated mechanotransduction regulates the metabolic response of chondrocytes to dynamic loading. *Proceedings of the National Academy of Sciences of the United States of America, 111*(4), 1316–1321. http://dx.doi.org/10.1073/pnas.1319569111.

Opsahl, L. R., & Webb, W. W. (1994). Lipid-glass adhesion in giga-sealed patch-clamped membranes. *Biophysical Journal, 66*(1), 75–79. http://dx.doi.org/10.1016/S0006-3495(94)80752-0.

Patel, A. J., Honoré, E., Maingret, F., Lesage, F., Fink, M., Duprat, F., & Lazdunski, M. (1998). A mammalian two pore domain mechano-gated S-like K+ channel. *The EMBO Journal, 17*(15), 4283–4290. http://dx.doi.org/10.1093/emboj/17.15.4283.

Pathak, M. M., Nourse, J. L., Tran, T., Hwe, J., Arulmoli, J., Le, D. T., … Tombola, F. (2014). Stretch-activated ion channel Piezo1 directs lineage choice in human neural stem cells. *Proceedings of the National Academy of Sciences of the United States of America, 111*(45), 16148–16153. http://dx.doi.org/10.1073/pnas.1409802111.

Perozo, E., Cortes, D. M., Sompornpisut, P., Kloda, A., & Martinac, B. (2002). Open channel structure of MscL and the gating mechanism of mechanosensitive channels. *Nature, 418*(6901), 942–948. http://dx.doi.org/10.1038/nature00992.

Perozo, E., Kloda, A., Cortes, D. M., & Martinac, B. (2002). Physical principles underlying the transduction of bilayer deformation forces during mechanosensitive channel gating. *Nature Structural & Molecular Biology, 9*(9), 696–703. http://dx.doi.org/10.1038/nsb827. pii:nsb827.

Pivetti, C. D., Yen, M. R., Miller, S., Busch, W., Tseng, Y. H., Booth, I. R., & Saier, M. H., Jr. (2003). Two families of mechanosensitive channel proteins. *Microbiology and Molecular Biology Reviews, 67*(1), 66–85. table of contents.

Pliotas, C., Dahl, A. C., Rasmussen, T., Mahendran, K. R., Smith, T. K., Marius, P., … Naismith, J. H. (2015). The role of lipids in mechanosensation. *Nature Structural & Molecular Biology, 22*(12), 991–998. http://dx.doi.org/10.1038/nsmb.3120.

Poole, K., Herget, R., Lapatsina, L., Ngo, H. D., & Lewin, G. R. (2014). Tuning Piezo ion channels to detect molecular-scale movements relevant for fine touch. *Nature Communications, 5*, 3520. http://dx.doi.org/10.1038/ncomms4520.

Prager-Khoutorsky, M., Khoutorsky, A., & Bourque, C. W. (2014). Unique interweaved microtubule scaffold mediates osmosensory transduction via physical interaction with TRPV1. *Neuron, 83*(4), 866–878. http://dx.doi.org/10.1016/j.neuron.2014.07.023.

Prole, D. L., & Taylor, C. W. (2013). Identification and analysis of putative homologues of mechanosensitive channels in pathogenic protozoa. *PLoS One, 8*(6), e66068. http://dx.doi.org/10.1371/journal.pone.0066068.

Qi, Y., Andolfi, L., Frattini, F., Mayer, F., Lazzarino, M., & Hu, J. (2015). Membrane stiffening by STOML3 facilitates mechanosensation in sensory neurons. *Nature Communications, 6*, 8512. http://dx.doi.org/10.1038/ncomms9512.

Rajendran, L., & Simons, K. (2005). Lipid rafts and membrane dynamics. *Journal of Cell Science, 118*(Pt 6), 1099–1102. http://dx.doi.org/10.1242/jcs.01681.

Ranade, S. S., Syeda, R., & Patapoutian, A. (2015). Mechanically activated ion channels. *Neuron, 87*(6), 1162–1179. http://dx.doi.org/10.1016/j.neuron.2015.08.032.

Ranade, S. S., Woo, S. H., Dubin, A. E., Moshourab, R. A., Wetzel, C., Petrus, M., … Patapoutian, A. (2014). Piezo2 is the major transducer of mechanical forces for touch sensation in mice. *Nature, 516*(7529), 121–125. http://dx.doi.org/10.1038/nature13980.

Retailleau, K., Arhatte, M., Demolombe, S., Jodar, M., Baudrie, V., Offermanns, S., … Duprat, F. (2016). Smooth muscle filamin A is a major determinant of conduit artery structure and function at the adult stage. *Pflugers Archiv*. http://dx.doi.org/10.1007/s00424-016-1813-x.

Retailleau, K., Duprat, F., Arhatte, M., Ranade, S. S., Peyronnet, R., Martins, J. R., … Honore, E. (2015). Piezo1 in smooth muscle cells is involved in hypertension-dependent arterial remodeling. *Cell Reports, 13*(6), 1161–1171. http://dx.doi.org/10.1016/j.celrep.2015.09.072.

Sachs, F. (2010). Stretch-activated ion channels: What are they? *Physiology, 25*(1), 50–56. http://dx.doi.org/10.1152/physiol.00042.2009.

Sachs, F. (2015). Mechanical transduction by ion channels: A cautionary tale. *World Journal of Neurology, 5*(3), 74–87.

Schiffhauer, E. S., Luo, T., Mohan, K., Srivastava, V., Qian, X., Griffis, E. R., … Robinson, D. N. (2016). Mechanoaccumulative elements of the mammalian actin cytoskeleton. *Current Biology*. http://dx.doi.org/10.1016/j.cub.2016.04.007.

Shaikh, S., Cox, C. D., Nomura, T., & Martinac, B. (2014). Energetics of gating MscS by membrane tension in azolectin liposomes and giant spheroplasts. *Channels (Austin), 8*(3).

Spagnoli, C., Beyder, A., Besch, S., & Sachs, F. (2008). Atomic force microscopy analysis of cell volume regulation. *Physical Review E: Statistical, Nonlinear, and Soft Matter Physics, 78*(3 Pt 1), 031916.

Sukharev, S. (2002). Purification of the small mechanosensitive channel of *Escherichia coli* (MscS): The subunit structure, conduction, and gating characteristics in liposomes. *Biophysical Journal, 83*(1), 290–298. http://dx.doi.org/10.1016/S0006-3495(02)75169-2.

Sukharev, S. I., Blount, P., Martinac, B., Blattner, F. R., & Kung, C. (1994). A large-conductance mechanosensitive channel in *E. coli* encoded by mscL alone. *Nature, 368*(6468), 265–268. http://dx.doi.org/10.1038/368265a0.

Sukharev, S. I., Martinac, B., Arshavsky, V. Y., & Kung, C. (1993). Two types of mechanosensitive channels in the *Escherichia coli* cell envelope: Solubilization and functional reconstitution. *Biophysical Journal, 65*(1), 177–183. http://dx.doi.org/10.1016/S0006-3495(93)81044-0.

Syeda, R., Xu, J., Dubin, A. E., Coste, B., Mathur, J., Huynh, T., … Patapoutian, A. (2015). Chemical activation of the mechanotransduction channel Piezo1. *Elife, 4*. http://dx.doi.org/10.7554/eLife.07369.

Teng, J., Loukin, S., Anishkin, A., & Kung, C. (2015). The force-from-lipid (FFL) principle of mechanosensitivity, at large and in elements. *Pflugers Archiv*. http://dx.doi.org/10.1007/s00424-014-1530-2.

Teng, J., Loukin, S. H., Anishkin, A., & Kung, C. (2015). L596-W733 bond between the start of the S4-S5 linker and the TRP box stabilizes the closed state of TRPV4 channel. *Proceedings of the National Academy of Sciences of the United States of America, 112*(11), 3386–3391. http://dx.doi.org/10.1073/pnas.1502366112.

Treptow, W., & Klein, M. L. (2010). The membrane-bound state of K2P potassium channels. *Journal of the American Chemical Society, 132*(23), 8145–8151. http://dx.doi.org/10.1021/ja102191s.

Wang, N., Butler, J. P., & Ingber, D. E. (1993). Mechanotransduction across the cell surface and through the cytoskeleton. *Science, 260*(5111), 1124–1127.
Watanabe, H., Vriens, J., Prenen, J., Droogmans, G., Voets, T., & Nilius, B. (2003). Anandamide and arachidonic acid use epoxyeicosatrienoic acids to activate TRPV4 channels. *Nature, 424*(6947), 434–438. http://dx.doi.org/10.1038/nature01807.
Woo, S. H., Ranade, S., Weyer, A. D., Dubin, A. E., Baba, Y., Qiu, Z., … Patapoutian, A. (2014). Piezo2 is required for Merkel-cell mechanotransduction. *Nature, 509*(7502), 622–626. http://dx.doi.org/10.1038/nature13251.
Yoo, J., & Cui, Q. (2009). Curvature generation and pressure profile modulation in membrane by lysolipids: Insights from coarse-grained simulations. *Biophysical Journal, 97*(8), 2267–2276. http://dx.doi.org/10.1016/j.bpj.2009.07.051.
Yoshimura, K., Nomura, T., & Sokabe, M. (2004). Loss-of-function mutations at the rim of the funnel of mechanosensitive channel MscL. *Biophysical Journal, 86*(4), 2113–2120. http://dx.doi.org/10.1016/S0006-3495(04)74270-8.
Zhang, W., Cheng, L. E., Kittelmann, M., Li, J., Petkovic, M., Cheng, T., … Jan, Y. N. (2015). Ankyrin repeats convey force to gate the NOMPC mechanotransduction channel. *Cell, 162*(6), 1391–1403. http://dx.doi.org/10.1016/j.cell.2015.08.024.
Zhao, Q., Wu, K., Geng, J., Chi, S., Wang, Y., Zhi, P., … Xiao, B. (2016). Ion permeation and mechanotransduction mechanisms of mechanosensitive piezo channels. *Neuron, 89*(6), 1248–1263. http://dx.doi.org/10.1016/j.neuron.2016.01.046.
Zheng, J. (2013). Molecular mechanism of TRP channels. *Comprehensive Physiology, 3*(1), 221–242. http://dx.doi.org/10.1002/cphy.c120001.
Zhong, D., & Blount, P. (2014). Electrostatics at the membrane define MscL channel mechanosensitivity and kinetics. *The FASEB Journal, 28*(12), 5234–5241. http://dx.doi.org/10.1096/fj.14-259309.
Zhong, D., Yang, L. M., & Blount, P. (2014). Dynamics of protein-protein interactions at the MscL periplasmic-lipid interface. *Biophysical Journal, 106*(2), 375–381. http://dx.doi.org/10.1016/j.bpj.2013.12.006.
Zimmerberg, J., & Kozlov, M. M. (2006). How proteins produce cellular membrane curvature. *Nature Reviews Molecular Cell Biology,* 7(1), 9–19. http://dx.doi.org/10.1038/nrm1784.

CHAPTER FOUR

Genetic Diseases of PIEZO1 and PIEZO2 Dysfunction

S.L. Alper
Beth Israel Deaconess Medical Center and Harvard Medical School, Boston, MA, United States
E-mail: salper@bidmc.harvard.edu

Contents

Abstract

Mutations in the genes encoding the mechanosensitive cation channels PIEZO1 and PIEZO2 are responsible for multiple hereditary human diseases. Loss-of-function mutations in the human *PIEZO1* gene cause autosomal recessive congenital lymphatic dysplasia. Gain-of-function mutations in the human *PIEZO1* gene cause the autosomal dominant hemolytic anemia, hereditary xerocytosis (also known as dehydrated stomatocytosis). Loss-of-function mutations in the human *PIEZO2* gene cause an autosomal recessive syndrome of muscular atrophy with perinatal respiratory distress, arthrogryposis, and scoliosis. Gain-of-function mutations in the human *PIEZO2* gene cause three clinical types of autosomal dominant distal arthrogryposis. This chapter will review the hereditary diseases caused by mutations in the *PIEZO* genes and will discuss additional physiological systems in which PIEZO channel dysfunction may contribute to human disease pathophysiology.

Current Topics in Membranes, Volume 79
ISSN 1063-5823
http://dx.doi.org/10.1016/bs.ctm.2017.01.001

1. INTRODUCTION

Cells are subjected to mechanical forces and are equipped with force sensors that include mechanosensitive receptors (such as integrins) and mechanosensitive ion channels. Mechanosensitive cation channels were first described as responding to application of negative pressure within a patch-clamp pipette or in response to cell swelling or stretch (see Guharay & Sachs, 1984; Lane, McBride, & Hamill, 1991; Morris, 1990; Yang & Sachs, 1989 as examples of early studies). The first molecular identification of the mechanosensitive cation channel polypeptides PIEZO1 and PIEZO2 of mouse by Coste et al. (2010) employed siRNA knockdown analysis of candidate polytopic transmembrane proteins. PIEZO proteins are expressed widely among metazoans and are the largest transmembrane proteins yet identified, with human PIEZO1 and PIEZO2 encompassing 2521 and 2752 amino acid residues, respectively. The two human polypeptide gene products exhibit 43% amino acid identity and 59% similarity, with highest sequence identity in the C-terminal third of the polypeptides.

PIEZO proteins function as nonspecific monovalent cation channels with lower permeability to divalent cations (Gnanasambandam, Bae, Gottlieb, & Sachs, 2015), characterized by rapid activation in response to mechanical displacement of the cell surface membrane from above (Bae, Sachs, & Gottlieb, 2011; Coste et al., 2010) or from below (Poole, Herget, Lapatsina, Ngo, & Lewin, 2014), with rapid subsequent inactivation (Coste et al., 2010). The rapid inactivation distinguishes PIEZO channels from the recently described, structurally unrelated tentonin-3/TMEM150c cation channels, which inactivate 8- to 10-fold more slowly than PIEZO1 (Hong et al., 2016), but share with most PIEZO proteins their pharmacological sensitivity to the nonspecific inhibitors Gd^{3+} and ruthenium red and to the more specific inhibitor peptide, *Grammostola spatulata* mechanotoxin-4 (GsMTx4) (Bae et al., 2011; Hong et al., 2016; Suchyna et al., 2000). Intrinsic bilayer stretch or distortion is sufficient for mechanical activation of the cation channel activity of PIEZO polypeptides (Cox et al., 2016; Syeda et al., 2016), although channel activity is subject to (at least) indirect modulation by cytoskeletal and scaffolding proteins inside cells (Gottlieb, Bae, & Sachs, 2012) and likely by ligands of extracellular matrix as well. PIEZO1 can also be activated by the novel drug ligand, Yoda1, in the absence of concurrent mechanical stress (Cahalan et al., 2015; Syeda et al., 2015, 2016).

The structure of mouse PIEZO1 has been determined by cryo-electron microscopic single particle analysis at 4.8 Å resolution (Fig. 1). The structure reveals a homotrimeric propeller structure in which each protomer has ~14 transmembrane spans (Ge et al., 2015) (despite hydropathy profile estimates

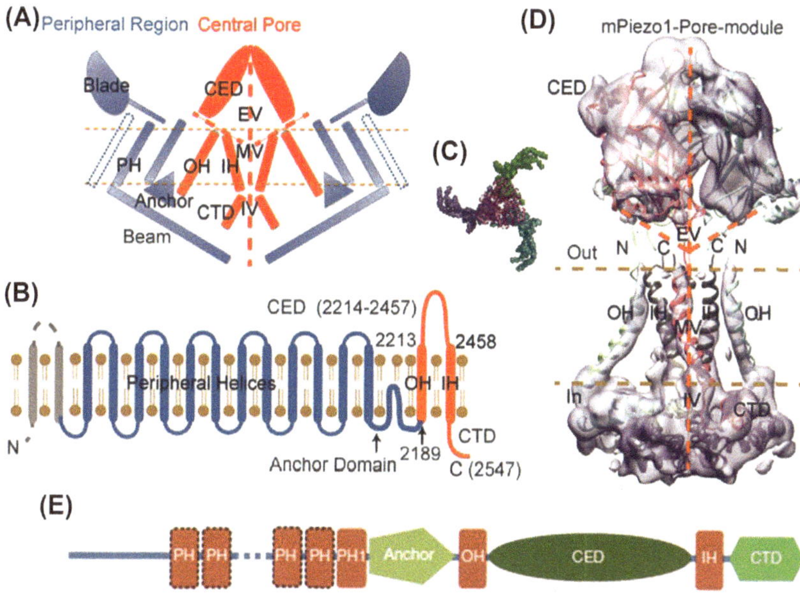

Figure 1 Schematic of mouse PIEZO1 structure based on single particle cryo-electron microscopic (cryo-EM) imaging at 4.8 Å resolution (Ge et al., 2015), based on PDB structures 4RAX and 3JAC. (A) Schematic diagram of the architecture of mouse PIEZO1 central pore region (two subunits of the homotrimer) encompassing CTD (C-terminal extracellular domain), IH (inner helix), CED (extracellular domain) and OH (outer helix), and the peripheral mechanotransduction region encompassing the anchor domain, peripheral helices (PH), and blades. (B) Schematic of mouse PIEZO1 2D topographical disposition in the lipid bilayer. (C) PyMOL rendering of PIEZO1 trimer viewed from above showing central pore region and three peripheral regions with blades proposed to serve as mechanosensors. (D) Transparent surface representation of the cryo-EM density map of the hypothesized central pore region depicted in panel A. (E) Linear depiction of C-terminal subdomain structure of mouse PIEZO1. *Panels A, B, and D are modified from Zhao, Q., Wu, K., Geng, J., Chi, S., Wang, Y., Zhi, P., …, Xiao, B. (2016). Ion permeation and mechanotransduction mechanisms of mechanosensitive piezo channels.* Neuron, *89(6), 1248–1263. http://dx.doi.org/10.1016/j.neuron.2016.01.046; C and E are modified from Akyuz, N., & Holt, J.R. (2016). Plug-N-Play: Mechanotransduction goes modular.* Neuron, *89(6), 1128–1130. http://dx.doi.org/10.1016/j.neuron.2016.02.041 and Ge, J., Li, W., Zhao, Q., Li, N., Chen, M., Zhi, P., …, Yang, M. (2015). Architecture of the mammalian mechanosensitive Piezo1 channel.* Nature, *527(7576), 64–69. http://dx.doi.org/10.1038/nature15247, respectively.*

of >30 transmembrane spans). The conducting pore has been proposed to comprise the final two transmembrane spans (OH and IH) (Coste et al., 2015) with the final span (IH) of each protomer within the trimer contributing the pore-lining residues (Zhao et al., 2016) (Fig. 1). The large, modular, mechanosensing domain comprises subdomains described as anchor, beam, blade, and peripheral helices (Fig. 1) and confers mechanosensitive activity upon engineered fusion to a mechanically insensitive, trimeric ASIC cation channel (Zhao et al., 2016).

Despite the above impressive advances in PIEZO1 structure determination, the state of anti-PIEZO1 antibodies remains problematic. At least one antibody has been validated for immunoblot by knockdown and knockout experiments (Li et al., 2014). However, this antibody also has extensive immunoblot reactivity with non-PIEZO1 polypeptides and has not been validated as an immunocytochemical or immunohistological reagent, since its immunostaining signal is retained in knockout mice despite the ability of competitor peptide to reduce signal. Thus, reports of PIEZO1 immunolocalization should be carefully evaluated for appropriate control studies. In contrast, at least one antibody to PIEZO2 has been validated for immunolocalization studies by demonstration of loss of reactivity in knockdown and knockout tissue, as well as colocalization with PIEZO2-GFP fusion protein (Woo et al., 2014).

2. PIEZO1 LOSS-OF-FUNCTION: CONGENITAL LYMPHATIC DYSPLASIA

PIEZO1 loss-of-function mutations cause autosomal recessive generalized lymphatic dysplasia (Fotiou et al., 2015; Lukacs et al., 2015). The lymphatic system is an independent circulatory system, distinct from the vascular endothelium-lined blood circulation that contributes to regulation of tissue turgor. Lymphatic endothelial cells are characterized by distinct surface markers and regulated by at least partially distinct transcription factors. The lymph transported by the lymphatic system is a protein-rich extracellular fluid transporting antigens and antigen-presenting cells to lymph nodes as part of the immune recognition and tolerance systems. Backflow of lymph is prevented by a system of valves. The initial lymphatics are characterized by a leaky endothelium. The endothelia of larger collecting lymphatic vessels have tighter intercellular junctions, and are invested with lymphatic smooth muscle cells. These contractile cells promote lymph flow toward the thoracic duct, which empties into the central venous circulation. Lymphatic contractions are sensitive to physical distention associated with Ca^{2+} signaling under the regulation of VEGF-C, endothelin, nitric oxide, and other ligands (Padera, Meijer, & Munn, 2016).

Failure of lymphatics to drain interstitial fluid from the extracellular space back into the blood leads to lymphedema, which can lead to local inflammation, promoting adipose hypertrophy, and progressive fibrosis and deformity. Acquired lymphedema reflecting obstruction secondary to infection, trauma, malignancy, surgery, radiation, or trauma affects nearly 1% of the population. In contrast, primary lymphedema is rare, found in about 0.01% of those under 20 years of age. Mutations in nearly 30 genes have been associated with intrinsic or syndromic lymphatic abnormalities (Padera et al., 2016). One of these syndromes, Hennekam lymphangiectasia-lymphedema syndrome, has been associated in ~50% of cases with loss-of-function mutations in the genes *CCBE1* and *FAT4*. The secreted protein CCBE1 (collagen and calcium-binding EGF domain-containing protein 1) is an essential part of VEGF-C/VEGFR-3 signaling in lymphatics (Le Guen et al., 2014) and is the mammalian homolog of the zebrafish protein "full of fluid." Mutations in VEGF-C and VEGFR-3 cause Nonne-Milroy lymphedema and Milroy-like disease. The Hennekam gene product FAT4 is an atypical cadherin of nearly 5000 aa in length genomic knockout of which dysregulates planar cell polarity, producing renal cystogenesis in mice and flies (Saburi et al., 2008).

Two recent reports have described homozygous or compound heterozygous mutations in *PIEZO1* in families with nonimmune hydrops fetalis secondary to generalized lymphatic dysplasia (Table 1), totaling 12 mutant alleles (Fotiou et al., 2015; Lukacs et al., 2015).

These mutations are predicted or documented as loss-of-function mutations. Fifteen percent of nonimmune hydrops fetalis is attributed to lymphatic dysplasia. The PIEZO1 mutation-associated cases exhibited full body edema and severe facial swelling, including periorbital edema, retrognathia, flat facial profile, elevated palate, gingival hypertrophy, and microstomia. Most patients also exhibited intestinal lymphangiectasia, growth retardation, seizures, microcephaly, and/or intellectual disability. Patients from unrelated families subjected to lymphoscintigraphy exhibited deep rerouting, evidenced by popliteal lymph node uptake and superficial rerouting through the skin, and suggesting deficient superficial lymphatic collector function (Fotiou et al., 2015). The cosegregating homozygous and compound heterozygous *PIEZO1* mutations in these families included nonsense, missense, and splice donor site mutations, all predicted to encode loss-of-function.

Table 1 PIEZO1 missense mutations reported in association with recessive generalized lymphatic dysplasia

Nucleotides	Exon	Amino acids	Domain,[a] protein level	Protein activity	References
c.G2263T	17	E755X[b]	PH, undetected in RBC	Unreported	Fotiou et al. (2015)
c.G4888T	36	E1630X	PH, undetected in RBC	Unreported	Fotiou et al. (2015)
c.6085G>C	42	G2029R	Anchor, ↓surf expr	↓Current	Lukacs et al. (2015)
c.3455+1G>A[c]	24i	S1153Wfs21★ splice donor	PH, retained 80nt Int24	↓RBC YODA resp in cmpd het RBC	Lukacs et al. (2015)
c.G6511T	45	V2171F	Anchor-OH jct, ↓polypeptide	Unreported	Fotiou et al. (2015)
c.3796+1G>A[d]	26i	Splice donor	PH, ↓ polypeptide	Unreported	Fotiou et al. (2015)
c.C6682T	46	Q2228X	CED, undetected in RBC	Unreported	Fotiou et al. (2015)
c.7289T	50	P2430L	IH, unreported	Unreported	Fotiou et al. (2015)
c.1669+1G>A[e]	13i	27★ neocodons	PH, terminates in Int13	Unreported	Fotiou et al. (2015)
c.C7366T	51	R2456C	IH-CTD jct, unreported	Unreported	Fotiou et al. (2015)
c.C7374G	51	F2458L	IH-CTD jct, unreported	Unreported	Fotiou et al. (2015)
c2815A	21	L939M	PH, unreported	Unreported	Fotiou et al. (2015)

CED, extracellular domain; *CTD*, C-terminal extracellular domain; *PH*, peripheral helices.
Each contiguous shaded or unshaded area represents one or two mutations found in a single patient; patients with two mutations are compound heterozygotes.
[a]Domains as described in Fig. 1 and Ge et al. (2015).
[b]X indicates truncation mutations shown to encode unstable proteins.
[c]Intron 24 mutation disrupts splice donor consensus sequence at Ex24–25 splice junction ("24i").
[d]Intron 26 mutation disrupts splice donor consensus sequence at Ex 26–27 splice junction, generating a transcript lacking Ex26 and encoding an unstable polypeptide.
[e]Intron 13 mutation disrupts splice donor consensus sequence at Ex13–14 splice junction, with retention of 27 codons from int13.

The compound heterozygous *PIEZO1* mutations from one family, encoding PIEZO1 polypeptide variants G2029R and S1153Wfs21* were subjected to functional analysis (Lukacs et al., 2015). In this family the older affected sibling exhibited hydrops fetalis and chylothrorax requiring extracorporeal membrane oxygenation, and at age 3 years maintained chronic pleural effusions with persistent lymphedema of legs, torso, and face. The younger sibling had less severe lymphedema at birth, developed pleural effusions by 3 months of age, and had a milder subsequent clinical course. Neither child in this family exhibited impaired intellectual function.

Both *PIEZO1* mutations in this family were shown to be loss-of-function mutants. Endogenous PIEZO1 in erythrocytes from the heterozygous parents mediated normal elevations of intracellular Ca^{2+} ($[Ca^{2+}]_i$) in response to negative pressure pulses applied through an attached pipette tip and in response to the activator ligand Yoda1. In contrast, erythrocytes from both affected compound heterozygotes showed no $[Ca^{2+}]_i$ response to negative pressure and either absent or greatly impaired response to Yoda1. HEK-293 cells overexpressing PIEZO1 G2029R similarly exhibited minimal $[Ca^{2+}]_i$ elevation elicited by Yoda1. Mutant G2029R also exhibited minimal amplitude currents elicited by progressively increased mechanical displacement by piezo-controlled blunt probe, in parallel with severely decreased surface expression of ecto-epitope-tagged channel polypeptide (Lukacs et al., 2015). Patients of some families were not anemic and exhibited normal hematological indices, including mean corpuscular volume (MCV) (Lukacs et al., 2015). Other patients variably exhibited occasional stomatocytes or mild to marked spherocytosis, with ~50% of patients showing one or more of borderline decreases in mean corpuscular hemoglobin concentration (MCHC), slight reticulocytosis and slightly decreased total hemoglobin, and borderline increased serum lactate dehydrogenase, suggesting mild or subclinical hemolysis (Fotiou et al., 2015).

All patients with generalized lymphatic dysplasia in association with PIEZO1 mutations differed from Hennekam syndrome and other lymphedema syndromes in the absence of intestinal lymphangiectasia, dysmorphic features, learning disabilities or seizures, microcephaly, chorioretinopathy, and hypotrichosis (Fotiou et al., 2015; Lukacs et al., 2015). PIEZO1 deficiency thus appears to represent a clinically distinct form of generalized lymphatic dysplasia. PIEZO1 may plausibly contribute to a signaling pathway active during development of the lymphatic endothelial system, with contributions from other Mendelian lymphatic dysplasia genes including *VEGF-C*, *VEGFR-3*, and *CCBE* and the congenital chylothorax gene, *ITGA9*, encoding α9 integrin (Ma et al., 2008). Interestingly, PIEZO1 was defined as an activator of integrin function (McHugh et al., 2010; McHugh, Murdoch, Haslett, & Sethi, 2012) before its definition as a mechanosensitive cation channel.

3. PIEZO1 GAIN-OF-FUNCTION: HEREDITARY XEROCYTOSIS/DEHYDRATED STOMATOCYTOSIS

Autosomal dominant hereditary xerocytosis (HX), also known as dehydrated stomatocytosis (DHSt), is a hereditary hemolytic anemia of normocytic or macrocytic type and with variable degrees of compensation. Reticulocytosis and jaundice can be severe, mild, or only rarely detectable. HX has been considered a rare disease [prevalence of ~1−50,000 (Andolfo, Russo, Gambale, & Iolascon, 2016)], but molecular genetic diagnostics applied to patients with mild clinical presentations or with asymptomatic increases in MCHC suggest a substantially higher prevalence. Most cases of HX are caused by mutations in *PIEZO1* (Albuisson et al., 2013; Andolfo, Alper, De Franceschi, et al., 2013; Zarychanski et al., 2012), but mutations in the porpyrin transporter gene *ABCB6* (Andolfo, Alper, Delaunay, et al., 2013; Andolfo, Russo, Manna, et al., 2016; Bawazir et al., 2014) and the Ca^{2+}-activated K^+ channel gene *KCNN4* also cause HX (Andolfo et al., 2015; Rapetti-Mauss et al., 2015; Zarychanski et al., 2012). Those HX-associated PIEZO1 mutations functionally investigated to date have uniformly exhibited gain-of-function properties (Table 2).

A large genome-wide association study (GWAS) has associated a rare *PIEZO1* variant with MCHC (Astle et al., 2016). HX red cells are characterized by the cardinal properties of erythrocyte dehydration: increased MCHC, resistance to osmotic lysis, and increased red cell density (Andolfo, Russo, Gambale, et al., 2016). HX red cells in some patients can exhibit the highest known mean density, greater than that of the more heterogeneous red cells of sickle cell disease. HX red cells can be of normal or increased MCV, with total hemoglobin sufficiently elevated to maintain increased MCHC. Intracellular K^+ content is decreased and intracellular Na^+ content is increased, despite frequently upregulated Na^+, K^+-ATPase activity. These changes reflect increased baseline nonspecific cation leak across the plasma membrane of the HX red cell. Such cation leak can also be associated with clinical pseudohyperkalemia (Badens & Guizouarn, 2016; Glogowska & Gallagher, 2015) (apparent but artifactual elevation of serum [K^+] not

Table 2 PIEZO1 missense mutations reported in association with dominant dehydrated stomatocytosis/hereditary xerocytosis

Nucleotides	Exon	Amino acids	Domain	Functional phenotype	References
c.G2344A	18	G782S	PH	Not tested	Andolfo, Alper, De Franceschi, et al. (2013)
c.G2423A	18	R808Q	PH	Not tested	Andolfo, Alper, De Franceschi, et al. (2013)
c.G4073C	29	R1358P	PH	Slowed inact	Albuisson et al. (2013)
6nt ins	32	R1462ins 2aa	PH	Unreported	Begtrup et al. (2013)
c.G5591A in cis with	39	R1864H in cis with	PH	Unreported	Andolfo, Russo, Gambale, et al. (2016)
c.7473-7478dup6	51	2492EL2493dup	CTD	Unreported	Andolfo, Russo, Manna, et al. (2016)
c.5863T	41	R1955C	PH	Unreported	Yang, Voelkel, Lezon-Geyda, Schulz, and Gallagher (2017)
c.C6008A	42	A2003D	PH	Not tested	Andolfo, Alper, De Franceschi, et al. (2013)
c.G6058A	42	A2020T	PH	Slowed inact	Albuisson et al. (2013)
c.C6059T	42;	A2020V	PH	Not tested	Andolfo, Alper, De Franceschi, et al. (2013)
c.C3350T	24	S1117L	PH	Not tested	Andolfo, Alper, De Franceschi, et al. (2013)
c.C6067G	42	L2023V	PH	Slowed inact	Risinger et al. (2014)
c.6239-6256dup18	43	2080−2085dup	Anchor	Dyserythropoiesis; current unreported	Paessler and Hartung (2015)
c.C6380T	44	T2127M	Anchor	Slowed inact	Albuisson et al. (2013); Andolfo, Alper, De Franceschi, et al. (2013)
c.6496−6507del12	45	2169delK	Anchor	Not tested	Andolfo, Alper, De Franceschi, et al. (2013)
c.T6674G	46	M2225R	CED	Unreported; slowed inact	Zarychanski et al. (2012); Albuisson et al. (2013)

(Continued)

Table 2 PIEZO1 missense mutations reported in association with dominant dehydrated stomatocytosis/hereditary xerocytosis—cont'd

Nucleotides	Exon	Amino acids	Domain	Functional phenotype	References
c.G6905A	47	R2302H found in β-thal pt	CED	Unreported	Bakeer et al. (2014)
c.C7367A	51	R2456H	IH-CTD jn	Unreported; slowed inact	Zarychanski et al. (2012); Albuisson et al. (2013); Andolfo, Alper, De Franceschi, et al. (2013); Shmukler et al. (2014)
c.C7426T	51	R2476C found in symptomatic neonatal sickle cell pt	CTD	Unreported	Bento et al. (2016)
c.G7463A	51	R2488Q	CTD	Incr RBC current	Andolfo, Alper, De Franceschi, et al. (2013); Archer et al. (2014)
c.G7463A c.G2152A	51; 16	R2488Q G718S	CTD PH	Incr current Unreported	Andolfo, Alper, De Franceschi, et al. (2013)
c.7473–7478dup6	51	2492–2493dup	CTD	Unreported	Andolfo, Russo, Manna, et al. (2016)
c.7488–7489ins6nt	51	E2496ELE	CTD	Slowed inact; unreported	Andolfo, Alper, De Franceschi, et al. (2013); Imashuku et al. (2016)

CED, extracellular domain; *CTD*, C-terminal extracellular domain; *IH*, inner helix; *jn*, junction; *PH*, peripheral helices.

present in the patient but appearing in the blood only after its postvenisection storage at room temperature or in the cold, usually within 4–24 h).

The autosomal dominant missense mutations in PIEZO1 found in affected members of HX families were overexpressed in HEK-293 cells and shown to encode mechanically sensitive cation channel activity remarkable for delayed inactivation, usually prolonged two- to six-fold. Overexpression in HEK-293 cells of PIEZO1 cDNA containing two HX gain-of-function mutations led to near-abrogation of inactivation of cation currents induced either by mechanical displacement or osmotic swelling. In addition, the red cells of HX patients have been shown to exhibit constitutive cation channel activity in on-cell patches without application of incremental negative hydrostatic pressure. This channel activity is sensitive to inhibition by the PIEZO inhibitor, GsMTx4 (Vandorpe et al., 2010). However, this channel activity's non-inactivating kinetics and its dependence on extracellular ATP (unpublished data) suggest that PIEZO1 activation likely triggers downstream signaling pathways including ATP release and P2 receptor activation.

Similar cation leaks have been described by net and unidirectional flux studies in overhydrated stomatocytosis and hereditary spherocytosis (Andolfo, Russo, Gambale, et al., 2016) caused by mutations in different genes. However, the mechanisms by which the increased cell Na^+ content and decreased cell K^+ content observed in both types of anemia lead to red cell overhydration (decreased MCHC) in overhydrated spherocytosis but to red cell dehydration (increased MCHC) in xerocytosis remain unclear. Hyperactivation of PIEZO1 is predicted to elevate intracellular $[Ca^{2+}]_i$ in HX red cells, leading to overactivity of the intermediate conductance Ca^{2+}-activated K^+ channel KCNN4. Overactive KCNN4, in concert with red cell Cl^- channels [likely including TMEM16A/ANO1 (Skals et al., 2010)], then leads to red cell shrinkage and dehydration. Indeed, HX can also be caused by gain-of-function missense mutations in KCNN4 (Andolfo et al., 2015; Glogowska, Lezon-Geyda, Maksimova, Schulz, & Gallagher, 2015; Rapetti-Mauss et al., 2015).

Clinical disease severity in HX usually differs to as great a degree among family members as among families with distinct mutations. This intrafamily variability probably represents polymorphic variation in numerous, genetically unlinked risk modifier genes. Among these are the many additional modifiers of red cell volume and density. In addition, *PIEZO1* is among the many erythroid gene products that undergo regulated intron retention of increasing frequency and prevalence during progression of erythroid differentiation (Pimentel et al., 2016). Such intron retention would be predicted to decrease levels of functional PIEZO1 in mature red cells. A hypomorphic PIEZO1 variant or a risk modifier variant leading to decreased "physiological" intron retention in PIEZO1 mRNA might then increase red cell PIEZO1 abundance, potentially exacerbating clinical manifestations of HX.

Activation of PIEZO1 elevates erythrocyte intracellular Ca^{2+} (Cahalan et al., 2015) and appears to be required for shear stress–induced release of ATP from red cells (Cinar et al., 2015). However, red cells from one HX patient with an (undescribed) PIEZO1 mutation exhibited shear-induced elevations of intracellular $[Ca^{2+}]$ and of ATP release of only ~50% their magnitudes in normal red cells (Cinar et al., 2015). These results contrast with the increases in both signals predicted by the simplest model of PIEZO1 gain-of-function, as illustrated by the loss of a component of shear-induced endothelial ATP release in the vascular endothelial cell-specific *Piezo1* knockout mouse (S. Wang et al., 2016). A possible role of PIEZO1 in the hypoxia-induced ATP release from red cells governed by deoxyhemoglobin binding to Band 3/SLC4A1, the major integral membrane protein of the red cell (Chu et al., 2016), remains untested.

Splenectomy is an important component of treatment for patients with severe hereditary hemolytic anemia due to spherocytosis. However, splenectomy is contraindicated in HX, due to increased susceptibility to life-threatening thrombotic events such as pulmonary embolism or more gradual development of pulmonary hypertension (Andolfo, Russo, Gambale, et al., 2016; Stewart et al., 1996). This important clinical distinction may be related to the apparent expression of PIEZO1 in platelets (Wright, Amisten, Goodall, & Mahaut-Smith, 2016), reflecting inhibition of flow-stimulated thrombus formation by the PIEZO1 antagonists Gd^{3+} and GsMTx4 (Ilkan et al., 2016).

Another distinguishing feature of HX is the high degree of systemic iron overload found in patients, out of proportion to that in patients with comparable levels of hemolysis attributable to other causes. Although the iron overload is independent of the more frequent genetic causes of autosomal dominant hemochromatosis, it can be exacerbated by their additional presence. The degree of iron overload indeed can correlate with clinical severity of HX, whereas coincident systemic iron deficiency can ameliorate HX clinical severity (Orvain et al., 2016). Patients not previously diagnosed with HX have presented to medical attention with heart failure secondary to cardiac hemochromatosis (Assis et al., 2013) that must be treated with iron chelation therapy. HX-associated iron deposition can also precipitate diabetes mellitus secondary to pancreatic hemochromatosis (Imashuku et al., 2016), as well as impairing hepatic and renal function.

The mechanism of tissue iron accumulation in HX remains unknown, as is any interaction with or regulation of PIEZO1 by the iron transporters DMT1/SLC11A2, ferroportin/SLC40A1, the ferroportin downregulatory ligand hepcidin, or the hepcidin regulator, erythroferrone. Although PIEZO1 permeability to Mn^{2+} is very low (Gnanasambandam et al., 2015), it remains possible that HX mutation-associated delayed inactivation of PIEZO1 might itself allow increased cellular Fe^{2+} uptake, thus contributing to progressive hemochromatosis.

HX associated with missense mutations in PIEZO1 is also variably associated with transient perinatal edema of variable severity that subsides within 3–6 months after birth (Andolfo, Alper, De Franceschi, et al., 2013; Grootenboer et al., 1998). The appearance and severity of edema is inconsistent among different families with the same PIEZO1 mutation and even among affected members of the same family. The mechanistic relationship between the completely penetrant, persistent lymphatic insufficiency associated with PIEZO1 loss-of-function mutations and the incompletely penetrant, transient, apparent lymphatic insufficiency associated with PIEZO1 gain-of-function mutations remains unexplained. Histologic, cell biological, and transcriptomic comparisons of these two clinically similar conditions associated with either gain or loss of PIEZO1 function are awaited.

4. PIEZO1 LOSS-OF-FUNCTION PHENOTYPES IN ANIMAL MODELS

The human phenotypes of PIEZO1 loss-of-function are less severe than those in the mouse, in which global *Piezo1* knockout led to disordered vasculogenesis evident by e9.5 in yolk sac vasculature and to midgestational embryonic lethality (Ranade et al., 2014). Investigations of lymphangiogenesis in the global mouse knockout have not yet been reported, but mouse *Piezo1* knockout in vascular endothelial cells led to loss of the normal shear stress–induced increases in endothelial Ca^{2+} signaling and ion current. PIEZO1 knockout fetal endothelial cells also failed to display normal cellular alignment along the axis of the shear vector, by a mechanism partially reflecting decreased calpain activation. Moreover, in the aged heterozygous mouse in situ, endothelial cells of large cerebral arteries failed to normally align their long axes along the vector of intraluminal blood flow (Li et al., 2014). In addition, knockdown of human *PIEZO1* in HUVEC cells led to loss of shear stress–induced endothelial cell reorientation and of stress fiber alignment and elongation along the axis of flow (Ranade et al., 2014). Moreover, repression of PIEZO1 biosynthesis by miR-103a was accompanied by reduced endothelial cell viability and in vitro capillary tube formation (Huang, Li, Chen, Zhang, & Shi, 2016). The atheroprotective influence of vascular endothelial cell orientation along the axis of blood flow suggests that PIEZO1 hypomorphic variants might be risk modifiers of atherogenesis.

The role of PIEZO1 in mouse erythrocytes was addressed in a tissue-specific knockout mouse in which the *Vav1* promoter (active in early hemato- and granulopoiesis) drove Cre excision of *Piezo1* exons 20–23 in nearly all erythroid and granuloid precursors, resulting in absence of detectable polypeptide in red cells (Cahalan et al., 2015). The non-anemic knockout mice had mildly overhydrated circulating red cells of normal discoid morphology without stomatocytes. However, MCV was increased by nearly 10%, MCHC was decreased by nearly 6%, and cells exhibited increased osmotic fragility and increased size, as reported by flow cytometry forward scatter. Reduced plasma haptoglobin suggested elevated hemolysis. The increased volume of PIEZO1-deficient mouse red cells was consistent with, but more prominent than, the incompletely penetrant overhydrated phenotype in red cells of most patients with generalized lymphoid dysplasia secondary to *PIEZO1* loss-of-function mutations (Fotiou et al., 2015; Lukacs et al., 2015).

Genomic inactivation of the *piezo1* gene in zebrafish had no apparent effect on nucleated red cell phenotype (Shmukler et al., 2015), but antisense morpholino-oligonucleotide knockdown of the Piezo1 polypeptide produced a severe hemolytic anemia with enlarged and sometimes dysmorphic red cells (Faucherre, Kissa, Nargeot, Mangoni, & Jopling, 2014). A similar phenotype discrepancy was observed between the zebrafish tailfin dysgenesis induced by morpholino knockdown of *piezo1* (Eisenhoffer et al., 2012) and the normal fin development observed in fish harboring genomic inactivation of *piezo1* (Shmukler et al., 2015). Such differences in phenotype have been noted previously among numerous comparisons of genomic knockout and oligonucleotide knockdown in zebrafish (Kok et al., 2015) and may reflect compensatory regulation or adaptive second-site mutations during development in the genomic knockout fish (Rossi et al., 2015).

As in mouse red cells, flow-induced ATP release from vascular endothelial cells requires endothelial expression of PIEZO1 (S. Wang et al., 2016). Endothelial specific knockout of PIEZO1 rendered endothelial cells deficient in release of both nitric oxide and ATP in response either to shear stress or to the PIEZO agonist Yoda1. Pharmacological induction of endothelium-limited, Cre-mediated inactivation of the *PIEZO1* gene in 2- to 3-month-old mice led within days to systemic hypertension persisting throughout the 2-week observation period (S. Wang et al., 2016), but longer-term effects on blood pressure were not reported. In addition, vascular smooth muscle-limited, Cre-mediated *PIEZO1* inactivation in mice led to multiple changes in arterial adaptation to hypertensive stimuli (Retailleau et al., 2015). Stretch-activated cation channel activity was abrogated in vascular myocytes without altering myogenic tone of intact arterioles. However, arterial remodeling in response to hypertension induced by angiotensin II infusion, or by DOCA/salt administration in the setting of uninephrectomy, was converted from the normal hypertrophic response to one of reduced arterial wall thickness, medial cell number, and arterial lumenal diameter, without endothelial involvement. PIEZO1 activity in vascular myocytes was found to be under the negative regulation of the actin cytoskeletal crosslinker, filamin A. The amplitude of arterial PIEZO1-dependent stretch-activated currents induced by increased hydrostatic pressure in the absence of filamin A was higher than in its (wild type) presence (Retailleau et al., 2015). However, despite these links of PIEZO1 to blood pressure regulation in mice, hypertension was not reported in the several adolescent and young adult patients with generalized lymphatic dysplasia secondary to biallelic loss-of-function mutations in PIEZO1 (Fotiou et al., 2015), and *PIEZO1* variants have not to date been identified as candidate risk modifiers in GWAS studies of human hypertensive subjects.

The PIEZO1 promoter is active throughout the mouse kidney at postnatal day 5, but predominantly in inner medulla in the adult kidney (Martins et al., 2016). Cell-attached basolateral membrane patches on epithelial cells of isolated inner medullary collecting ducts from mice with KspCre-directed genetic inactivation of *PIEZO1* limited to distal nephron segments revealed loss of stretch-activated cation channel activity elicited by negative pipette pressure. Strikingly, this PIEZO1-dependent activity was non-inactivating and only slowly deactivating (Martins et al., 2016), in remarkable contrast to the behavior of recombinant PIEZO1 from human, mouse, or fly expressed in HEK-293 cells. The absence of distal nephron PIEZO1 did not alter baseline mouse urine osmolality or urinary concentration induced by water restriction. However, absence of PIEZO1 delayed the urinary dilution following reintroduction of water to the mice. Moreover, the vasopressin-independent reduction in urinary concentration induced by fasting was abolished (Martins et al., 2016).

Epithelial cell basolateral membranes in isolated perfused mouse proximal tubules exhibit similar stretch-activated cation channel activity sensitive to GsMTx4 and ruthenium red, and remarkable for lack of inactivation and very slow deactivation. This activity in immortalized mouse proximal tubular cells was decreased dramatically by PIEZO1 siRNA (Peyronnet et al., 2013). The putative PIEZO1 activity was also suppressed by overexpression in the same proximal tubular cells of polycystin-2 (PC2), with still stronger suppression by PC2 mutant 740X lacking its ER retention signal. PC2-associated suppression of recombinant PIEZO1 activity required its >200 aa N-terminal cytosolic domain of PC2 and was associated with the ability of recombinant PC2 to bind directly to PIEZO1 without decreasing its surface expression. These data strongly suggest a contributory role for PIEZO1 in human polycystic kidney disease (Peyronnet et al., 2013), but more definitive evidence supporting this hypothesis is currently lacking.

Urinary bladder is a tissue subjected to repeated cycles of stretch and relaxation. Preliminary evidence for PIEZO1 expression has been documented in urothelium of mouse and human. Primary mouse urothelial cells responded to mechanical stretch with elevated cytosolic Ca^{2+} and ATP release, and both were reduced by GsMTx4 and by PIEZO1 knockdown (Miyamoto et al., 2014). In female rats, a surgical model of partial bladder obstruction increased total bladder PIEZO1 mRNA 2.5-fold over 7 days, in parallel with appearance of bladder hypertrophy and nonvoiding bladder contractions in the absence of elevated baseline intravesical pressure. In situ hybridization and colocalization revealed that PIEZO1 mRNA increased in detrusor and suburothelial smooth muscle layers of the bladder, whereas the high urothelial PIEZO1 mRNA levels and lesser suburothelial levels remained statistically unchanged (Michishita, Yano, Tomita, Matsuzaki, & Kasahara, 2016).

PIEZO1 has been shown to play a role in epithelial confinement sensing (Eisenhoffer et al., 2012; Gudipaty & Rosenblatt, 2016; Hung et al., 2016). The transition from unconfined or subconfluent to confined or highly confluent environments has been shown in a model cell type to require PIEZO1-mediated Ca^{2+} signaling, leading to PDE1 phosphodiesterase activation and suppression of protein kinase A activity. Cell confinement itself can activate PIEZO1 (Hung et al., 2016). In intact tissues, confined, confluent epithelial cell layers must provide and maintain their function as compartmental barriers while sustaining (in different settings) high rates of division, cell differentiation, cell injury and repair, or cell elimination from the monolayer and death by anoikis. Disruption of apical cell extrusion can interfere with physiological death pathways while promoting loss of contact inhibition and formation of cell masses. Developing zebrafish embryos subjected to *piezo1* knockdown or Gd^{3+} exposure exhibit impaired cell extrusion leading to growth of cell masses at sites (such as the dorsal fin) developmentally specified for extrusion-dependent organogenesis (Eisenhoffer et al., 2012). However, this process can be governed by multiple pathways, insofar as genomic knockout of *piezo1* in zebrafish produces no grossly apparent developmental phenotype (Kok et al., 2015).

The microtubule and actin cytoskeletal regulatory scaffold protein adenomatous polyposis coli (APC) undergoes C-terminal truncating mutation in 80% of colon cancers. C-terminally truncated APC in cell culture and zebrafish epidermis inhibits normal apical cell extrusion and promotes basal cell extrusion of the type believed to promote invasion and metastasis. The discovery through exome sequencing of *PIEZO1* mutations in familial colorectal adenomatous polyposis unexplained by mutations in *APC* or in the other known causal tumor suppressors (Spier et al., 2016) encourages the speculation that even haploinsufficiency of *PIEZO1* might promote polyp formation through impairment of apical extrusion or disordered tumor angiogenesis and/or lymphangiogenesis, with possible later contribution to the inevitable progression to carcinoma. The speculation is supported by the frequent downregulation of PIEZO1 in small cell lung cancer cell lines and by the increased cell invasiveness into Matrigel in the setting of PIEZO1 knockdown (McHugh et al., 2012). It is further supported by the role of R-Ras in integrin inactivation by PIEZO1 (McHugh et al., 2010), since R-Ras is a critical regulator of microvascular integrity in tumors and in the retina (Sawada, Li, & Komatsu, 2015; Vahatupa et al., 2016) through a VE-cadherin-dependent mechanism.

5. ADDITIONAL PATHOLOGICAL ASSOCIATIONS OF HUMAN PIEZO1

PIEZO1 has been identified by yeast two-hybrid screen and co-immunoprecipitation as a binding partner of trefoil factor 1 (TFF1) (Yang et al., 2014), a stomach-specific tumor-suppressor gene. Further analysis suggested that the C-terminal 537 amino acid residues of PIEZO1 interacted with TFF1, but not with related polypeptides TFF2 or TFF3. PIEZO1 knockdown in a gastric tumor cell line reduced both TFF1-stimulated cell migration and β1 integrin expression, similar to the effects of PIEZO1 knockdown on increased lung tumor cell migration in vitro and metastasis in vivo (McHugh et al., 2012). Interestingly, four *PIEZO1* missense variants predicted as "pathogenic" have been identified as candidate risk modifier alleles among the 30% of colonic adenomatosis patients in whom the familial disease is not explained by mutations in the previously defined causal genes *APC, MUTYH, POLE1*, or *POLD1* (Spier et al., 2016). This association will be discussed later.

No PIEZO1-based human blood group or transplantation antigen has been yet defined. However, strong PIEZO1 phosphorylation at two putatively intracellular residues has been detected within 2 min of *Plasmodium falciparum* merozoite attachment to the human red cell surface in the presence of inhibitors of invasion (Zuccala et al., 2016), suggesting possible merozoite recognition either of PIEZO1 itself or of an upstream signaling receptor. Both phosphorylation sites were among the 23 sites detected in mouse PIEZO1-GST fusion protein overexpressed in HEK-293 cells (Coste et al., 2015).

Two patients with compound heterozygous forms of the recessive lysosomal storage disease Mucolipidosis Type IVA, a deficiency of the enzyme N-acetylgalactosamine-6-sulfatase (NGALTS), are also likely haploinsufficient for PIEZO1. Each patient has one deletion allele of *NGALTS* extending into the neighboring *PIEZO1* gene to encompass the *PIEZO1* promoter, exon 1, and part of intron 1 (Caciotti et al., 2015). Any resulting mutant PIEZO1 polypeptide product is predicted to initiate at or downstream of PIEZO1 Met268, with high likelihood of loss-of-function.

The ExAC database of ~60,000 nominally healthy individuals reports (as of late 2016) 15 subjects with heterozygous predicted loss-of-function mutations, including frameshift termination, termination, and splice-site inactivation, in addition to 1-2 inframe indel mutations of 1-2 codons in length and numerous single missense substitutions of unknown consequence. Although *PIEZO1* haploid sufficiency is not known to be pathogenic, it may constitute a risk variant for other disorders, especially those characterized by mutant gene products that normally interact directly or functionally with PIEZO1.

6. PIEZO2 LOSS-OF-FUNCTION DISEASE: RECESSIVE SYNDROME OF DISTAL ARTHROGRYPOSIS, SCOLIOSIS, MUSCLE ATROPHY, PROPRIOCEPTION DEFICITS, AND TRANSIENT NEONATAL RESPIRATORY INSUFFICIENCY

Whereas the global knockout of mouse *PIEZO2* is perinatal lethal (Dubin et al., 2012; Nonomura et al., 2016), biallelic loss of function of *PIEZO2* in humans leads to a congenital and progressive arthrogryposis syndrome (multiple congenital contractures) of variable clinical phenotype that allows survival to adulthood, without currently evident limitation to life span. The *PIEZO2* loss-of-function mutations implicated to date in the arthrogryposis syndrome are found throughout the length of the protein and are dominated by frameshift/termination mutations (Table 3).

One instance of multiexon deletion and one missense mutation are each accompanied by decreased mRNA levels (Chesler et al., 2016; McMillin et al., 2014). Arthrogryposis has been associated with mutations in over 400 individual genes, including those encoding components of the neuromuscular junction, gene products involved in myelination and survival

Table 3 PIEZO2 missense mutations associated with scoliosis with peripheral sensory dysfunction

Nucleotides	Exon	Amino acids	Domain	Functional phenotype	References
c.1550–1552del3 ins4	13	S517Tfs48*	PH	Unreported	McMillin et al. (2014)
c.2708C>G	18	S903X	PH	Unreported	Mahmud et al. (2016)
c.3019–29del	20	P1007Lfs3*	PH	Unreported	McMillin et al. (2014)
c.4723C>T	32	R1575X	PH	Unreported	Chesler et al. (2016)
c.5053C>T	35	R1685X	PH	Unreported	Chesler et al. (2016)
c.5054C>G	35	R1685P	PH	Unreported	Chesler et al. (2016)
c.5621delT	37	L1874Rfs5*	PH	Unreported	McMillin et al. (2014)
c.493?–917? (deletion of Exons 6 and 7)		fs*	PH	Absence of protein	McMillin et al. (2014)

PH, peripheral helices.

of motor neurons, components of skeletal myocytes, components of extracellular matrix, and others (Hall, 2014).

The *PIEZO2* mutation–associated arthrogryposis syndrome includes muscular atrophy predominantly of lower limbs, kyphoscoliosis, and distal sensory neuropathy characterized chiefly by defective proprioception (Chesler et al., 2016; Mahmud et al., 2016; McMillin et al., 2014) without radiologically evident brain or spinal cord abnormality. Kyphoscoliosis and congenital contractures of the feet are present in nearly all patients, and congenital contractures of the wrist or hands (including duck bill deformity) are present in most. Muscle weakness and atrophy tends to be more prominent distally, but some patients exhibit truncal weakness with reduced head control and shoulder girdle strength, and impaired ambulation. A minority of patients exhibit masked facies without ocular phenotype, dysarthria, or delayed or reduced communication. The syndrome is often accompanied by transient perinatal respiratory insufficiency with shallow breathing, resolving within a few days after birth. Versions of each of these clinical components of the human arthrogryposis syndrome have been reported among available global and targeted Piezo2 knockout mouse models.

The sensory deficits of two unrelated patients with biallelic PIEZO2 loss-of-function mutations were studied in detail (Chesler et al., 2016). These patients had similar developmental histories that included delayed walking, early impairment of fine motor skills, and delayed mastery of activities of daily living. Neurological examination revealed each patient to have lost discriminative and light touch perception and vibratory perception on glabrous skin, whereas vibratory sensation and detection of and pain threshold for pressure, pinprick, light touch, and warm temperature were normal for hairy skin. Functional magnetic resonance imaging revealed reduced or absent brain activation on light brush stimulation of glabrous skin. Most tendon reflexes were absent. Each patient had profound proprioception deficits manifest as marked ataxia, dysmetria, stationary balance, and pseudoathetosis, exacerbated by loss of visual cues. Patients were unable to identify direction of joint movements across an arc of less than 20 degree. Sensory nerve action potentials were of reduced magnitude and borderline low-to-normal velocity, but motor nerve conduction studies were normal, without evidence of denervation.

The above deficits of PIEZO2-deficient patients are consistent with mouse PIEZO2 localization defined by immunofluorescence in PIEZO2-GFP knock-in mice (Woo et al., 2014), and with the phenotypes of mice with sensory neuron-targeted Cre-mediated knockout of *Piezo2* in Merkel cells (Woo et al., 2014), in sensory endings of proprioceptor neurons innervating muscle spindles and Golgi tendon organs (Woo et al., 2015), and in peripheral low threshold mechanosensory neurons of skin. However, in contrast to humans with PIEZO2 loss-of-function, loss of mouse PIEZO2 function abolished light touch sensation in hairy as well as in glabrous skin, in parallel with loss of rapidly inactivating mechanosensory currents in these cells (Ranade et al., 2014). PIEZO2 knockout in mice did not affect mechanical allodynia or nociception, just as compound heterozygous PIEZO2 loss-of-function mutations in human patients did not affect mechanical pain sensation.

7. PIEZO2 GAIN-OF-FUNCTION DISEASE: DOMINANT DISTAL ARTHROGRYPOSIS TYPES 3 (GORDON SYNDROME) AND 5, AND MARDEN–WALKER SYNDROME

PIEZO2 gain-of-function mutations have been found to cause a group of clinically similar syndromes of distal arthrogryposis formerly known as Gordon syndrome (distal arthrogryposis type 3), distal arthrogryposis type 5, and Marden–Walker syndrome (Coste et al., 2013; McMillin et al., 2014). The reported patients with PIEZO2 mutations have presented with generalized arthrogryposis with normal flexion but limited extension, and chronic degenerative problems of multiple joints. Clinical features distinguishing the dominant arthrogryposis associated with PIEZO2 gain-of-function mutations from the recessive arthrogryposis syndrome associated with biallelic PIEZO2 loss-of-function mutations include the presence at high frequency of ophthalmoplegia and/or ptosis, and a high frequency of moderate-to-severe restrictive lung disease associated with late-onset increases in resting muscle tone. Additional differences between the dominant and recessive syndromes include occasional presence of cerebellar malformations (absent from the recessive disease), lower frequency of scoliosis, and the absence of the dysarthria sometimes present in the recessive disease. The late onset reduction in resting muscle tone may reflect overactivation of lung stretch responses by PIEZO2-expressing vagal neurons (Nonomura et al., 2016).

Only two distal arthrogryposis mutations have been physiologically characterized for gain-of-function by recombinant expression in HEK-293 cells (Coste et al., 2013). Arthrogryposis-associated mutant PIEZO2 polypeptides I802F and E2727del (Table 4) exhibited mechanical displacement-activated

Table 4 PIEZO2 missense mutations associated with dominant distal arthrogryposes type 3 (Gordon syndrome), type 5, and Marden–Walker syndrome

Nucleotides	Exon	Amino acids	Disease	Domain	Functional phenotype	References
c.2134A>G	15	M712V	DA5	PH	Unreported	Okubo et al. (2015)
c.2136G>T	15	M712I	DA5	PH	Unreported	Almanza, Buckingham, Shively, McMillin, & Bamshad (2014)
c.2404A>T	17	I802F	DA5	PH	Faster recovery from inact	Coste et al. (2013)
c.2993T>C	20	M998T	DA5	PH	Unreported	McMillin et al. (2014)
c4456G>C	30	A1486P	DA5	PH	Unreported	Okubo et al. (2015)
c.6662C>T	43	T2221I	DA5	PH	Unreported	Coste et al. (2013); McMillin et al. (2014)
c.6668C>T	43	S2223L	DA5	PH	Unreported	McMillin et al. (2014)
c.7067C>T	45	T2356M	DA5	Anchor	Unreported	McMillin et al. (2014)
c.8057G>A	52	R2686H	DA3	IH-CTD jct	Unreported	McMillin et al. (2014)
c.8056C>T	52	R2686C	DA5-MWS	IH-CTD jct	Unreported	McMillin et al. (2014)
c.8153G>T	52	R2718L	DA5	CTD	Unreported	McMillin et al. (2014)
c.8153G>C	52	R2718P	DA5	CTD	Unreported	McMillin et al. (2014)
c.8179–8181del3	52	E2727del	DA5	CTD	Slower onset, faster recovery from inact	Coste et al. (2013); McMillin et al. (2014)
c.8208delA	52	Y2737Ifs7*	DA5	CTD	Unreported	McMillin et al. (2014)
c.8215T>C	52	S2739P	DA5	CTD	Unreported	McMillin et al. (2014)
c.8238–8245del8	52	W2746X	DA3	CTD	Unreported	McMillin et al. (2014)

CTD, C-terminal extracellular domain; *IH*, inner helix; *jct*, junction; *PH*, peripheral helices.

currents resembling those of wild-type PIEZO2 both in peak magnitude and in linear current–voltage relationships of near-zero reversal potential.

Whereas I802F inactivation kinetics was indistinguishable from wild-type kinetics, mechanically activated E2727del currents displayed slower inactivation at all test potentials. However, mechanically activated cells expressing both PIEZO2 mutants I802F and E2727del recovered from inactivation more than twice as rapidly than did wild-type PIEZO2 currents, suggesting increased responsiveness to repeated physiological signals in PIEZO2-expressing tissues such as Merkel cells, dorsal root ganglia sensory neurons, and Golgi tendon organ neurons (Coste et al., 2013). These properties of mechanical hyperresponsiveness to pre- and postnatal developmental forces and to stresses of gravity and movement are believed to account for the varied manifestations of distal arthrogryposis types 3 and 5 and the type 5-like Marden–Walker syndrome, and for the partial clinical overlap with the numerous other genetic causes of distal arthrogryposis.

Of particular interest among these is the congenital distal arthrogryposis in the dominantly inherited syndrome CLIFAHDD (congenital contractures of the limbs and face with hypotonia and developmental delay) associated with mutations in the neuronal cation channel NALCN (Chong et al., 2015), a channel not known to be mechanosensitive. CLIFAHDD is associated with both gain-of-function channel mutations and dominant negative (antimorphic) NALCN mutations with loss-of-function phenotype (Bend et al., 2016). However, the NALCN null mutations associated with the phenotypically related recessive disease IHPRF (infantile hypotonia and psychomotor retardation and characteristic facies) (Al-Sayed et al., 2013) encode classical loss-of-function mutants, and distal arthrogryposis has not been observed in IHPRF (Bend et al., 2016).

8. PIEZO2 LOSS-OF-FUNCTION PHENOTYPES IN MOUSE MODELS

PIEZO2 is required for control of breathing in mice. Embryonic lung development appears to proceed normally in the global PIEZO2 knockout mouse, and 6 h postnatally the pulmonary wet/dry weight ratio is unchanged, consistent with normal alveolar fluid clearance after birth. However, shortly thereafter lung wet weight begins to increase, and the mice die by 24 h postnatally (Nonomura et al., 2016). PIEZO2 expression was detected in the P2Y1+ vagal sensory neurons of the nodose ganglion, known to participate in the Hering–Breuer inspiratory reflex that protects the lungs from overexpansion by inducing transient apnea. PIEZO2 was also present in neurons of the jugular ganglion and dorsal root ganglia innervating the lungs, as well as in pulmonary neuroepithelial cell bodies contacted by afferents from PIEZO2-expressing neurons (Nonomura et al., 2016).

Selective knockout of PIEZO2 in neurons of the jugular and thoracic dorsal root ganglia, but not in nodose ganglion, led to neonatal respiratory distress and death. However, selective nodose ganglion knockout of PIEZO2 allowed survival to adulthood. Artificial lung inflation in these mice demonstrated impairment of the Hering–Breuer inflation reflex, with increased tidal volume. Vagal neuron-specific expression of photosensitive channel rhodopsin allowed induction of exhalation state apnea by focal illumination, with chronic reduction of respiratory rate. PIEZO2 knockout in most vagal, spinal sensory, and dorsal root ganglion neurons led to reduced vagal nerve firing in response to lung inflation. These mice exhibited increased tidal volume (as in vagotomized animals), with prolongation of expiratory airflow and blunting of the Hering–Breuer reflex, without impairment of chemosensor function (Nonomura et al., 2016). Thus, PIEZO2 plays a role in sensing lung expansion during cyclical respiration.

PIEZO2 also contributes to optimal hearing in mice. Mouse PIEZO2 is expressed in the apical membrane of cochlear and vestibular hair cells near junctional complexes with their supporting cells (Wu et al., 2017), but is not an essential component of the auditory mechanotransduction cation channel of outer hair cell stereocilia. However, PIEZO2 is required for the reverse polarity currents observed during normal hair cell development and following experimental stereocilia tip-link breakage. These PIEZO2-dependent currents are also observed in hair cells genetically deficient in the TMC1/TMC2 complex believed to contribute to the physiological forward polarity currents of mature stereocilia. Thus, PIEZO2 has been proposed as a strong candidate pore-forming subunit of the channel mediating reverse polarity currents of developing, genetically modified, or injured hair cells. *Piezo2* knockout throughout the inner ear led to 15-db threshold elevation of auditory brain stem response and to mild distortion of auditory signal, in the absence of vestibular defect. Outer hair cell electromotility as measured by nonlinear capacitance was unaltered, suggesting unimpaired prestin-mediated auditory signal amplification. Lack of PIEZO2 in primary cultured outer hair cells was associated with accelerated in vitro maturation of forward polarity sensory transduction currents. Since recombinant PIEZO2 expressed in HEK-293 cells is inhibited by cytosolic Ca^{2+},

PIEZO2 activation in hair cells may reflect the reduced cytosolic [Ca^{2+}] following inactivation of the constitutive forward polarity sensory transduction current. PIEZO2 may thus contribute either to beneficial repair of damaged hair cells or in their elimination during aging (Wu et al., 2017). However, the role of PIEZO2 in human hearing remains unclear.

PIEZO2 and PIEZO1 are both important contributors to normal function of mouse articular cartilage. Chondrocytes are the only cells in cartilage and are responsible for cartilage biosynthesis and remodeling. The cartilage of diarthrodial joints is subjected to episodically high stress and strain required for weight bearing during all types of body movement. Both PIEZO1 and PIEZO2 mRNAs are expressed at high abundance in primary articular chondrocytes of mouse and pig. Native, mechanically activated, actin cytoskeleton-dependent chondrocyte currents of high magnitude and slow (or absent) inactivation require expression of both PIEZO1 and PIEZO2. Pharmacological inhibition or knockdown of either PIEZO1 or PIEZO2 inhibited mechanically induced Ca^{2+} transients in chondrocytes (Lee et al., 2014). High strain mechanical injury of chondrocytes led to cell death that was reduced in frequency by PIEZO channel inhibitor GsMTx4. The role of PIEZO channels in the chondrocyte response to high strain injury may complement the role of osmosensor TRPV4 in the response of chrondrocytes to lower magnitude mechanical stresses or stresses of other types (O'Conor et al., 2016).

PIEZO2 may contribute to regulation of mouse gut motility. Isolated gut enterochromaffin cells (ECs) and EC within intact jejunal mucosa release serotonin in response to stretch (F. Wang et al., 2016). Enterically secreted serotonin controls or modulates gut motility, sensation, and fluid secretion. PIEZO2 mRNA is expressed in human EC, and PIEZO2 protein is detected in nearly 90% of mouse EC throughout the length of the small intestine. Mouse EC expresses mechanically activated currents with pharmacological sensitivity indistinguishable from that of recombinant PIEZO2. PIEZO2 inhibitors block stretch-activated serotonin release and serotonin-induced increase in gut epithelial short-circuit current (F. Wang et al., 2016). Interestingly, EC cells share with Merkel cells a developmental lineage requiring the transcription factor gene *Atoh1* (F. Wang et al., 2016).

PIEZO2 likely contributes to some types of nociception in mice. STOML3 depletion in heterologous systems decreases mechanosensitivity of both PIEZO1 and PIEZO2. *Stoml3* knockout mice do not exhibit wild type-like attenuation of tactile allodynia by systemic cholesterol depletion (Qi et al., 2015). Targeting STOML3-cholesterol association might control chronic pain in part through modulation of PIEZO channel activity. Injection of PIEZO2 antisense RNA attenuated mechanically stimulated pain in both the peripheral neuropathy model of oxaliplatin exposure and in the endothelium-dependent neuropathy model of endothelin-1 topical injection (Ferrari, Bogen, Green, & Levine, 2015).

Osmotic swelling of primary afferent dorsal root ganglion neurons converts rapidly adapting, mechanically activated currents to slowly adapting currents, similar to the effect of osmotic swelling of HEK-293 cells expressing PIEZO2. This switch is abolished by the actin cytoskeletal disruptor, cytochalasin D. In the intact rat, hind paw withdrawal upon injection of hypotonic solution is abolished by coadministration of the nonspecific PIEZO2 inhibitor Gd^{3+} or of cytochalasin D (Jia, Ikeda, Ling, Viatchenko-Karpinski, & Gu, 2016).

Cytoplasmic injection of GTP or GTPγS in mouse dorsal root ganglion neurons also converts the rapid adaptation of mechanically activated PIEZO2 channels to a pattern of slow adaptation (Jia, Ikeda, Ling, & Gu, 2013). The GTPase activating protein EPAC1 activates both small G proteins RAP1 and RAS in dorsal root ganglion neurons. In capsaicin-sensitive dorsal root ganglion neurons of rat, only RAS mediates ability of EPAC1 to increase neuronal excitability and promote hypersensitivity in models of chronic inflammation (Shariati, Thompson, Nicol, & Vasko, 2016). In contrast, in mouse dorsal root ganglion neurons RAP1 uniquely mediates the mechanical hyperalgesic effects of EPAC1 that are partially mediated by PIEZO2 and that can be inhibited by GRK2-mediated phosphorylation of EPAC1 at Ser108 (Singhmar et al., 2016).

Neuropathic pain increases EPAC1 mRNA in mice, and nerve damage-induced allodynia is decreased in EPAC1 knockout mice. The EPAC agonist 8-CPT-cAMP sensitizes mechanically activated currents both in dorsal root ganglion neurons and in PIEZO2-expressing HEK-293 cells. In vivo 8-CPT injection induces prolonged allodynia prevented by EPAC1 knockdown and attenuated by knockdown of PIEZO2, (which also elevated the light touch threshold without changing electrical excitability of sensory nerve fibers). Thus, PIEZO2 can signal through EPAC1 to contribute to mechanical allodynia and neuropathic pain (Eijkelkamp et al., 2013).

9. ADDITIONAL HUMAN GENETIC ASSOCIATIONS WITH PIEZO2

The PIEZO2 interactome of mouse dorsal root ganglion somatosensory neurons, as determined by mass spectroscopy (Narayanan et al., 2016) without further functional examination, includes several candidate binding proteins of potential pathological interest. Examples are the arthrogryposis multiplex congenita gene product MTMR1 and the Charcot–Marie–Tooth disease gene products MTMR2 and MTMR5. Subtypes of Charcot–Marie–Tooth exhibit phenotypic overlap with some arthrogryposes. MTMR1 and MTMR2 exhibit phosphatidylinositol phosphatase activity, and PIP2 depletion can inhibit PIEZO2 activity in cells (Borbiro, Badheka, & Rohacs, 2015).

Several candidate binding proteins suggest more speculative pathological links. Thus, the putative PIEZO2-binding protein COL4A3BP is an ER-Golgi ceramide transporter in which a single heterozygous de novo phosphorylation site mutation predicted to be upregulatory has been found in three unrelated individuals. The mutation is associated with global developmental delay syndromes that include craniofacial dysmorphism and variable oligohydramnios, oculomotor problems, and focal hypotonia (Consortium et al., 2015). Ceramide is a precursor of sphingosine-1-phosphate (S1P), a regulator of confinement sensing and monolayer cell extrusion processes coregulated by PIEZO1 (Gudipaty & Rosenblatt, 2016). Moreover, the ratio of ceramide to sphingomyelin regulates membrane tension (Lopez-Montero, Velez, & Devaux, 2007) in a way that might alter the activation threshold of PIEZO channels.

Mutation of another putative PIEZO2-interacting protein, SAMHD1, can be associated with Aicardi-Goutieres syndrome, which can be associated with scoliosis and, rarely, spondyloenchondrodysplasia. Additional intriguing, putative PIEZO2 interactors include potential modulators of PIEZO2 regulatory pathways, including cyclic nucleotide phosphodiesterase PDE11A and Ca^{2+}-calmodulin kinase 2 CAMK2 kinase subunits alpha, beta, gamma, and delta, as well as several classes of K^+ channel polypeptides (Narayanan et al., 2016).

The ExAC-14 genomic variant database generated from exomes of nominally healthy individuals includes (as of late 2016) 14 predicted loss-of-function mutations in one allele of the *PIEZO2* gene (frameshift-terminator mutations, termination mutations, and consensus splice-site inactivating mutations). In addition, a single in-frame short indel and numerous missense mutations of uncertain pathogenicity have been identified. Although haploid sufficiency of *PIEZO2* is not known to be pathogenic, it may serve as a risk modifier in other disorders, especially those of gene products interacting directly or functionally with PIEZO2.

10. CONCLUSIONS AND FUTURE PROSPECTS

Mutations in the genes encoding stretch-activated cation channels PIEZO1 and PIEZO2 are associated with both dominant gain-of-function diseases and recessive loss-of-function diseases. A common characteristic of PIEZO1- and PIEZO2-associated diseases is the overlap in clinical phenotypes of gain-of-function and loss-of-function mutations. This overlap suggests that both too little and too much mechanical sensitivity can be pathogenic, with either disorder leading to overlapping clinical and pathological manifestations. Moreover, multiple cell types may exhibit a limited repertoire of the full spectrum of phenotypic responses to altered mechanosensitivity, influenced in severity and duration by developmental state or chronicity of illness. Understanding these overlapping phenotypic responses to apparently opposite physiological dysfunctions of PIEZO channels will likely be important in development of therapeutics designed to correct both groups of mechanosensitivity disorders. Comparable considerations apply to the clinically similar, overlapping disorders associated with gain-of-function and loss-of-function mutations in the *NALCN* cation channel gene.

Diseases associated with PIEZO channel gain-of-function mutations are plausible targets for treatment with pharmacological inhibitors. The peptide inhibitor GsMTx4 blocks both PIEZO1- and PIEZO2-associated currents (Bae et al., 2011). GsMTx4 can be effective and nontoxic in rodents in either *d*- or *l*-forms (Suchyna et al., 2004; Wang, Ma, Sachs, Li, & Suchyna, 2016), and so has been postulated to target the PIEZO channel—membrane lipid interface. However, no small molecule PIEZO1 inhibitor of even moderate specificity has yet been reported. Modulators of STOML3 may also serve to downregulate gain-of-function PIEZO mutants (Poole et al., 2014; Wetzel et al., 2017).

A limited set of PIEZO1 loss-of-function missense mutations may retain sensitivity to the only defined small molecule PIEZO1 activator, Yoda1 (Syeda et al., 2015), or to congeners likely under development. Whether Yoda1 recognizes the channel polypeptide or the channel—lipid bilayer interface remains unreported. No small molecule pharmacological activator of PIEZO2 has yet been reported.

The minority of PIEZO loss-of-function mutations secondary to premature in-frame terminator mutations might respond to treatment with read-through drugs such as ataluren (Roy et al., 2016) or aminoglycoside antibiotics. Some PIEZO loss-of-function mutations may be misfolding mutations, for which design and synthesis of small molecule refolding and/or stabilizing chaperonin ligands may be possible. However, the great length of PIEZO polypeptides may pose challenges to the secretory and membrane protein biosynthesis and targeting apparatus. The TANGO complex has been shown to be important in the biosynthesis and secretion of certain collagen subtypes and other proteins too large or too extended in conformation to fit easily into conventional, size-limited, COPII-coated vesicles for export from the endoplasmic reticulum (Malhotra & Erlmann, 2011; Saito & Katada, 2015). Interestingly, among the candidate proteins of the PIEZO2 mass spectroscopic interactome in dorsal root ganglion neurons are the TANGO complex proteins TANGO1 and TANGO-like. Thus, biosynthesis and trafficking of the putative PIEZO trimers of ~800–900 kDa or more in mass may be regulated by the TANGO chaperonin system, which in complex with PIEZO polypeptides could constitute an additional target for pharmacological intervention.

Achievement of pharmacoselectivity between PIEZO1 and PIEZO2 has already been demonstrated feasible by the development of Yoda1, which activates PIEZO1 but not PIEZO2 (Syeda et al., 2015). Another precedent for future development of pharmacoselective agents is the contrast between the sensitivity of mouse PIEZO1 and insensitivity of the single *Drosophila melanogaster* Piezo polypeptide to the pore blocker, ruthenium red (Coste et al., 2012).

Inhibitors of PIEZO channels may prove useful in treatment of sickle cell disease (Cahalan et al., 2015; Vandorpe et al., 2010), HX (Andolfo, Alper, De Franceschi, et al., 2013; Zarychanski et al., 2012), and other cation-leak hemolytic anemias (Andolfo, Russo, Gambale, et al., 2016). They may also hold promise for treatment of some types of muscular dystrophy (Suchyna & Sachs, 2007), for cardiac arrhythmias and ischemia-reperfusion injury (J. Wang et al., 2016), and in some types of osteoarthritis (Lee et al., 2014). PIEZO agonists might have clinical uses in chronic obstructive lung disease and acute respiratory distress syndromes, sleep-apnea, and some types of sudden infant death syndrome (Nonomura et al., 2016). PIEZO modulators may find therapeutic uses in treatment of some forms of irritable bowel syndrome and gastrointestinal dysmotility disorders (F. Wang et al., 2016). The expression of PIEZO1 and PIEZO2 in astrocytes of the optic nerve head has led to the speculation that PIEZO modulators might be therapeutic in the setting of glaucoma or other disorders of elevated intraocular pressure (Choi, Sun, & Jakobs, 2015; Tehrani et al., 2016).

The current development of recombinant PIEZO expression systems in multiple cell types and in cellular and reconstituted lipid vesicles should accelerate our understanding of the pathophysiology of hereditary and acquired diseases of the PIEZO cation channels. Moreover, these methods will also accelerate the high-throughput search for new small molecules, peptides, antibodies, and aptamers with properties of specific high-affinity antagonists and agonists of PIEZO1 and PIEZO2 and to allow their testing for efficacy in cellular and animal models of disease.

REFERENCES

Akyuz, N., & Holt, J. R. (2016). Plug-N-Play: Mechanotransduction goes modular. *Neuron, 89*(6), 1128–1130. http://dx.doi.org/10.1016/j.neuron.2016.02.041.

Al-Sayed, M. D., Al-Zaidan, H., Albakheet, A., Hakami, H., Kenana, R., Al-Yafee, Y., … Kaya, N. (2013). Mutations in NALCN cause an autosomal-recessive syndrome with severe hypotonia, speech impairment, and cognitive delay. *American Journal of Human Genetics, 93*(4), 721–726. http://dx.doi.org/10.1016/j.ajhg.2013.08.001.

Albuisson, J., Murthy, S. E., Bandell, M., Coste, B., Louis-Dit-Picard, H., Mathur, J., … Patapoutian, A. (2013). Dehydrated hereditary stomatocytosis linked to gain-of-function mutations in mechanically activated PIEZO1 ion channels. *Nature Communications, 4*, 1884. http://dx.doi.org/10.1038/ncomms2899.

Almanza, E., Buckingham, K., Shively, K., McMillin, M., & Bamshad, M. (2014). *Screening for mutations in* PIEZO2 *that cause Distal Arthrogryposis Type 5*. Univ. Washington GenOM Project ALVA 2014.

Andolfo, I., Alper, S. L., De Franceschi, L., Auriemma, C., Russo, R., De Falco, L., … Iolascon, A. (2013). Multiple clinical forms of dehydrated hereditary stomatocytosis arise from mutations in PIEZO1. *Blood, 121*(19), 3925–3935. http://dx.doi.org/10.1182/blood-2013-02-482489. S3921–3912.

Andolfo, I., Alper, S. L., Delaunay, J., Auriemma, C., Russo, R., Asci, R., … Iolascon, A. (2013). Missense mutations in the ABCB6 transporter cause dominant familial pseudohyperkalemia. *American Journal of Hematology, 88*(1), 66–72. http://dx.doi.org/10.1002/ajh.23357.

Andolfo, I., Russo, R., Gambale, A., & Iolascon, A. (2016). New insights on hereditary erythrocyte membrane defects. *Haematologica, 101*(11), 1284–1294. http://dx.doi.org/10.3324/haematol.2016.142463.

Andolfo, I., Russo, R., Manna, F., De Rosa, G., Gambale, A., Zouwail, S., … Iolascon, A. (2016). Functional characterization of novel ABCB6 mutations and their clinical implications in familial pseudohyperkalemia. *Haematologica, 101*(8), 909–917. http://dx.doi.org/10.3324/haematol.2016.142372.

Andolfo, I., Russo, R., Manna, F., Shmukler, B. E., Gambale, A., Vitiello, G., … Iolascon, A. (2015). Novel Gardos channel mutations linked to dehydrated hereditary stomatocytosis (xerocytosis). *American Journal of Hematology, 90*(10), 921–926. http://dx.doi.org/10.1002/ajh.24117.

Archer, N. M., Shmukler, B. E., Andolfo, I., Vandorpe, D. H., Gnanasambandam, R., Higgins, J. M., … Nathan, D. G. (2014). Hereditary xerocytosis revisited. *American Journal of Hematology, 89*, 1142–1146. http://dx.doi.org/10.1002/ajh.23799.

Assis, R. A., Kassab, C., Seguro, F. S., Costa, F. F., Silveira, P. A., Wood, J., & Hamerschlak, N. (2013). Iron overload in a teenager with xerocytosis: The importance of nuclear magnetic resonance imaging. *Einstein, 11*, 528–532.

Astle, W. J., Elding, H., Jiang, T., Allen, D., Ruklisa, D., Mann, A. L., … Soranzo, N. (2016). The allelic landscape of human blood cell trait variation and links to common complex disease. *Cell, 167*(5), 1415–1429. http://dx.doi.org/10.1016/j.cell.2016.10.042. e1419.
Badens, C., & Guizouarn, H. (2016). Advances in understanding the pathogenesis of the red cell volume disorders. *British Journal of Haematology, 174*(5), 674–685. http://dx.doi.org/10.1111/bjh.14197.
Bae, C., Sachs, F., & Gottlieb, P. A. (2011). The mechanosensitive ion channel Piezo1 is inhibited by the peptide GsMTx4. *Biochemistry, 50*(29), 6295–6300. http://dx.doi.org/10.1021/bi200770q.
Bakeer, N., Luo, H. Y., Grimley, M., Risinger, M., Manganaris, P., Begrup, A. H., … Kalfa, T. (2014). Hemoglobin cincinnati: A novel beta globin gene mutation causing dominant beta-thalassemia [abst]. *Pediatric Blood & Cancer, 61*, S53.
Bawazir, W. M., Flatt, J. F., Wallis, J. P., Rendon, A., Cardigan, R. A., New, H. V., … Bruce, L. J. (2014). Familial pseudohyperkalemia in blood donors: A novel mutation with implications for transfusion practice. *Transfusion, 54*(12), 3043–3050. http://dx.doi.org/10.1111/trf.12757.
Begtrup, A. H., Dagaonkar, N., Pushkaran, S., Giger, K. M., Husami, A., Kissell, D., … Kalfa, T. A. (2013). Development of a comprehensive rapid next-generation sequencing assay for the diagnosis of inherited hemolytic anemia [abst]. *Blood, 122*, 949.
Bend, E. G., Si, Y., Stevenson, D. A., Bayrak-Toydemir, P., Newcomb, T. M., Jorgensen, E. M., & Swoboda, K. J. (2016). NALCN channelopathies: Distinguishing gain-of-function and loss-of-function mutations. *Neurology, 87*(11), 1131–1139. http://dx.doi.org/10.1212/WNL.0000000000003095.
Bento, C., Magalhaes Maia, T., Oliveira, A. C., Relvas, L., Almeida, H., Pereira, J., … Ribeiro, L. (2016). Separating the wheat from the chaff: Congenital hemolytic anemia study with a targeted next generation sequencing panel [abst]. *Haematologica, 101*, 615.
Borbiro, I., Badheka, D., & Rohacs, T. (2015). Activation of TRPV1 channels inhibits mechanosensitive Piezo channel activity by depleting membrane phosphoinositides. *Science Signaling, 8*(363), ra15. http://dx.doi.org/10.1126/scisignal.2005667.
Caciotti, A., Tonin, R., Rigoldi, M., Ferri, L., Catarzi, S., Cavicchi, C., … Morrone, A. (2015). Optimizing the molecular diagnosis of GALNS: Novel methods to define and characterize Morquio-A syndrome-associated mutations. *Human Mutation, 36*(3), 357–368. http://dx.doi.org/10.1002/humu.22751.
Cahalan, S. M., Lukacs, V., Ranade, S. S., Chien, S., Bandell, M., & Patapoutian, A. (2015). Piezo1 links mechanical forces to red blood cell volume. *eLife, 4*. http://dx.doi.org/10.7554/eLife.07370.
Chesler, A. T., Szczot, M., Bharucha-Goebel, D., Ceko, M., Donkervoort, S., Laubacher, C., … Bonnemann, C. G. (2016). The role of PIEZO2 in human mechanosensation. *The New England Journal of Medicine, 375*(14), 1355–1364. http://dx.doi.org/10.1056/NEJMoa1602812.
Choi, H. J., Sun, D., & Jakobs, T. C. (2015). Astrocytes in the optic nerve head express putative mechanosensitive channels. *Molecular vision, 21*, 749–766.
Chong, J. X., McMillin, M. J., Shively, K. M., Beck, A. E., Marvin, C. T., Armenteros, J. R., … Bamshad, M. J. (2015). De novo mutations in NALCN cause a syndrome characterized by congenital contractures of the limbs and face, hypotonia, and developmental delay. *American Journal of Human Genetics, 96*(3), 462–473. http://dx.doi.org/10.1016/j.ajhg.2015.01.003.
Chu, H., McKenna, M. M., Krump, N. A., Zheng, S., Mendelsohn, L., Thein, S. L., … Low, P. S. (2016). Reversible binding of hemoglobin to band 3 constitutes the molecular switch that mediates O_2 regulation of erythrocyte properties. *Blood, 128*(23), 2708–2716. http://dx.doi.org/10.1182/blood-2016-01-692079.

Cinar, E., Zhou, S., DeCourcey, J., Wang, Y., Waugh, R. E., & Wan, J. (2015). Piezo1 regulates mechanotransductive release of ATP from human RBCs. *Proceedings of the National Academy of Sciences of the United States of America, 112*(38), 11783–11788. http://dx.doi.org/10.1073/pnas.1507309112.

Consortium Uk K, Walter, K., Min, J. L., Huang, J., Crooks, L., Memari, Y., … Soranzo, N. (2015). The UK10K project identifies rare variants in health and disease. *Nature, 526*(7571), 82–90. http://dx.doi.org/10.1038/nature14962.

Coste, B., Houge, G., Murray, M. F., Stitziel, N., Bandell, M., Giovanni, M. A., … Patapoutian, A. (2013). Gain-of-function mutations in the mechanically activated ion channel PIEZO2 cause a subtype of distal arthrogryposis. *Proceedings of the National Academy of Sciences of the United States of America, 110*(12), 4667–4672. http://dx.doi.org/10.1073/pnas.1221400110.

Coste, B., Mathur, J., Schmidt, M., Earley, T. J., Ranade, S., Petrus, M. J., … Patapoutian, A. (2010). Piezo1 and Piezo2 are essential components of distinct mechanically activated cation channels. *Science, 330*(6000), 55–60. http://dx.doi.org/10.1126/science.1193270.

Coste, B., Murthy, S. E., Mathur, J., Schmidt, M., Mechioukhi, Y., Delmas, P., & Patapoutian, A. (2015). Piezo1 ion channel pore properties are dictated by C-terminal region. *Nature Communications, 6*, 7223. http://dx.doi.org/10.1038/ncomms8223.

Coste, B., Xiao, B., Santos, J. S., Syeda, R., Grandl, J., Spencer, K. S., … Patapoutian, A. (2012). Piezo proteins are pore-forming subunits of mechanically activated channels. *Nature, 483*(7388), 176–181. http://dx.doi.org/10.1038/nature10812.

Cox, C. D., Bae, C., Ziegler, L., Hartley, S., Nikolova-Krstevski, V., Rohde, P. R., … Martinac, B. (2016). Removal of the mechanoprotective influence of the cytoskeleton reveals PIEZO1 is gated by bilayer tension. *Nature Communications, 7*, 10366. http://dx.doi.org/10.1038/ncomms10366.

Dubin, A. E., Schmidt, M., Mathur, J., Petrus, M. J., Xiao, B., Coste, B., & Patapoutian, A. (2012). Inflammatory signals enhance piezo2-mediated mechanosensitive currents. *Cell Reports, 2*(3), 511–517. http://dx.doi.org/10.1016/j.celrep.2012.07.014.

Eijkelkamp, N., Linley, J. E., Torres, J. M., Bee, L., Dickenson, A. H., Gringhuis, M., … Wood, J. N. (2013). A role for Piezo2 in EPAC1-dependent mechanical allodynia. *Nature Communications, 4*, 1682. http://dx.doi.org/10.1038/ncomms2673.

Eisenhoffer, G. T., Loftus, P. D., Yoshigi, M., Otsuna, H., Chien, C. B., Morcos, P. A., & Rosenblatt, J. (2012). Crowding induces live cell extrusion to maintain homeostatic cell numbers in epithelia. *Nature, 484*(7395), 546–549. http://dx.doi.org/10.1038/nature10999.

Faucherre, A., Kissa, K., Nargeot, J., Mangoni, M. E., & Jopling, C. (2014). Piezo1 plays a role in erythrocyte volume homeostasis. *Haematologica, 99*(1), 70–75. http://dx.doi.org/10.3324/haematol.2013.086090.

Ferrari, L. F., Bogen, O., Green, P., & Levine, J. D. (2015). Contribution of Piezo2 to endothelium-dependent pain. *Molecular Pain, 11*, 65. http://dx.doi.org/10.1186/s12990-015-0068-4.

Fotiou, E., Martin-Almedina, S., Simpson, M. A., Lin, S., Gordon, K., Brice, G., … Ostergaard, P. (2015). Novel mutations in PIEZO1 cause an autosomal recessive generalized lymphatic dysplasia with non-immune hydrops fetalis. *Nature Communications, 6*, 8085. http://dx.doi.org/10.1038/ncomms9085.

Ge, J., Li, W., Zhao, Q., Li, N., Chen, M., Zhi, P., … Yang, M. (2015). Architecture of the mammalian mechanosensitive Piezo1 channel. *Nature, 527*(7576), 64–69. http://dx.doi.org/10.1038/nature15247.

Glogowska, E., & Gallagher, P. G. (2015). Disorders of erythrocyte volume homeostasis. *International Journal of Hematology, 37*(Suppl. 1), 85–91. http://dx.doi.org/10.1111/ijlh.12357.

Glogowska, E., Lezon-Geyda, K., Maksimova, Y., Schulz, V. P., & Gallagher, P. G. (2015). Mutations in the Gardos channel (KCNN4) are associated with hereditary xerocytosis. *Blood, 126*(11), 1281–1284. http://dx.doi.org/10.1182/blood-2015-07-657957.

Gnanasambandam, R., Bae, C., Gottlieb, P. A., & Sachs, F. (2015). Ionic selectivity and permeation properties of human PIEZO1 channels. *PLoS One, 10*(5), e0125503. http://dx.doi.org/10.1371/journal.pone.0125503.

Gottlieb, P. A., Bae, C., & Sachs, F. (2012). Gating the mechanical channel Piezo1: A comparison between whole-cell and patch recording. *Channels (Austin), 6*(4), 282–289. http://dx.doi.org/10.4161/chan.21064.

Grootenboer, S., Schischmanoff, P. O., Cynober, T., Rodrigue, J. C., Delaunay, J., Tchernia, G., & Dommergues, J. P. (1998). A genetic syndrome associating dehydrated hereditary stomatocytosis, pseudohyperkalaemia and perinatal oedema. *British Journal of Haematology, 103*(2), 383–386.

Gudipaty, S. A., & Rosenblatt, J. (2016). Epithelial cell extrusion: pathways and pathologies. *Seminars in Cell & Developmental Biology.* http://dx.doi.org/10.1016/j.semcdb.2016.05.010.

Guharay, F., & Sachs, F. (1984). Stretch-activated single ion channel currents in tissue-cultured embryonic chick skeletal muscle. *Journal of Physiology, 352*, 685–701.

Hall, J. G. (2014). Arthrogryposis (multiple congenital contractures): Diagnostic approach to etiology, classification, genetics, and general principles. *European Journal of Medical Genetics, 57*(8), 464–472. http://dx.doi.org/10.1016/j.ejmg.2014.03.008.

Hong, G. S., Lee, B., Wee, J., Chun, H., Kim, H., Jung, J., … Oh, U. (2016). Tentonin 3/TMEM150c confers distinct mechanosensitive currents in dorsal-root ganglion neurons with proprioceptive function. *Neuron, 91*(1), 107–118. http://dx.doi.org/10.1016/j.neuron.2016.05.029.

Huang, L., Li, L., Chen, X., Zhang, H., & Shi, Z. (2016). MiR-103a targeting Piezo1 is involved in acute myocardial infarction through regulating endothelium function. *Cardiology Journal.* http://dx.doi.org/10.5603/CJ.a2016.0056.

Hung, W. C., Yang, J. R., Yankaskas, C. L., Wong, B. S., Wu, P. H., Pardo-Pastor, C., … Konstantopoulos, K. (2016). Confinement sensing and signal optimization via piezo1/PKA and Myosin II pathways. *Cell Reports, 15*(7), 1430–1441. http://dx.doi.org/10.1016/j.celrep.2016.04.035.

Ilkan, Z., Wright, J. R., Francescut, L., Goodall, A. H., & Mahaut-Smith, M. P. (2016). Thrombus formation under flow is inhibited by the mechanosensitive cation channel blockers GsMTx-4 peptide and gadolinium chloride. *Circulation, 132*, A11593.

Imashuku, S., Muramatsu, H., Sugihara, T., Okuno, Y., Wang, X., Yoshida, K., … Kanno, H. (2016). PIEZO1 gene mutation in a Japanese family with hereditary high phosphatidylcholine hemolytic anemia and hemochromatosis-induced diabetes mellitus. *International Journal of Hematology, 104*(1), 125–129. http://dx.doi.org/10.1007/s12185-016-1970-x.

Jia, Z., Ikeda, R., Ling, J., & Gu, J. G. (2013). GTP-dependent run-up of Piezo2-type mechanically activated currents in rat dorsal root ganglion neurons. *Molecular Brain, 6*, 57. http://dx.doi.org/10.1186/1756-6606-6-57.

Jia, Z., Ikeda, R., Ling, J., Viatchenko-Karpinski, V., & Gu, J. G. (2016). Regulation of Piezo2 mechanotransduction by static plasma membrane tension in primary afferent neurons. *Journal of Biological Chemistry, 291*(17), 9087–9104. http://dx.doi.org/10.1074/jbc.M115.692384.

Kok, F. O., Shin, M., Ni, C. W., Gupta, A., Grosse, A. S., van Impel, A., … Lawson, N. D. (2015). Reverse genetic screening reveals poor correlation between morpholino-induced and mutant phenotypes in zebrafish. *Developmental Cell, 32*(1), 97–108. http://dx.doi.org/10.1016/j.devcel.2014.11.018.

Lane, J. W., McBride, D. W., Jr., & Hamill, O. P. (1991). Amiloride block of the mechanosensitive cation channel in *Xenopus* oocytes. *Journal of Physiology, 441*, 347–366.

Le Guen, L., Karpanen, T., Schulte, D., Harris, N. C., Koltowska, K., Roukens, G., … Hogan, B. M. (2014). Ccbe1 regulates Vegfc-mediated induction of Vegfr3 signaling during embryonic lymphangiogenesis. *Development, 141*(6), 1239–1249. http://dx.doi.org/10.1242/dev.100495.

Lee, W., Leddy, H. A., Chen, Y., Lee, S. H., Zelenski, N. A., McNulty, A. L., … Liedtke, W. B. (2014). Synergy between Piezo1 and Piezo2 channels confers high-strain mechanosensitivity to articular cartilage. *Proceedings of the National Academy of Sciences of the United States of America, 111*(47), E5114–E5122. http://dx.doi.org/10.1073/pnas.1414298111.

Li, J., Hou, B., Tumova, S., Muraki, K., Bruns, A., Ludlow, M. J., … Beech, D. J. (2014). Piezo1 integration of vascular architecture with physiological force. *Nature, 515*(7526), 279–282. http://dx.doi.org/10.1038/nature13701.

Lopez-Montero, I., Velez, M., & Devaux, P. F. (2007). Surface tension induced by sphingomyelin to ceramide conversion in lipid membranes. *Biochimica et Biophysica Acta, 1768*(3), 553–561. http://dx.doi.org/10.1016/j.bbamem.2007.01.001.

Lukacs, V., Mathur, J., Mao, R., Bayrak-Toydemir, P., Procter, M., Cahalan, S. M., … Krock, B. L. (2015). Impaired PIEZO1 function in patients with a novel autosomal recessive congenital lymphatic dysplasia. *Nature Communications, 6*, 8329. http://dx.doi.org/10.1038/ncomms9329.

Ma, G. C., Liu, C. S., Chang, S. P., Yeh, K. T., Ke, Y. Y., Chen, T. H., … Chen, M. (2008). A recurrent ITGA9 missense mutation in human fetuses with severe chylothorax: Possible correlation with poor response to fetal therapy. *Prenatal Diagnosis, 28*(11), 1057–1063. http://dx.doi.org/10.1002/pd.2130.

Mahmud, A. A., Nahid, N. A., Nassif, C., Sayeed, M. S., Ahmed, M. U., Parveen, M., … Michaud, J. L. (2016). Loss of the proprioception and touch sensation channel PIEZO2 in siblings with a progressive form of contractures. *Clinical Genetics.* http://dx.doi.org/10.1111/cge.12850.

Malhotra, V., & Erlmann, P. (2011). Protein export at the ER: Loading big collagens into COPII carriers. *EMBO Journal, 30*(17), 3475–3480. http://dx.doi.org/10.1038/emboj.2011.255.

Martins, J. R., Penton, D., Peyronnet, R., Arhatte, M., Moro, C., Picard, N., … Demolombe, S. (2016). Piezo1-dependent regulation of urinary osmolarity. *Pflugers Arch, 468*(7), 1197–1206. http://dx.doi.org/10.1007/s00424-016-1811-z.

McHugh, B. J., Buttery, R., Lad, Y., Banks, S., Haslett, C., & Sethi, T. (2010). Integrin activation by Fam38A uses a novel mechanism of R-Ras targeting to the endoplasmic reticulum. *Journal of Cell Science, 123*(Pt 1), 51–61. http://dx.doi.org/10.1242/jcs.056424.

McHugh, B. J., Murdoch, A., Haslett, C., & Sethi, T. (2012). Loss of the integrin-activating transmembrane protein Fam38A (Piezo1) promotes a switch to a reduced integrin-dependent mode of cell migration. *PLoS One,* 7(7), e40346. http://dx.doi.org/10.1371/journal.pone.0040346.

McMillin, M. J., Beck, A. E., Chong, J. X., Shively, K. M., Buckingham, K. J., Gildersleeve, H. I., … Bamshad, M. J. (2014). Mutations in PIEZO2 cause Gordon syndrome, Marden-Walker syndrome, and distal arthrogryposis type 5. *American Journal of Human Genetics, 94*(5), 734–744. http://dx.doi.org/10.1016/j.ajhg.2014.03.015.

Michishita, M., Yano, K., Tomita, K. I., Matsuzaki, O., & Kasahara, K. I. (2016). Piezo1 expression increases in rat bladder after partial bladder outlet obstruction. *Life Sciences, 166*, 1–7. http://dx.doi.org/10.1016/j.lfs.2016.10.017.

Miyamoto, T., Mochizuki, T., Nakagomi, H., Kira, S., Watanabe, M., Takayama, Y., … Tominaga, M. (2014). Functional role for Piezo1 in stretch-evoked Ca(2)(+) influx and ATP release in urothelial cell cultures. *Journal of Biological Chemistry, 289*(23), 16565–16575. http://dx.doi.org/10.1074/jbc.M113.528638.

Morris, C. E. (1990). Mechanosensitive ion channels. *Journal of Membrane Biology, 113*(2), 93–107.

Narayanan, P., Sondermann, J., Rouwette, T., Karaca, S., Urlaub, H., Mitkovski, M., … Schmidt, M. (2016). Native Piezo2 interactomics identifies pericentrin as a novel regulator of Piezo2 in somatosensory neurons. *Journal of Proteome Research, 15*(8), 2676–2687. http://dx.doi.org/10.1021/acs.jproteome.6b00235.

Nonomura, K., Woo, S. H., Chang, R. B., Gillich, A., Qiu, Z., Francisco, A. G., … Patapoutian, A. (January 12, 2017). Piezo2 senses airway stretch and mediates lung inflation-induced apnoea. *Nature, 541*(7636), 176–181. http://dx.doi.org/10.1038/nature20793.

O'Conor, C. J., Ramalingam, S., Zelenski, N. A., Benefield, H. C., Rigo, I., Little, D., … Guilak, F. (2016). Cartilage-specific knockout of the mechanosensory ion channel TRPV4 decreases age-related osteoarthritis. *Scientific Reports, 6*, 29053. http://dx.doi.org/10.1038/srep29053.

Okubo, M., Fujita, A., Saito, Y., Komaki, H., Ishiyama, A., Takeshita, E., … Sasaki, M. (2015). A family of distal arthrogryposis type 5 due to a novel *PIEZO2* mutation. *American Journal of Medical Genetics Part A, 167A*, 1100–1106. http://dx.doi.org/10.1002/ajmg.a.36881.

Orvain, C., Da Costa, L., van Wijk, R., Pissard, S., Picard, V., Mansour-Hendili, L., … Aguilar-Martinez, P. (2016). Inherited and acquired modifiers of iron status may dramatically affect the phenotypic expression of dehydrated hereditary stomatocytosis. *Blood, 128*, 1281.

Padera, T. P., Meijer, E. F., & Munn, L. L. (2016). The lymphatic system in disease processes and cancer progression. *Annual Review of Biomedical Engineering, 18*, 125–158. http://dx.doi.org/10.1146/annurev-bioeng-112315-031200.

Paessler, M., & Hartung, H. (2015). Dehydrated hereditary stomatocytosis masquerading as MDS [abst]. *Blood, 125*, 1841.

Peyronnet, R., Martins, J. R., Duprat, F., Demolombe, S., Arhatte, M., Jodar, M., … Patel, A. (2013). Piezo1-dependent stretch-activated channels are inhibited by polycystin-2 in renal tubular epithelial cells. *EMBO Reports, 14*(12), 1143–1148. http://dx.doi.org/10.1038/embor.2013.170.

Pimentel, H., Parra, M., Gee, S. L., Mohandas, N., Pachter, L., & Conboy, J. G. (2016). A dynamic intron retention program enriched in RNA processing genes regulates gene expression during terminal erythropoiesis. *Nucleic Acids Research, 44*(2), 838–851. http://dx.doi.org/10.1093/nar/gkv1168.

Poole, K., Herget, R., Lapatsina, L., Ngo, H. D., & Lewin, G. R. (2014). Tuning Piezo ion channels to detect molecular-scale movements relevant for fine touch. *Nature Communications, 5*, 3520. http://dx.doi.org/10.1038/ncomms4520.

Qi, Y., Andolfi, L., Frattini, F., Mayer, F., Lazzarino, M., & Hu, J. (2015). Membrane stiffening by STOML3 facilitates mechanosensation in sensory neurons. *Nature Communications, 6*, 8512. http://dx.doi.org/10.1038/ncomms9512.

Ranade, S. S., Qiu, Z., Woo, S. H., Hur, S. S., Murthy, S. E., Cahalan, S. M., … Patapoutian, A. (2014). Piezo1, a mechanically activated ion channel, is required for vascular development in mice. *Proceedings of the National Academy of Sciences of the United States of America, 111*(28), 10347–10352. http://dx.doi.org/10.1073/pnas.1409233111.

Rapetti-Mauss, R., Lacoste, C., Picard, V., Guitton, C., Lombard, E., Loosveld, M., … Badens, C. (2015). A mutation in the Gardos channel is associated with hereditary xerocytosis. *Blood, 126*(11), 1273–1280. http://dx.doi.org/10.1182/blood-2015-04-642496.

Retailleau, K., Duprat, F., Arhatte, M., Ranade, S. S., Peyronnet, R., Martins, J. R., … Honore, E. (2015). Piezo1 in smooth muscle cells is involved in hypertension-dependent arterial remodeling. *Cell Reports, 13*(6), 1161–1171. http://dx.doi.org/10.1016/j.celrep.2015.09.072.

Risinger, M. A., Glogowska, E., Begtrup, A. H., Dagaonkar, N., Chonat, S., Joiner, C. H., … Gallagher, P. G. (2014). The novel PIEZO1 mutation p.L2023V is causal for hereditary xerocytosis resulting in delayed channel inactivation and a dehydrated red blood cell phenotype [abst]. *Blood, 124*, 741.

Rossi, A., Kontarakis, Z., Gerri, C., Nolte, H., Holper, S., Kruger, M., & Stainier, D. Y. (2015). Genetic compensation induced by deleterious mutations but not gene knockdowns. *Nature, 524*(7564), 230–233. http://dx.doi.org/10.1038/nature14580.

Roy, B., Friesen, W. J., Tomizawa, Y., Leszyk, J. D., Zhuo, J., Johnson, B., … Jacobson, A. (2016). Ataluren stimulates ribosomal selection of near-cognate tRNAs to promote nonsense suppression. *Proceedings of the National Academy of Sciences of the United States of America, 113*(44), 12508–12513. http://dx.doi.org/10.1073/pnas.1605336113.

Saburi, S., Hester, I., Fischer, E., Pontoglio, M., Eremina, V., Gessler, M., … McNeill, H. (2008). Loss of Fat4 disrupts PCP signaling and oriented cell division and leads to cystic kidney disease. *Nature Genetics, 40*(8), 1010–1015. http://dx.doi.org/10.1038/ng.179.

Saito, K., & Katada, T. (2015). Mechanisms for exporting large-sized cargoes from the endoplasmic reticulum. *Cellular and Molecular Life Sciences, 72*(19), 3709–3720. http://dx.doi.org/10.1007/s00018-015-1952-9.

Sawada, J., Li, F., & Komatsu, M. (2015). R-Ras protein inhibits autophosphorylation of vascular endothelial growth factor receptor 2 in endothelial cells and suppresses receptor activation in tumor vasculature. *Journal of Biological Chemistry, 290*(13), 8133–8145. http://dx.doi.org/10.1074/jbc.M114.591511.

Shariati, B., Thompson, E. L., Nicol, G. D., & Vasko, M. R. (2016). Epac activation sensitizes rat sensory neurons through activation of Ras. *Molecular and Cellular Neurosciences, 70*, 54–67. http://dx.doi.org/10.1016/j.mcn.2015.11.005.

Shmukler, B. E., Huston, N. C., Thon, J. N., Ni, C. W., Kourkoulis, G., Lawson, N. D., … Alper, S. L. (2015). Homozygous knockout of the piezo1 gene in the zebrafish is not associated with anemia. *Haematologica, 100*(12), e483–485. http://dx.doi.org/10.3324/haematol.2015.132449.

Shmukler, B. E., Vandorpe, D. H., Rivera, A., Auerbach, M., Brugnara, C., & Alper, S. L. (2014). Dehydrated stomatocytic anemia due to the heterozygous mutation R2456H in the mechanosensitive cation channel PIEZO1: A case report. *Blood Cells, Molecules and Diseases, 52*, 53–54. http://dx.doi.org/10.1016/j.bcmd.2013.07.015.

Singhmar, P., Huo, X., Eijkelkamp, N., Berciano, S. R., Baameur, F., Mei, F. C., … Kavelaars, A. (2016). Critical role for Epac1 in inflammatory pain controlled by GRK2-mediated phosphorylation of Epac1. *Proceedings of the National Academy of Sciences of the United States of America, 113*(11), 3036–3041. http://dx.doi.org/10.1073/pnas.1516036113.

Skals, M., Jensen, U. B., Ousingsawat, J., Kunzelmann, K., Leipziger, J., & Praetorius, H. A. (2010). *Escherichia coli* alpha-hemolysin triggers shrinkage of erythrocytes via K(Ca)3.1 and TMEM16A channels with subsequent phosphatidylserine exposure. *Journal of Biological Chemistry, 285*(20), 15557–15565. http://dx.doi.org/10.1074/jbc.M109.082578.

Spier, I., Kerick, M., Drichel, D., Horpaopan, S., Altmuller, J., Laner, A., … Aretz, S. (2016). Exome sequencing identifies potential novel candidate genes in patients with unexplained colorectal adenomatous polyposis. *Familial Cancer, 15*(2), 281–288. http://dx.doi.org/10.1007/s10689-016-9870-z.

Stewart, G. W., Amess, J. A., Eber, S. W., Kingswood, C., Lane, P. A., Smith, B. D., & Mentzer, W. C. (1996). Thrombo-embolic disease after splenectomy for hereditary stomatocytosis. *British Journal of Haematology, 93*(2), 303–310.

Suchyna, T. M., Johnson, J. H., Hamer, K., Leykam, J. F., Gage, D. A., Clemo, H. F., … Sachs, F. (2000). Identification of a peptide toxin from *Grammostola spatulata* spider venom that blocks cation-selective stretch-activated channels. *Journal of General Physiology, 115*(5), 583–598.

Suchyna, T. M., & Sachs, F. (2007). Mechanosensitive channel properties and membrane mechanics in mouse dystrophic myotubes. *Journal of Physiology, 581*(Pt 1), 369–387. http://dx.doi.org/10.1113/jphysiol.2006.125021.

Suchyna, T. M., Tape, S. E., Koeppe, R. E., 2nd, Andersen, O. S., Sachs, F., & Gottlieb, P. A. (2004). Bilayer-dependent inhibition of mechanosensitive channels by neuroactive peptide enantiomers. *Nature, 430*(6996), 235–240. http://dx.doi.org/10.1038/nature02743.

Syeda, R., Florendo, M. N., Cox, C. D., Kefauver, J. M., Santos, J. S., Martinac, B., & Patapoutian, A. (2016). Piezo1 channels are Inherently mechanosensitive. *Cell Reports, 17*(7), 1739–1746. http://dx.doi.org/10.1016/j.celrep.2016.10.033.

Syeda, R., Xu, J., Dubin, A. E., Coste, B., Mathur, J., Huynh, T., … Patapoutian, A. (2015). Chemical activation of the mechanotransduction channel Piezo1. *eLife, 4*. http://dx.doi.org/10.7554/eLife.07369.

Tehrani, S., Davis, L., Cepurna, W. O., Choe, T. E., Lozano, D. C., Monfared, A., … Morrison, J. C. (2016). Astrocyte structural and molecular response to elevated intraocular pressure occurs rapidly and precedes axonal tubulin rearrangement within the optic nerve head in a rat model. *PLoS One, 11*(11), e0167364. http://dx.doi.org/10.1371/journal.pone.0167364.

Vahatupa, M., Prince, S., Vataja, S., Mertimo, T., Kataja, M., Kinnunen, K., … Uusitalo-Jarvinen, H. (2016). Lack of R-Ras leads to increased vascular permeability in ischemic retinopathy. *Investigative Ophthalmology & Visual Science, 57*(11), 4898–4909. http://dx.doi.org/10.1167/iovs.16-19212.

Vandorpe, D. H., Xu, C., Shmukler, B. E., Otterbein, L. E., Trudel, M., Sachs, F., … Alper, S. L. (2010). Hypoxia activates a Ca^{2+}-permeable cation conductance sensitive to carbon monoxide and to GsMTx-4 in human and mouse sickle erythrocytes. *PLoS One, 5*(1), e8732. http://dx.doi.org/10.1371/journal.pone.0008732.

Wang, F., Knutson, K., Alcaino, C., Linden, D. R., Gibbons, S. J., Kashyap, P., … Beyder, A. (2016). Mechanosensitive ion channel Piezo2 is important for enterochromaffin cell response to mechanical forces. *Journal of Physiology*. http://dx.doi.org/10.1113/JP272718.

Wang, J., Ma, Y., Sachs, F., Li, J., & Suchyna, T. M. (2016). GsMTx4-D is a cardioprotectant against myocardial infarction during ischemia and reperfusion. *Journal of Molecular and Cellular Cardiology, 98*, 83–94. http://dx.doi.org/10.1016/j.yjmcc.2016.07.005.

Wang, S., Chennupati, R., Kaur, H., Iring, A., Wettschureck, N., & Offermanns, S. (2016). Endothelial cation channel PIEZO1 controls blood pressure by mediating flow-induced ATP release. *Journal of Clinical Investigation, 126*(12), 4527–4536. http://dx.doi.org/10.1172/JCI87343.

Wetzel, C., Pifferi, S., Picci, C., Gok, C., Hoffmann, D., Bali, K. K., … Lewin, G. R. (2017). *Nature Neuroscience, 20*, 209–218. http://dx.doi.org/10.1038/nn.4454.

Woo, S. H., Lukacs, V., de Nooij, J. C., Zaytseva, D., Criddle, C. R., Francisco, A., … Patapoutian, A. (2015). Piezo2 is the principal mechanotransduction channel for proprioception. *Nature Neuroscience, 18*(12), 1756–1762. http://dx.doi.org/10.1038/nn.4162.

Woo, S. H., Ranade, S., Weyer, A. D., Dubin, A. E., Baba, Y., Qiu, Z., … Patapoutian, A. (2014). Piezo2 is required for Merkel-cell mechanotransduction. *Nature, 509*(7502), 622–626. http://dx.doi.org/10.1038/nature13251.

Wright, J. R., Amisten, S., Goodall, A. H., & Mahaut-Smith, M. P. (2016). Transcriptomic analysis of the ion channelome of human platelets and megakaryocytic cell lines. *Thrombosis and Haemostasis, 116*(2), 272–284. http://dx.doi.org/10.1160/TH15-11-0891.

Wu, Z., Grillet, N., Zhao, B., Cunningham, C., Harkins-Perry, S., Coste, B., … Muller, U. (2017). Mechanosensory hair cells express two molecularly distinct mechanotransduction channels. *Nature Neuroscience, 20*, 24–35. http://dx.doi.org/10.1038/nn.4449.

Yang, E., Voelkel, E., Lezon-Geyda, K., Schulz, V., & Gallagher, P. (2017). Hemoglobin C trait accentuating erythrocyte dehydration in hereditary xerocytosis [abst]. *Pediatric Blood & Cancer*, 64 (in press).

Yang, X. C., & Sachs, F. (1989). Block of stretch-activated ion channels in Xenopus oocytes by gadolinium and calcium ions. *Science, 243*(4894 Pt 1), 1068–1071.

Yang, X. N., Lu, Y. P., Liu, J. J., Huang, J. K., Liu, Y. P., Xiao, C. X., … Guleng, B. (2014). Piezo1 is as a novel trefoil factor family 1 binding protein that promotes gastric cancer cell mobility in vitro. *Digestive Diseases and Sciences, 59*(7), 1428–1435. http://dx.doi.org/10.1007/s10620-014-3044-3.

Zarychanski, R., Schulz, V. P., Houston, B. L., Maksimova, Y., Houston, D. S., Smith, B., … Gallagher, P. G. (2012). Mutations in the mechanotransduction protein PIEZO1 are associated with hereditary xerocytosis. *Blood, 120*(9), 1908–1915. http://dx.doi.org/10.1182/blood-2012-04-422253.

Zhao, Q., Wu, K., Geng, J., Chi, S., Wang, Y., Zhi, P., … Xiao, B. (2016). Ion permeation and mechanotransduction mechanisms of mechanosensitive piezo channels. *Neuron, 89*(6), 1248–1263. http://dx.doi.org/10.1016/j.neuron.2016.01.046.

Zuccala, E. S., Satchwell, T. J., Angrisano, F., Tan, Y. H., Wilson, M. C., Heesom, K. J., & Baum, J. (2016). Quantitative phospho-proteomics reveals the Plasmodium merozoite triggers pre-invasion host kinase modification of the red cell cytoskeleton. *Scientific Reports, 6*, 19766. http://dx.doi.org/10.1038/srep19766.

CHAPTER FIVE

The Structural Basis for Sensing by the Piezo1 Protein

W. Li, N. Gao and M. Yang[1]
Tsinghua University, Beijing, China
[1]Corresponding author: E-mail: maojunyang@tsinghua.edu.cn

Contents

Abstract

Mechanotransduction is one of the processes by which cells sense and convert mechanical stimuli into biological signals. Experimental data from various species have revealed crucial roles for mechanotransduction in organ development and a plethora of physiological activities. Piezo proteins have recently been identified as the long-sought-after mechanically activated cation channels in eukaryotes. The architecture of mouse Piezo1 (mPiezo1) channel determined by cryoelectron microscopic single-particle analysis at medium resolution yielded important insights into the mechanical force sensing mechanism. mPiezo1 is found to form a trimeric propeller-like structure with the extracellular domains resembling three distal blades and a central cap. The transmembrane region consists of a central pore module that likely determines the ion-conducting properties of mPiezo1, and three peripheral wings formed by arrays of paired transmembrane helices. Compared with the central pore module, the three distal blades display considerably larger flexibility. In the intracellular region, three long beam-like domains (~80 Å in length) support the whole transmembrane region and connect the mobile peripheral regions to the central pore module. This unique design suggests that the trimeric mPiezo1 may mechanistically function in similar principles as how propellers sense and transduce force to control the ion

Current Topics in Membranes, Volume 79
ISSN 1063-5823
http://dx.doi.org/10.1016/bs.ctm.2016.10.001

conductivity. This review summarizes the current knowledge on the structure and proposes possible gating mechanisms of mPiezo1.

1. INTRODUCTION

1.1 Function of mechanosensitive ion channels

Mechanosensitive (MS) channels are the membrane pores that sense mechanical forces. On activation they conduct specific ions based on the ion concentration gradient across membrane. MS channels have diverse biological and physiological functions regulating many biological processes in different organisms. The most extensively studied MS channels are from bacteria, which protect the organisms from severe hypoosmotic shocks (Sukharev, Blount, Martinac, Blattner, & Kung, 1994). There are two major channels, MscL (the channel of large conductance) and MscS (the channel of small conductance), with minor contributions from MscS homologs, MscK and YbdG. Loss of both MscL and MscS compromises the physical integrity of *Escherichia coli* cell during hypoosmotic shock; the further loss of MscK and YbdG lowers the threshold at which cell death occurs in response to a rapid lowering of osmolarity. In multicellular plants and animals, MS channels are responsible for sensing not only external mechanical forces such as gravity, touch and hearing, but also internal mechanical forces from cell interactions, and therefore plays a variety of roles in cell growth, volume regulation, migration, proliferation, blood pressure control, tension detection, and pain sensation, which are of vital importance in proper growth, development, and function of various organs (Gillespie & Walker, 2001; Haswell, Phillips, & Rees, 2011; McHugh, Murdoch, Haslett, & Sethi, 2012; Miyamoto et al., 2014; Sachs, 2010; Vogel, 2006).

The existence of ion channels directly activated by mechanical force among mammalian cells was first demonstrated by Corey and Hudspeth in the late 1970s. They developed an accurate system to record the transepithelial microphonic potential on hair cells from the *bullfrog sacculus* when activated by well-defined mechanical stimuli, showing that there exists a 40-μs delay after a fast stimuli and that the response current can be maintained over many milliseconds. This suggested the existence of a transmembrane channel because, if the measured response is only associated with extracellular charge displacement or with otolithic membrane, the current should be transient. This latency also has moderate temperature dependence, which excludes diffusion from contributing to the delay. Furthermore, addition of the ototoxic antibiotic streptomycin or removal of Ca^{2+} can block the response (Corey & Hudspeth, 1979). A series of subsequent studies performed in various mammalian cell types including cultured mesangial cells (Craelius, el-Sherif, & Palant, 1989), skeletal myotubes (Franco & Lansman, 1990; McBride & Hamil, 1992), putative aortic baroreceptor neurons (Kraske et al., 1998), dorsal root ganglion sensory neurons (Delmas, Hao, & Rodat-Despoix, 2011; McCarter, Reichling, & Levine, 1999), inner-ear hair cells (Beurg, Evans, Hackney, & Fettiplace, 2006), and rat pulmonary arterial smooth muscle cells (Delmas et al., 2011) revealed that mechanically activated cation channels in the cell membrane play key roles in converting mechanical stimuli into biological activities such as touch, hearing and blood pressure regulation, a process that is termed mechano-transduction (Chalfie, 2009).

Since many MS ion channels also have other roles than sensing the mechanical force, the possibility that some known channels in other cellular pathways can participate in mechanotransduction is difficult to rule out. Some criteria of identifying MS ion channels are proposed: (1) activated by stretch or pressure; (2) have rapid channel kinetics; (3) current kinetics are dependent on the pressure amplitude; (4) the gating mechanism is associated with conformational changes. In obedience with most of these criteria, many new classes of MS ion channels are identified in both invertebrates and vertebrates recently, including the TRP channel family (Christensen & Corey, 2007), the DEG/ENaC family (Delmas & Coste, 2013), Kv channels (Hao et al., 2013; Tabarean & Morris, 2002), TMC proteins (Kawashima et al., 2015), and Piezo proteins (Coste et al., 2010). In these studies, cells with different channel complements exhibit their own unique pattern of response to specific kinds of mechanical stimuli.

1.2 Identification of Piezo proteins

In 2010, Coste et al. in the Scripps Research Institute identified two mouse proteins of uncharacterized functions that exhibited properties of mechano-transducers when introduced into cells. They named those two proteins with highly similar sequences as Piezo1 and Piezo2, with "piezo" meaning pressure or squeezing in Greek. Both Piezo1 and Piezo2 can independently induce cationic nonselective currents when mechanically activated, and these currents display comparably fast kinetics of inactivation when overexpressed in heterologous systems. Consistent with the later-demonstrated physiological roles, Piezo1 and Piezo2 are widely expressed in eukaryotic MS tissues (Coste et al., 2010). Piezo1 and Piezo2 are subsequently confirmed as the pore-forming subunits of the long-sought mechanically activated cation channels in mammals (Kim, Coste, Chadha, Cook, & Patapoutian, 2012). Once these proteins were identified, it greatly facilitated comprehensive studies on the channels using genetic, biochemical, and biophysical approaches, reinforcing their functional significance as crucial mechanotransducers in various biological processes (Honore, Martins, Penton, Patel, & Demolombe, 2015). In flies, a single Piezo gene exists and it is involved in mechanical nociception (Kim et al., 2012). Piezo2 mediates gentle touch sensation in vertebrates, including fish (Faucherre, Nargeot, Mangoni, & Jopling, 2013), birds (Schneider et al., 2014), rodents (Ikeda et al., 2014; Maksimovic et al., 2014; Ranade, Woo, et al., 2014; Woo et al., 2014), and human (Schrenk-Siemens et al., 2015). Piezo1 appears to play a more broad role in different aspects of physiology, such as sensing shear stress of blood flow for proper blood vessel development (Li et al., 2014; Ranade, Qiu, et al., 2014), regulating red blood cell function (Faucherre, Kissa, Nargeot, Mangoni, & Jopling, 2014), sensing the extension of bladder (Miyamoto et al., 2014), and controlling cell migration and differentiation (McHugh et al., 2010; Pathak et al., 2014). In human, mutations of Piezo1 or Piezo2 genes have been linked to many genetic diseases, such as dehydrated hereditary stomatocytosis (DHSt) (Albuisson et al., 2013; Andolfo et al., 2013; Bae, Gnanasambandam, Nicolai, Sachs, & Gottlieb, 2013; Beneteau et al., 2014; Sandberg, Nybo, Birgens, & Frederiksen, 2014; Shmukler et al., 2014; Zarychanski et al., 2012), distal arthrogryposis type 5 (Coste et al., 2013), Gordon syndrome, and Marden—Walker syndrome (McMillin et al., 2014), indicating their pathophysiological relevance and potential therapeutic targets.

Piezo proteins thus represent an evolutionally conserved and physiologically important prototypic class of mechanically activated cation channels. They do not possess significant sequence homology to any known classes of ion channels including voltage- or ligand-gated channels, TRP channels, prokaryotic mechanosensitive channels (MSCs), or eukaryotic MS K2P channels. Mammalian Piezo proteins contain over 2500 residues with more than 30 predicted transmembrane segments (TM) (Coste et al., 2010), and they homo-oligomerize to form channel complexes (Kim et al., 2012). These features are quite uncommon, which make Piezo as the membrane channel protein with the largest number of TM segments to our knowledge. Their distinct sequences and predicted topology raised intriguing questions on how these huge channel machines assemble and function, e.g., how many copies of protomers are required to assembly the homo-oligomerized channel complex. At the structural level, such issues that as to the real number of TM segments in Piezo proteins beyond predicted topology or whether the channel complex constitutes a single ion-conduction pore or whether each protomer has its own pore (Bae et al., 2013). Biophysical questions related to how Piezo channels are specifically gated by forces are critical for understanding the conversion of force into an electrical signal. Many of these issues can be addressed once a detailed structure of the full length Piezo channel complex is provided.

Recent advancements of cryoelectron microscopy (cryo-EM) single particle technique allows structural determination of traditionally challenging samples (Bai, McMullan, & Scheres, 2015; Cheng, 2015; Cheng, Grigorieff, Penczek, & Walz, 2015). With this cutting-edge technique, we reported the first structure of the full-length mPiezo1 channel featuring a trimeric three-blade-propeller-like architecture (Ge et al., 2015). This review describes the recent findings on the structure of mPiezo1 and structural insights into the ion-conducting and gating mechanisms of mPiezo1 and discusses the remaining questions on these channels.

2. STRUCTURE OF MOUSE PIEZO1

2.1 Overall structure of mouse Piezo1

Piezo1 formed a three-blade, propeller-shaped architecture, with distinct regions resembling the typical structural components as a propeller, including three blades and a central cap (Fig. 1).

The transmembrane region, which could be readily located on averaged particle images contains three extended and twisted arrays of transmembrane helices (Fig. 1). Beyond the transmembrane helical array, three thick distal blades are arranged in a superhelical fashion and each blade also has a helicoidal surface (Fig. 1A). Viewed from the top, the diameter and the axial height of the structure are ~200 and 155 Å, respectively (Fig. 1A and B). A single central cap sits above the surface of the transmembrane core with a gap (~8 Å) in between (Fig. 1A and B). Furthermore, a tightly packed region, likely to be a compact soluble domain, is located on the opposite side of the cap, right below the transmembrane region (Fig. 1A and B). Three long, distinct density rods (~80 Å) exposed on the outer surface of the transmembrane region, hereafter termed beam, seem to connect the distal end of the transmembrane region and the blades mechanically to

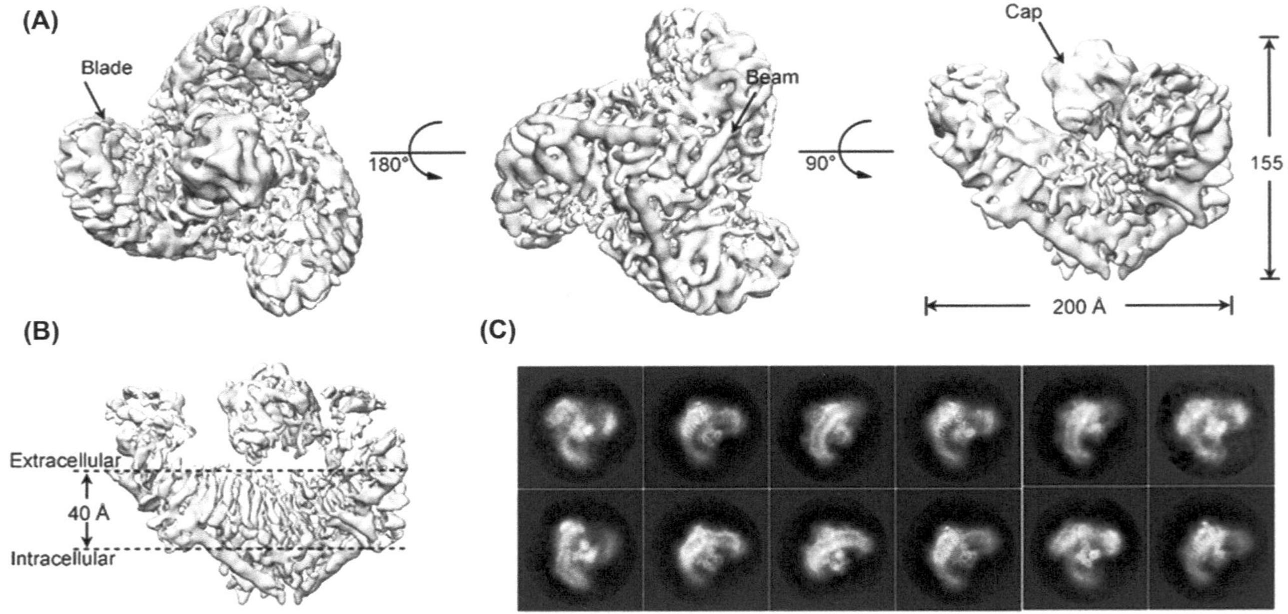

Figure 1 ***Overall structure of mouse Piezo1 (mPiezo1) determined by single particle cryoelectron microscopy.*** (A) Top (left), bottom (middle), and side (right) views of unsharpened map (at 5σ contour level) of mPiezo1 with distinct regions labeled and indicated by the *black arrows*. The height and width of the structure are shown in the right panel. (B) Side view of the sharpened map (6σ contour level) of mPiezo1 filtered to a resolution of 4.8 Å. The transmembrane region is indicated. (C) Representative 2D class averages of different views show fine features of the trimeric complex.

the center of the trimeric complex at the bottom face. The diameter of the density rod suggests that the beam is composed of a two-stranded coiled coil (Fig. 1B). In summary, mPiezo1 forms a homo-trimer displaying an overall structure that resembles a typical three-bladed propeller.

2.2 Topology of mouse Piezo1

One major question about the structure is which side of the mPiezo1 faces the intracellular side. The density map revealed that the cap is separated from the transmembrane core indicates that it is likely to be a soluble region. A topological prediction model suggests that residues from 2210 to 2457 (termed the C-terminal extracellular domain, CED) constitute a large extracellular loop followed by the last transmembrane segment at the C terminus (Kamajaya, Kaiser, Lee, Reid, & Rees, 2014). Two-dimensional classification of deletion-mutant Piezo1$^{2219-2453}$ particles by negative-staining electron microscopy revealed that the central cap was absent in 2D class averages (Fig. 2A and B) confirming that this loop in the primary sequence indeed forms the cap.

Consequently, the high resolution crystal structure of the CED domain of mPiezo1 was resolved. The crystal structure of the CED (Piezo1$^{2214-2457}$) was similar to that of the same region in *Caenorhabditis elegans* Piezo that has been reported recently (Kamajaya et al., 2014). The amino (N) and C termini of the CED are on the same side and close to each other (Fig. 2C). Consistent with this finding, bioinformatics analysis, immunostaining to detect extracellular tags inserted in predicted loop regions, and detection of intracellular phosphorylation sites by mass spectrometry were used to characterize the transmembrane topology of mPiezo1, which came out to give a similar topological prediction that the CED is located between the last two transmembrane segments in the C-terminal region of Piezo1 (Coste et al., 2015; Kamajaya et al., 2014).

In both gel filtration and crystal lattice, the CED tends to form a homotrimer. A direct and rigid fitting of the crystallographic trimer of the CED into the cryo-EM density map resulted in a perfect match (Fig. 2D). These results demonstrate that the cap is formed by a CED trimer, further supporting the conclusion that the full-length Piezo1 forms a homotrimer.

Further confirmation of the topological location of the CED and the C terminus of Piezo1 was demonstrated by immunolabeling of various tagged proteins in cells. Immunolabeling of live HEK293 T cells expressing Piezo1 with a flag-tagged fused either in a flexible loop of the CED (after A2419) or at the C terminus of Piezo1 showed that the flag tag could be labeled on the

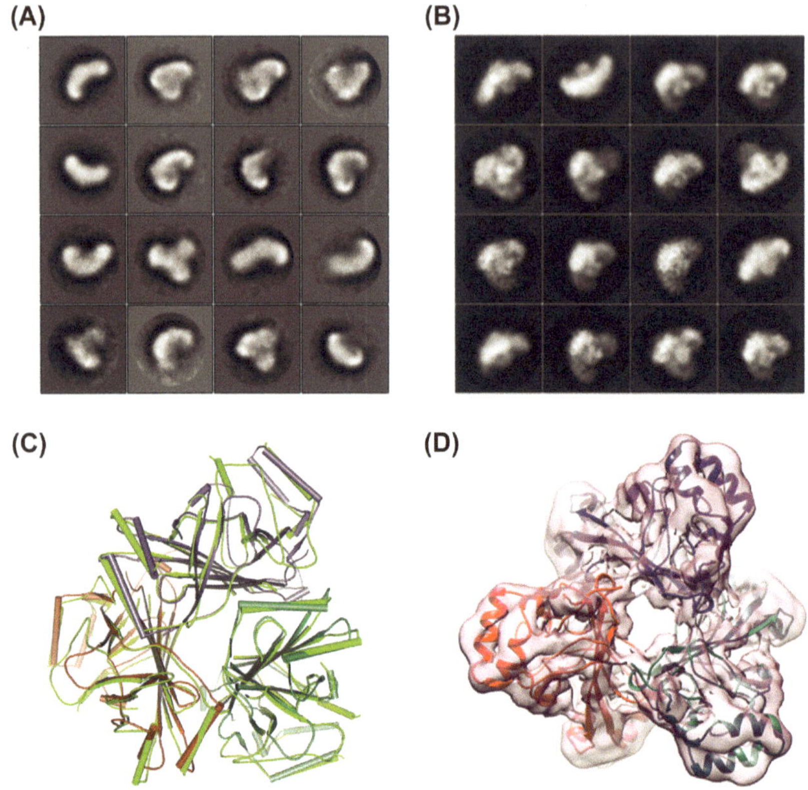

Figure 2 ***Topological localization of C-terminal extracellular domain (CED).*** (A) 2D class averages of negatively stained mouse Piezo1ΔCED particles. It is evident that the central cap domain is missing. (B) 2D class averages of negatively stained mouse Piezo1 particles (detergent: C12E10). (C) Crystal structure alignment of the trimeric CED of Piezo1 from mouse and *Caenorhabditis elegans*. The crystal structure of *C. elegans* CED is colored in green, while the trimeric CED from mouse is colored in three different colors (pink, cyan, and blue). (D) Transparent surface representation of segmented density map of the "cap" superimposed with the trimeric CED crystal structure. The trimeric CEDs are colored in red, cyan, and blue, respectively.

plasma membrane of live cells only when inserted in the CED and not at the C terminus (Fig. 3). These results demonstrated that the CED is an extracellular domain, whereas the C terminus is intracellular, consistent with a recent report that ion-permeation properties of mPiezo1 and dPiezo are conferred by C-terminal region (Coste et al., 2015). Consequently, both the central cap and the three blades are located at the extracellular side, whereas the beams are located at the intracellular side.

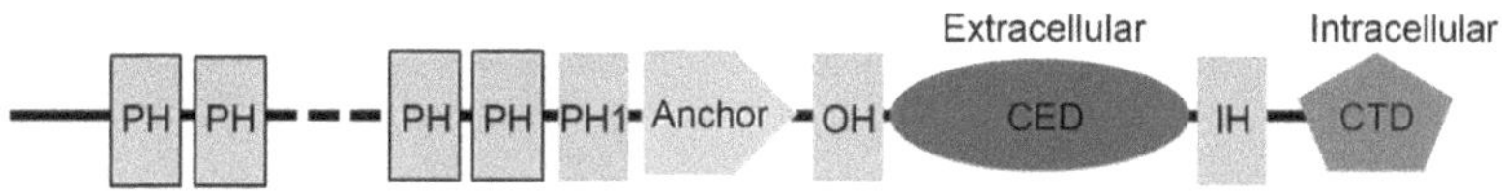

Figure 3 The model represents the topology of the C-terminal part of Piezo1. Different structural units are indicated. The CED is extracellular, while the CTD is intracellular. *CED*, C-terminal extracellular domain; *CTD*, C-terminal domain; *IH*, inner helix; *OH*, outer helix; *PH*, peripheral helix.

2.3 Transmembrane skeleton of mouse Piezo1

The most unique feature of Piezo proteins is their unprecedentedly large number of predicted TMs (30–40 TMs). Although the resolution of the density map is not sufficient for de novo sequence assignment, we could build an alanine backbone model (~502 residues) covering the apparent TM segments, including the "beam" and the CTD. Together with the 227 amino acids of the CED, each protomer contains 719 amino acids (~one-third of the total sequence) (Fig. 4A). The whole TM skeleton displays a three-winged arrangement, with each extended wing being slightly twisted (Fig. 4B).

Fourteen TMs could be readily recognized on each wing which is lower than the predicted number. This indicates that about half of the predicted TM helices are not buried in the membrane, contrasting the previously constructed topology model topology with 38 predicted transmembrane segments (Coste et al., 2012). The discrepancy of the number of TM is likely due to the fact that bioinformatics prediction heavily relies on primary amino acid sequences and lack of experimental evidence for many of the predicted extracellular loops. A recent work suggested that mPiezo1 has a minimum of 18 TM segments (Coste et al., 2015). In line with this, a single "blade" displays a comparable volume of the "cap" region that accommodates about 700 residues indicating that some of the previously predicted helices should be located at the distal extracellular regions.

Due to the lack of reliable density at the junction of the C-terminus of CED and the N-terminus of the IH, the three IHs are not assigned to each subunit and thus labeled as IH, IH′, and IH″.

For the sake of clarity in referring to known ion channel TM organization, the TM core constituting TMs are named as inner helix (IH) and outer helix (OH), and the peripheral TM array constituting TMs as peripheral helix (PH1–12) (Fig. 4). 12 PHs from the same protomer are roughly organized as six helical pairs extending from the central axis to the periphery

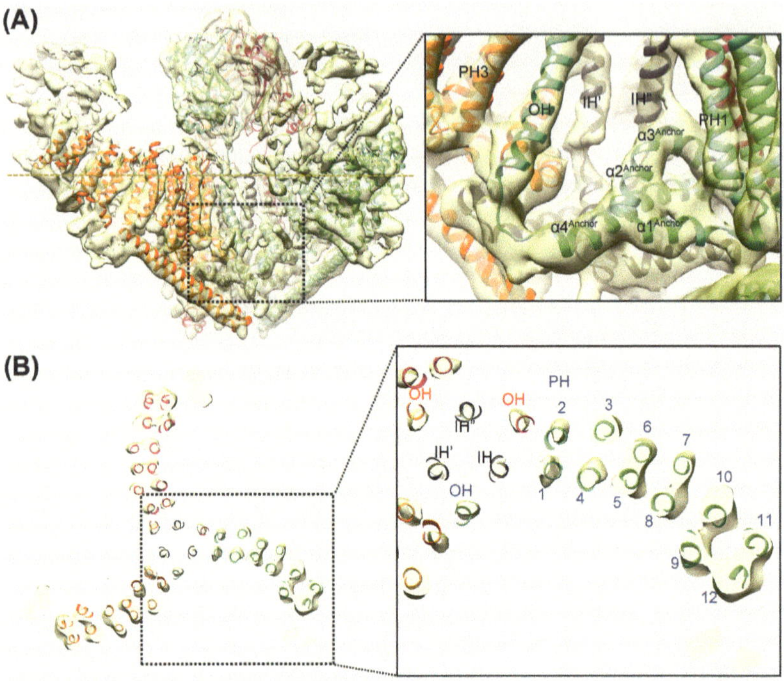

Figure 4 ***The transmembrane skeleton organization.*** (A) A side view of the mouse Piezo1 cryoelectron microscopy density map superimposed with polyalanine models of each protomer. Three protomers are colored differently. The boxed region is enlarged in the right panel to illustrate the anchor domain, the connecting helices (α1−α4) between the outer helix (OH) and peripheral helix 1 (PH1), resulting in the swap of the OH to a neighboring subunit. (B) A z-slice representation of the layer indicated by the yellow dashed line shown in (A) highlighting the overall organization of the TM skeleton. The boxed region is amplified in the right panel to illustrate the central TM core that consists of three inner helices (IH) and three OHs and a wing of the PH1−12.

of the complex (Fig. 4), and connected to the extracellular "blade." Interestingly, these six continuous pairs of helices are arranged in a much organized way (Fig. 4). The IH and OH connect to the C- and N-terminus of the CED, respectively (Fig. 4). However, the lack of reliable density at the junction point between the C-terminus of the CED and the IH makes the assignment of three IHs to each protomer unclear.

Interestingly, a unique hairpin structure has been observed at the interface of two adjacent subunits, which contains four helices, α1−α4, and penetrates into the inner leaflet of the membrane, with a longer helix

(α4) roughly parallel to the membrane (Fig. 4). This region has been assigned as the linker sequence between OH and PH1. This structure was termed as the "anchor" motif. Intriguingly, this arrangement essentially swaps the OH-CED-IH containing region of one subunit into the neighboring subunit in a clockwise direction (viewed from the "cap") leading to a clear separation of the three subunits on the 3D structure (Fig. 4). This helix-swapping arrangement might be critical for trimerization, stabilization, and the mechanical force transduction of the mPiezo1 complex. Alignment of Piezo sequences from various protozoan parasites identified a conserved sequence in this "anchor" region that might be involved in channel pore properties (Prole & Taylor, 2013). This short stretch of residues includes the PF(X2)E(X6)W motif, which is conserved among species (Coste et al., 2015; Prole & Taylor, 2013). In line with this prediction, it was shown that the E2133A mutation reduces Piezo1 single-channel conductance by about half (Coste et al., 2015). Similarly, the E2133K mutation dramatically decreased single-channel conductance, altered ionic selectivity, and reduced block by extracellular ruthenium red (Coste et al., 2015).

Strikingly, each wing of the TM region appears to stay on a coiled-coil "beam" exposed at the intracellular surface. The "beam" is about 80 Å in length and positioned in a 30-degree angle relative to the membrane plane (Fig. 4). It originates peripherally at the intracellular side of the PH7-8 pair and ends near the central axis of the channel complex, where it forms extensive interactions with the "anchor" and the CTD (Fig. 4). This unique organization suggests that the three "beam" motifs may not only support the whole TM skeleton, but also transmit conformational changes from peripheral TMs and the extracellular "blade" regions to the central region where the ion-conduction pore locates.

2.4 Ion-conducting pore of mouse Piezo1

The pore modules are crucial in determining the ion permeation and selectivity properties of particular channels. In the past few decades, efforts have been taken to identify key elements of the pore regions of different classes of ion channels. The center of the mPiezo1 structure within the membrane consists of six TM helices in a triangular arrangement (Fig. 5).

Three IHs, extended from the C-terminus of the extracellular "cap," are located at innermost and apparently line a central pore. Three OHs, extended from the N-terminus of the cap, enclose the three IHs (Fig. 5). This central region, including the IH–OH pairs, CED, and CTD comprises the pore-module of Piezo1 channels (Ge et al., 2015). The lack of side chain

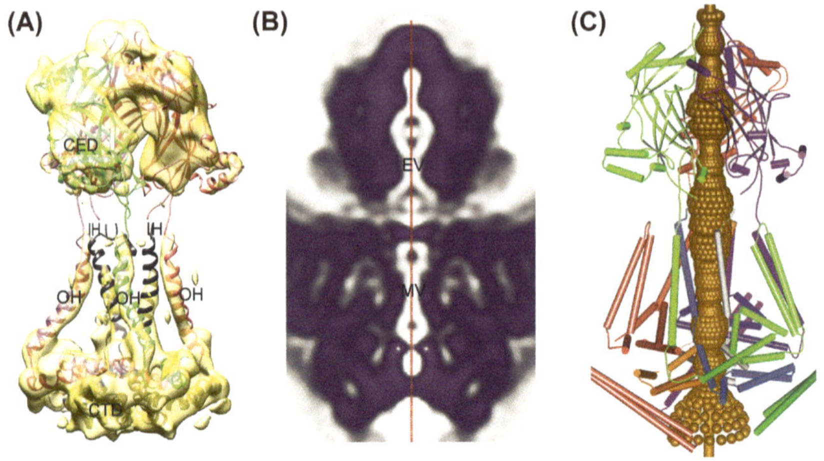

Figure 5 ***Putative ion-conducting pore.*** (A) Surface representation (transparent) of the segmented map for the CTD, transmembrane core, and cap regions superimposed with the model. (B) Central slice of the rotationally averaged density map highlighting a continuous central pore along the Z-axis. The EV within the "cap" domain, the MV enclosed by IHs and an intracellular vestibule in the CTD region are labeled. (C) Same as (A), but the model is displayed with the putative ion-conducting pore (in golden) superimposed. *CED*, C-terminal extracellular domain; *CTD*, C-terminal domain; *EV*, extracellular vestibule; *IH*, inner helix; *MV*, transmembrane vestibule; *OH*, outer helix.

information of the ion conduction pathway prevents accurate measurement of the radius of the pore and the key residues that gating the channel. However, apparent restriction sites are readily revealed indicating potential gate locations. The central slice of the rotationally averaged density map revealed a continuous central channel along the Z-axis including an EV within the "cap" region, a transmembrane vestibule (MV) enclosed by the three IHs and an intracellular vestibule (IV) which could be formed by the trimeric CTD (Fig. 5). The organization of the central TM core and the pore is strikingly reminiscent to trimeric P2X4 channels (Kawate, Michel, Birdsong, & Gouaux, 2009) and acid-sensing ion channels (ASIC) (Gonzales, Kawate, & Gouaux, 2009), which only possess two transmembrane helices and a large extracellular domain in each protomer. Thus the OH-CED-IH-CTD containing region was proposed to function as the pore module of Piezo channels.

Surprisingly, the result of one previous functional study is in contrast to this structurally indicated channel pore, which shows that human Piezo1 lacking the entire C-terminus starting from residue 2218 (the corresponding residue 2234 in mPiezo1) retains stretch-activated channel activities in a cell-attached recording mode, but the inactivation or deactivation behavior are severely slowed if not completely lost. It was claimed that the C-terminal region functions as an independent unit responsible for channel closure by inactivation or deactivation (Bae et al., 2013). However, many other functional studies are in favor of our model. Chimeras between mPiezo1 and dPiezo were used to show that ion-permeation properties are conferred by C-terminal region (Coste et al., 2015). Mutations of a glutamate residue E2133 affect unitary conductance and ion selectivity of the channel (Coste et al., 2015). The latest research has found that residues 2189–2547 of mPiezo1 determine the fundamental pore properties, including unitary conductance and pore blockage, as well as other channel properties such as current amplitude and inactivation kinetics (Zhao et al., 2016). Negatively charged residues along the ion-conducting pathway of cation channels are generally involved in determining pore properties including unitary conductance, ion selectivity, and pore blockage (Owsianik, Talavera, Voets, & Nilius, 2006). Simultaneously mutating E2495, E2496, and D2501 to lysine (K) (mPiezo1-E2495E2496D2501-KKK mutant) drastically reduced the single-channel conductance (10.3 ± 1.2 pS vs. 27.3 ± 1.8 pS of mPiezo1) (Zhao et al., 2016).

The structural information of Piezo1 was mainly speculated from predicted transmembrane topologies or presumed important structural motifs identified by remarkable amino acid mutations, until we solved the first full length structure of mPiezo1 protein with medium resolution. Our structure by and large confirmed one of the two predicted transmembrane topologies proposed by Coste et al., which claim 18 TMs in each protomer (Lewis & Grandl, 2015), whereas our structure suggests the existence at least 14 TMs in each protomer and no more than 18 TMs. The missing TMs may locate at the distal end of the transmembrane skeleton and actually interact with the blade region, but the low resolution at this area holds us back from identifying them. Those previously identified important amino acids mostly locates in the vital functional modules provided by our structure, which explains their remarkable influence on the MSC activity. In conclusion, our structure firstly identified the functional modules of Piezo1 and presented a relatively convincible model to elucidate the coupling mechanism between mechanical force sensing and nonselective cation channeling, but the detailed parameters about the channel pole and accurate structure of the blade region are still missing and require higher resolution structures.

3. POSSIBLE GATING MECHANISMS OF MOUSE PIEZO1

Structural data also found that the blade region is likely intrinsically flexible. Comparing different classes of the structures from symmetry-free 3D classification reveals several motion modes for the "blade" (Ge et al., 2015) (Fig. 6). 3D soft-edged masks were applied to different subregions (e.g., the "cap," the "blade") to improve the resolution. However, with all available means, these efforts significantly improved the densities of the "cap," but failed on the "blade."

The large conformational heterogeneity in the "blade" regions could be the main reason hampering us to obtain the high-resolution structure. The most significant one is that the rotational spacing between adjacent blades could vary between 100 and 140 degrees (Ge et al., 2015) (Fig. 6). Other less-pronounced but identifiable conformational variations include the tilt of the "blade" relatively to the plasma membrane plane and curvature changes on the helical surface (Fig. 6). These structural observations suggest

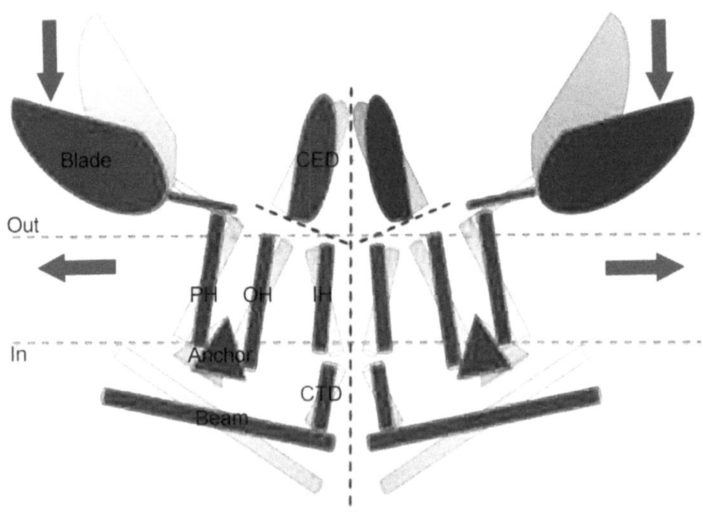

Figure 6 A proposed model of force-induced gating of Piezo channels. The light and dark gray models represent the closed and open state channels, respectively. The distinct domains of mPiezo1 structure are labeled and the *black dash lines* indicate the ion-conduction pathway. Presumably, force-induced motion (*arrows*) of the peripheral "blade" or PHs leads to conformational arrangement and gating of the channel. *CED*, C-terminal extracellular domain; *CTD*, C-terminal domain; *IH*, inner helix; *OH*, outer helix; *PH*, peripheral helix.

that the "blade" regions are highly flexible. Amanda H. et al. measured Piezo1 currents evoked by membrane stretch in three patch configurations, while simultaneously visualizing and measuring membrane geometry (Lewis & Grandl, 2015). They found that Piezo1 responds to lateral membrane tension with exquisite sensitivity as compared to other mechanically activated channels and that resting tension can drive channel inactivation, thereby tuning overall mechanical sensitivity of Piezo1. Piezo1 is activated by both convex and concave membrane curvature in multiple patch configurations (Lewis & Grandl, 2015). Mechanistically, the highly flexible "blade" regions might serve as sensors of force executed on the protein complex resulting in mechanical gating of the Piezo channel (Ge et al., 2015). Force-induced motion of the peripheral "blade" or PHs leads to conformational arrangement of the channel resulting in opening of the central pore via the coordinated action of the "beam," "anchor," and CTD domains together (Fig. 6).

Although the full length structure of Piezo1 was obtained, we still cannot exclude the possibility that some cofactors are required by Piezo1 to completely perform its function. Actually, some recent reports indeed discovered proteins that can coimmunoprecipitate with and regulated Piezo1, such as polycystin 2 (PC2) (Peyronnet et al., 2013), stomatin-like protein 3 (STOML3) (Poole, Herget, Lapatsina, Ngo, & Lewin, 2014), and Trefoil factor family 1 (TFF1) (Yang et al. 2014). PC2 is a member of TRP protein family, and in renal epithelial cells PC2 can interact with Piezo1 and inhibit its stretch-activated cationic channel activity, while this inhibition is under control of PC1/PC2 ratio. Deletion of the N-terminal of PC2 can prevent both its interaction with and inhibition of Piezo1. In contrast to PC2, STOML3 can dramatically lower the activation threshold for Piezo1 and Piezo2 currents down to ~10 nm TFF1 can interact with the C-terminal portion of the Piezo1 protein and its effect of promoting cell mobility can be attenuated by knockdown of Piezo1. Other than protein cofactors, some small molecules can also influence Piezo1 activity, including ruthenium red, gadolinium, GsMTx4, and streptomycin (Bae, Sachs, & Gottlieb, 2011; Coste et al., 2010), which can block Piezo1 activity, while the Yoda1 (Syeda et al., 2015) can enhance the activity of Piezo1. These functional studies strongly suggest that the pore-module of Piezo1 is not rigid and its flexibility can be exquisitely regulated by various protein cofactors or small molecules. Based on the huge volume of this channel complex, it is anticipated that different modules of Piezo1, consisting of blades, beams, peripheral TMs, anchors, OHs and IHs, CEDs and CTDs, can interact with different classes of proteins that participated in different physiological pathways, making Piezo1 a central pivot in integrating MS biological processes, such as organ development, angiogenesis, nervous system function, etc.

4. CONCLUSIONS AND FUTURE PROSPECTS

Since the discovery in 2010, numerous studies of Piezo proteins in many species and tissues have been performed to demonstrate their physiological importance. Recent finding on the first cryo-EM structure of full-length mPiezo1 establishes it as a three-blade-propeller-like homotrimer (Ge et al., 2015) consistent with previous biochemical data that they form homo-oligomers (Kim et al., 2012) and the observation that the CED alone also exist as trimers (Fig. 2). The possibility that Piezo1 may exist in other oligomeric states in living cells under some specific conditions, a scenario occurred in previous studies of other ion channels such as Orai channels (Hou, Pedi, Diver, & Long, 2012; Penna et al., 2008) has not been totally excluded. Intriguingly, unlike the family of voltage-gated channels with most of its members displaying fourfold symmetry, MA channels possess a variety of organizational symmetry spanning from two- to sevenfold. *Escherichia coli* and *Thermoanaerobacter tengcongensis* MscS channels fold as a membrane-spanning heptamer; human TRAAK creates a twofold symmetric K^+ channel; the MscL homolog from *Mycobacterium tuberculosis* is organized as a homopentamer; SaMscL(CDelta26) forms a tetrameric channel (Bass, Strop, Barclay, & Rees, 2002; Brohawn, del Marmol, & MacKinnon, 2012; Chang, Spencer, Lee, Barclay, & Rees, 1998; Liu, Gandhi, & Rees, 2009; Zhang, Wang, et al., 2012). All these states of oligomerization demonstrate the high structural diversity of MSCs.

The trimeric mPiezo1 channel complex comprises at least 42 TMs with each protomer of 14 TMs, making it the membrane complex with the largest number of TMs known so far. How would so many TMs organize into a stable and functional complex? It may rely on several possible mechanisms. Firstly, the 14 TMs in each protomer are organized in pair-wise fashion to form an ordered wing, an observation that might explain the formidable folding and membrane insertion of these TMs during mPiezo1 protein biosynthesis. Secondly, the three TM wings converge at the center and intertwine each other through a domain-swapping strategy by the "anchor" region (Fig. 4). Thirdly, three coiled-coil "beam" motifs potentially provide a strong support for the whole TM skeleton (Fig. 4). Finally, the extracellular trimeric "cap" and the extensive interactions between the "beam," "anchor," and CTDs may further stabilize the whole complex (Fig. 6).

Despite its unprecedented number of TMs, Piezo channels possess a relatively simple ion-conducting pore lined by three IHs without apparent pore helices and selectivity filter loops (Fig. 5) as commonly observed in potassium- (Doyle et al., 1998) or sodium-selective channels (Payandeh, Scheuer, Zheng, & Catterall, 2011; Zhang, Ren, et al., 2012), or nonselective cation channels such as TRPV1 (Liao, Cao, Julius, & Cheng, 2013) and RyR1 (Efremov, Leitner, Aebersold, & Raunser, 2015; Yan et al., 2015; Zalk et al., 2015), but apparently reminiscent to trimeric P2X4 (Kawate et al., 2009) and ASICs (Gonzales et al., 2009). Perhaps the rather moderate selectivity feature, only selecting cations over anions, is based on the comparatively simple structure of the pole module of Piezo1.

The 3D structure of mPiezo1 also provides clues for questions, such as how Piezo channels are specially gated by force and what is the molecular basis for their mechanosensitivity. Mouse Piezo1 can be activated by various forms of mechanical stimulation including poking, suction, and flow-induced shear stress (Coste et al., 2010; Li et al., 2014; Ranade, Qiu, et al., 2014), and exhibits much higher pressure sensitivity than prokaryotic MA channels such as MscL and MscS (Bae et al., 2013). The striking similarity of the structure to a three-blade-propeller leads us to speculate that mechanically activated Piezo channels could possibly function in similar principles as how a propeller senses and transduces force (Fig. 6). The extracellular protruding "blade" regions appear to be ideal to sense shear stress or other forms of mechanical stimulation. In line with this idea, 3D classification revealed that these domains are highly mobile and occupied in variable positions (Ge et al., 2015). The extending TM wing could also be involved in sensing membrane tensions. Motions of these peripheral domains could be transmitted to the central ion-conduction pore via the "beam" domains that display extensive interactions with the "anchor" and CTD domains (Fig. 6). However, an intrinsic mechanosensitivity of Piezo channels in lipid membrane has not been demonstrated as shown for Msc channels (Kung, Martinac, & Sukharev, 2010) or vertebrate two-pore domain K^+ (K2P) channels (Berrier et al., 2013; Brohawn, Su, & MacKinnon, 2014), and no specific force-sensing domains have precedentedly been established in other MSCs (Anishkin, Loukin, Teng, & Kung, 2014; Brohawn, Campbell, & MacKinnon, 2014; Kung et al., 2010). Thus, it remains possible that other extracellular or intracellular proteins may interact with and regulate Piezo channel function (Peyronnet et al., 2013; Poole et al., 2014). Nevertheless, given their sophisticated mechanosensitivity, Piezo channels are highly likely to possess dedicated force-sensing domains and mechanotransdunction mechanisms as indicated by the mPiezo1 structure. This hypothesis merits further investigations.

Other than the direct stretch—activated channel activity, Piezo proteins also participated in many other biological processes, most likely through indirect mechanisms. As reported by McHugh et al., Piezo1, once known as Fam38a, is involved in integrin activation and knockdown of Piezo1 can reduce cell adherence and promotes cell migration (McHugh et al., 2010, 2012). The structural basis of Piezo1 participating indirectly in biological processes other than the direct stretch—activated channel activity like the example earlier is still lacking and requires further structural investigations.

Our structure can also interpret the mechanism under some pathological conditions. Concluded from previous studies, more than 14 mutations in Piezo1 are identified to be linked to different clinical forms of diseases (Albuisson et al., 2013; Andolfo et al., 2013; Zarychanski et al., 2012), and most of these mutations are located in the C-terminal region of Piezo1 (from 2000 to 2500), which corresponds to the central pole-module in our structure, including CED, CTD, OH, IH, and anchor domains. These domains are crucial for MS activity, which provides the structural basis for them becoming hot spots of pathological mutation. T2127M mutation in human Piezo1 was reported to cause DHSt (Andolfo et al., 2013), and this site is mapped to the anchor domain in our structure, which functions in linking the PH segments to OH segments. From other groups, mutation of amino acids from 2225 to 2456, R2456H and R2488Q can also cause DHSt, with slightly different symptoms (Zarychanski et al., 2012). Although many pathological mutation sites can be mapped into the pole module of Piezo proteins, the detailed mechanism and the information necessary for drug development are still missing. A higher resolution structure is still in need to solve these problems.

All in all, the first cryo-EM structure of full-length Piezo1 has resolved a series of speculations, and meanwhile raised more interesting questions: How large is the ion conducting pore? Does three protomers of Piezo1 form a single pore or each protomer owns one pore? What is the exact number of the TMs of Piezo1? What is the subtle structure difference between Piezo1 and Piezo2, and how this structure difference determines their individual functions? How the MS activity of Piezo proteins participates in other physiological pathways such as cell proliferation and cell migration? All these questions warrant further structural studies of Piezo1 and Piezo2 with higher resolution. While previous studies reported that Piezo channels could be blocked by the nonspecific blockers ruthenium red, gadolinium, and streptomycin (Andolfo et al., 2013; Coste et al., 2010), a single compound called Yoda1 that selectively opens Piezo1 but not Piezo2 has been identified by screening over three million synthetic small molecules using a fluorescence cellular assay to monitor calcium influx through Piezo1 (Syeda et al., 2015). Applying the blockers and/or activators to structural analyses, Piezo1 could be pharmaceutically manipulated to reduce the conformational flexibility, to achieve a structure with higher resolution and then shed light on those intriguing questions stated earlier.

REFERENCES

Albuisson, J., Murthy, S. E., Bandell, M., Coste, B., Louis-Dit-Picard, H., Mathur, J., … Patapoutian, A. (2013). Dehydrated hereditary stomatocytosis linked to gain-of-function mutations in mechanically activated PIEZO1 ion channels. *Nature Communications, 4*, 1884. http://dx.doi.org/10.1038/ncomms2899.

Andolfo, I., Alper, S. L., De Franceschi, L., Auriemma, C., Russo, R., De Falco, L., … Iolascon, A. (2013). Multiple clinical forms of dehydrated hereditary stomatocytosis arise from mutations in PIEZO1. *Blood, 121*(19). http://dx.doi.org/10.1182/blood-2013-02-482489, 3925–3935, S3921–S3912.

Anishkin, A., Loukin, S. H., Teng, J., & Kung, C. (2014). Feeling the hidden mechanical forces in lipid bilayer is an original sense. *Proceedings of the National Academy of Sciences of the United States of America, 111*(22), 7898–7905. http://dx.doi.org/10.1073/pnas.1313364111.

Bae, C., Gnanasambandam, R., Nicolai, C., Sachs, F., & Gottlieb, P. A. (2013). Xerocytosis is caused by mutations that alter the kinetics of the mechanosensitive channel PIEZO1. *Proceedings of the National Academy of Sciences of the United States of America, 110*(12), E1162–E1168. http://dx.doi.org/10.1073/pnas.1219777110.

Bae, C., Sachs, F., & Gottlieb, P. A. (2011). The mechanosensitive ion channel Piezo1 is inhibited by the peptide GsMTx4. *Biochemistry, 50*(29), 6295–6300. http://dx.doi.org/10.1021/bi200770q.

Bai, X. C., McMullan, G., & Scheres, S. H. (2015). How cryo-EM is revolutionizing structural biology. *Trends in Biochemical Sciences, 40*(1), 49–57. http://dx.doi.org/10.1016/j.tibs.2014.10.005.

Bass, R. B., Strop, P., Barclay, M., & Rees, D. C. (2002). Crystal structure of *Escherichia coli* MscS, a voltage-modulated and mechanosensitive channel. *Science, 298*(5598), 1582–1587. http://dx.doi.org/10.1126/science.1077945.

Beneteau, C., Thierry, G., Blesson, S., Le Vaillant, C., Picard, V., Bene, M. C., … Le Caignec, C. (2014). Recurrent mutation in the PIEZO1 gene in two families of hereditary xerocytosis with fetal hydrops. *Clinical Genetics, 85*(3), 293–295. http://dx.doi.org/10.1111/cge.12147.

Berrier, C., Pozza, A., de Lacroix de Lavalette, A., Chardonnet, S., Mesneau, A., Jaxel, C., … Ghazi, A. (2013). The purified mechanosensitive channel TREK-1 is directly sensitive to membrane tension. *Journal of Biological Chemistry, 288*(38), 27307–27314. http://dx.doi.org/10.1074/jbc.M113.478321.

Beurg, M., Evans, M. G., Hackney, C. M., & Fettiplace, R. (2006). A large-conductance calcium-selective mechanotransducer channel in mammalian cochlear hair cells. *The Journal of Neuroscience, 26*(43), 10992–11000. http://dx.doi.org/10.1523/JNEUROSCI.2188-06.2006.

Brohawn, S. G., Campbell, E. B., & MacKinnon, R. (2014). Physical mechanism for gating and mechanosensitivity of the human TRAAK K^+ channel. *Nature, 516*(7529), 126–130. http://dx.doi.org/10.1038/nature14013.

Brohawn, S. G., del Marmol, J., & MacKinnon, R. (2012). Crystal structure of the human K2P TRAAK, a lipid- and mechano-sensitive K^+ ion channel. *Science, 335*(6067), 436–441. http://dx.doi.org/10.1126/science.1213808.

Brohawn, S. G., Su, Z., & MacKinnon, R. (2014). Mechanosensitivity is mediated directly by the lipid membrane in TRAAK and TREK1 K^+ channels. *Proceedings of the National Academy of Sciences of the United States of America, 111*(9), 3614–3619. http://dx.doi.org/10.1073/pnas.1320768111.

Chalfie, M. (2009). Neurosensory mechanotransduction. *Nature Reviews. Molecular Cell Biology, 10*(1), 44–52. http://dx.doi.org/10.1038/nrm2595.

Chang, G., Spencer, R. H., Lee, A. T., Barclay, M. T., & Rees, D. C. (1998). Structure of the MscL homolog from *Mycobacterium tuberculosis*: a gated mechanosensitive ion channel. *Science, 282*(5397), 2220–2226.

Cheng, Y. (2015). Single-particle cryo-EM at crystallographic resolution. *Cell, 161*(3), 450–457. http://dx.doi.org/10.1016/j.cell.2015.03.049.

Cheng, Y., Grigorieff, N., Penczek, P. A., & Walz, T. (2015). A primer to single-particle cryo-electron microscopy. *Cell, 161*(3), 438–449. http://dx.doi.org/10.1016/j.cell.2015.03.050.

Christensen, A. P., & Corey, D. P. (2007). TRP channels in mechanosensation: direct or indirect activation? *Nature Reviews. Neuroscience, 8*(7), 510–521. http://dx.doi.org/10.1038/nrn2149.

Corey, D. P., & Hudspeth, A. J. (1979). Ionic basis of the receptor potential in a vertebrate hair cell. *Nature, 281*(5733), 675–677.

Coste, B., Houge, G., Murray, M. F., Stitziel, N., Bandell, M., Giovanni, M. A., … Patapoutian, A. (2013). Gain-of-function mutations in the mechanically activated ion channel PIEZO2 cause a subtype of distal arthrogryposis. *Proceedings of the National Academy of Sciences of the United States of America, 110*(12), 4667–4672. http://dx.doi.org/10.1073/pnas.1221400110.

Coste, B., Mathur, J., Schmidt, M., Earley, T. J., Ranade, S., Petrus, M. J., … Patapoutian, A. (2010). Piezo1 and Piezo2 are essential components of distinct mechanically activated cation channels. *Science, 330*(6000), 55–60. http://dx.doi.org/10.1126/science.1193270.

Coste, B., Murthy, S. E., Mathur, J., Schmidt, M., Mechioukhi, Y., Delmas, P., & Patapoutian, A. (2015). Piezo1 ion channel pore properties are dictated by C-terminal region. *Nature Communications, 6*, 7223. http://dx.doi.org/10.1038/ncomms8223.

Coste, B., Xiao, B., Santos, J. S., Syeda, R., Grandl, J., Spencer, K. S., … Patapoutian, A. (2012). Piezo proteins are pore-forming subunits of mechanically activated channels. *Nature, 483*(7388), 176–181. http://dx.doi.org/10.1038/nature10812.

Craelius, W., el-Sherif, N., & Palant, C. E. (1989). Stretch-activated ion channels in cultured mesangial cells. *Biochemical and Biophysical Research Communications, 159*(2), 516–521.

Delmas, P., & Coste, B. (2013). Mechano-gated ion channels in sensory systems. *Cell, 155*(2), 278–284. http://dx.doi.org/10.1016/j.cell.2013.09.026.

Delmas, P., Hao, J., & Rodat-Despoix, L. (2011). Molecular mechanisms of mechanotransduction in mammalian sensory neurons. *Nature Reviews. Neuroscience, 12*(3), 139–153. http://dx.doi.org/10.1038/nrn2993.

Doyle, D. A., Morais Cabral, J., Pfuetzner, R. A., Kuo, A., Gulbis, J. M., Cohen, S. L., ... MacKinnon, R. (1998). The structure of the potassium channel: molecular basis of K^+ conduction and selectivity. *Science, 280*(5360), 69–77.

Efremov, R. G., Leitner, A., Aebersold, R., & Raunser, S. (2015). Architecture and conformational switch mechanism of the ryanodine receptor. *Nature, 517*(7532), 39–43. http://dx.doi.org/10.1038/nature13916.

Faucherre, A., Kissa, K., Nargeot, J., Mangoni, M. E., & Jopling, C. (2014). Piezo1 plays a role in erythrocyte volume homeostasis. *Haematologica, 99*(1), 70–75. http://dx.doi.org/10.3324/haematol.2013.086090.

Faucherre, A., Nargeot, J., Mangoni, M. E., & Jopling, C. (2013). *piezo2b* regulates vertebrate light touch response. *The Journal of Neuroscience, 33*(43), 17089–17094. http://dx.doi.org/10.1523/JNEUROSCI.0522-13.2013.

Franco, A., Jr., & Lansman, J. B. (1990). Calcium entry through stretch-inactivated ion channels in *mdx* myotubes. *Nature, 344*(6267), 670–673. http://dx.doi.org/10.1038/344670a0.

Ge, J., Li, W., Zhao, Q., Li, N., Chen, M., Zhi, P., ... Yang, M. (2015). Architecture of the mammalian mechanosensitive Piezo1 channel. *Nature, 527*(7576), 64–69. http://dx.doi.org/10.1038/nature15247.

Gillespie, P. G., & Walker, R. G. (2001). Molecular basis of mechanosensory transduction. *Nature, 413*(6852), 194–202. http://dx.doi.org/10.1038/35093011.

Gonzales, E. B., Kawate, T., & Gouaux, E. (2009). Pore architecture and ion sites in acid-sensing ion channels and P2X receptors. *Nature, 460*(7255), 599–604. http://dx.doi.org/10.1038/nature08218.

Hao, J., Zhang, Y., Ye, R., Zheng, Y., Zhao, Z., & Li, J. (2013). Mechanotransduction in cancer stem cells. *Cell Biology International, 37*(9), 888–891. http://dx.doi.org/10.1002/cbin.10111.

Haswell, E. S., Phillips, R., & Rees, D. C. (2011). Mechanosensitive channels: what can they do and how do they do it? *Structure, 19*(10), 1356–1369. http://dx.doi.org/10.1016/j.str.2011.09.005.

Honore, E., Martins, J. R., Penton, D., Patel, A., & Demolombe, S. (2015). The piezo mechanosensitive ion channels: may the force be with you! *Reviews of Physiology, Biochemistry and Pharmacology, 169*, 25–41. http://dx.doi.org/10.1007/112_2015_26.

Hou, X., Pedi, L., Diver, M. M., & Long, S. B. (2012). Crystal structure of the calcium release-activated calcium channel Orai. *Science, 338*(6112), 1308–1313. http://dx.doi.org/10.1126/science.1228757.

Ikeda, R., Cha, M., Ling, J., Jia, Z., Coyle, D., & Gu, J. G. (2014). Merkel cells transduce and encode tactile stimuli to drive Aβ-afferent impulses. *Cell, 157*(3), 664–675. http://dx.doi.org/10.1016/j.cell.2014.02.026.

Kamajaya, A., Kaiser, J. T., Lee, J., Reid, M., & Rees, D. C. (2014). The structure of a conserved piezo channel domain reveals a topologically distinct beta sandwich fold. *Structure, 22*(10), 1520–1527. http://dx.doi.org/10.1016/j.str.2014.08.009.

Kawashima, Y., Kurima, K., Pan, B., Griffith, A. J., & Holt, J. R. (2015). Transmembrane channel-like (TMC) genes are required for auditory and vestibular mechanosensation. *Pflugers Archiv, 467*(1), 85–94. http://dx.doi.org/10.1007/s00424-014-1582-3.

Kawate, T., Michel, J. C., Birdsong, W. T., & Gouaux, E. (2009). Crystal structure of the ATP-gated P2X(4) ion channel in the closed state. *Nature, 460*(7255), 592–598. http://dx.doi.org/10.1038/nature08198.

Kim, S. E., Coste, B., Chadha, A., Cook, B., & Patapoutian, A. (2012). The role of Drosophila Piezo in mechanical nociception. *Nature, 483*(7388), 209–212. http://dx.doi.org/10.1038/nature10801.

Kraske, S., Cunningham, J. T., Hajduczok, G., Chapleau, M. W., Abboud, F. M., & Wachtel, R. E. (1998). Mechanosensitive ion channels in putative aortic baroreceptor neurons. *American Journal of Physiology, 275*(4 Pt 2), H1497–H1501.

Kung, C., Martinac, B., & Sukharev, S. (2010). Mechanosensitive channels in microbes. *Annual Review of Microbiology, 64*, 313–329. http://dx.doi.org/10.1146/annurev.micro.112408.134106.

Lewis, A. H., & Grandl, J. (2015). Mechanical sensitivity of Piezo1 ion channels can be tuned by cellular membrane tension. *eLife, 4*. http://dx.doi.org/10.7554/eLife.12088.

Li, J., Hou, B., Tumova, S., Muraki, K., Bruns, A., Ludlow, M. J., … Beech, D. J. (2014). Piezo1 integration of vascular architecture with physiological force. *Nature, 515*(7526), 279–282. http://dx.doi.org/10.1038/nature13701.

Liao, M., Cao, E., Julius, D., & Cheng, Y. (2013). Structure of the TRPV1 ion channel determined by electron cryo-microscopy. *Nature, 504*(7478), 107–112. http://dx.doi.org/10.1038/nature12822.

Liu, Z., Gandhi, C. S., & Rees, D. C. (2009). Structure of a tetrameric MscL in an expanded intermediate state. *Nature, 461*(7260), 120–124. http://dx.doi.org/10.1038/nature08277.

Maksimovic, S., Nakatani, M., Baba, Y., Nelson, A. M., Marshall, K. L., Wellnitz, S. A., … Lumpkin, E. A. (2014). Epidermal Merkel cells are mechanosensory cells that tune mammalian touch receptors. *Nature, 509*(7502), 617–621. http://dx.doi.org/10.1038/nature13250.

McBride, D. W., Jr., & Hamill, O. P. (1992). Pressure-clamp: a method for rapid step perturbation of mechanosensitive channels. *Pflugers Archiv, 421*(6), 606–612.

McCarter, G. C., Reichling, D. B., & Levine, J. D. (1999). Mechanical transduction by rat dorsal root ganglion neurons in vitro. *Neuroscience Letters, 273*(3), 179–182.

McHugh, B. J., Buttery, R., Lad, Y., Banks, S., Haslett, C., & Sethi, T. (2010). Integrin activation by Fam38A uses a novel mechanism of R-Ras targeting to the endoplasmic reticulum. *Journal of Cell Science, 123*(Pt 1), 51–61. http://dx.doi.org/10.1242/jcs.056424.

McHugh, B. J., Murdoch, A., Haslett, C., & Sethi, T. (2012). Loss of the integrin-activating transmembrane protein Fam38A (Piezo1) promotes a switch to a reduced integrin-dependent mode of cell migration. *PLoS One, 7*(7), e40346. http://dx.doi.org/10.1371/journal.pone.0040346.

McMillin, M. J., Beck, A. E., Chong, J. X., Shively, K. M., Buckingham, K. J., Gildersleeve, H. I., … Bamshad, M. J. (2014). Mutations in PIEZO2 cause Gordon syndrome, Marden-Walker syndrome, and distal arthrogryposis type 5. *American Journal of Human Genetics, 94*(5), 734–744. http://dx.doi.org/10.1016/j.ajhg.2014.03.015.

Miyamoto, T., Mochizuki, T., Nakagomi, H., Kira, S., Watanabe, M., Takayama, Y., … Tominaga, M. (2014). Functional role for Piezo1 in stretch-evoked Ca(2)(+) influx and ATP release in urothelial cell cultures. *Journal of Biological Chemistry, 289*(23), 16565–16575. http://dx.doi.org/10.1074/jbc.M113.528638.

Owsianik, G., Talavera, K., Voets, T., & Nilius, B. (2006). Permeation and selectivity of TRP channels. *Annual Review of Physiology, 68*, 685–717. http://dx.doi.org/10.1146/annurev.physiol.68.040204.101406.

Pathak, M. M., Nourse, J. L., Tran, T., Hwe, J., Arulmoli, J., Le, D. T., … Tombola, F. (2014). Stretch-activated ion channel Piezo1 directs lineage choice in human neural stem cells. *Proceedings of the National Academy of Sciences of the United States of America, 111*(45), 16148–16153. http://dx.doi.org/10.1073/pnas.1409802111.

Payandeh, J., Scheuer, T., Zheng, N., & Catterall, W. A. (2011). The crystal structure of a voltage-gated sodium channel. *Nature, 475*(7356), 353–358. http://dx.doi.org/10.1038/nature10238.

Penna, A., Demuro, A., Yeromin, A. V., Zhang, S. L., Safrina, O., Parker, I., & Cahalan, M. D. (2008). The CRAC channel consists of a tetramer formed by Stim-induced dimerization of Orai dimers. *Nature, 456*(7218), 116–120. http://dx.doi.org/10.1038/nature07338.

Peyronnet, R., Martins, J. R., Duprat, F., Demolombe, S., Arhatte, M., Jodar, M., … Patel, A. (2013). Piezo1-dependent stretch-activated channels are inhibited by Polycystin-2 in renal tubular epithelial cells. *EMBO Reports, 14*(12), 1143–1148. http://dx.doi.org/10.1038/embor.2013.170.

Poole, K., Herget, R., Lapatsina, L., Ngo, H. D., & Lewin, G. R. (2014). Tuning Piezo ion channels to detect molecular-scale movements relevant for fine touch. *Nature Communications, 5*, 3520. http://dx.doi.org/10.1038/ncomms4520.

Prole, D. L., & Taylor, C. W. (2013). Identification and analysis of putative homologues of mechanosensitive channels in pathogenic protozoa. *PLoS One, 8*(6), e66068. http://dx.doi.org/10.1371/journal.pone.0066068.

Ranade, S. S., Qiu, Z., Woo, S. H., Hur, S. S., Murthy, S. E., Cahalan, S. M., … Patapoutian, A. (2014). Piezo1, a mechanically activated ion channel, is required for vascular development in mice. *Proceedings of the National Academy of Sciences of the United States of America, 111*(28), 10347–10352. http://dx.doi.org/10.1073/pnas.1409233111.

Ranade, S. S., Woo, S. H., Dubin, A. E., Moshourab, R. A., Wetzel, C., Petrus, M., … Patapoutian, A. (2014). Piezo2 is the major transducer of mechanical forces for touch sensation in mice. *Nature, 516*(7529), 121–125. http://dx.doi.org/10.1038/nature13980.

Sachs, F. (2010). Stretch-activated ion channels: What are they? *Physiology (Bethesda), 25*(1), 50–56. http://dx.doi.org/10.1152/physiol.00042.2009.

Sandberg, M. B., Nybo, M., Birgens, H., & Frederiksen, H. (2014). Hereditary xerocytosis and familial haemolysis due to mutation in the PIEZO1 gene: A simple diagnostic approach. *International Journal of Laboratory Hematology, 36*(4), e62–65. http://dx.doi.org/10.1111/ijlh.12172.

Schneider, E. R., Mastrotto, M., Laursen, W. J., Schulz, V. P., Goodman, J. B., Funk, O. H., … Bagriantsev, S. N. (2014). Neuronal mechanism for acute mechanosensitivity in tactile-foraging waterfowl. *Proceedings of the National Academy of Sciences of the United States of America, 111*(41), 14941–14946. http://dx.doi.org/10.1073/pnas.1413656111.

Schrenk-Siemens, K., Wende, H., Prato, V., Song, K., Rostock, C., Loewer, A., … Siemens, J. (2015). PIEZO2 is required for mechanotransduction in human stem cell-derived touch receptors. *Nature Neuroscience, 18*(1), 10–16. http://dx.doi.org/10.1038/nn.3894.

Shmukler, B. E., Vandorpe, D. H., Rivera, A., Auerbach, M., Brugnara, C., & Alper, S. L. (2014). Dehydrated stomatocytic anemia due to the heterozygous mutation R2456H in the mechanosensitive cation channel PIEZO1: a case report. *Blood Cells, Molecules & Diseases, 52*(1), 53–54. http://dx.doi.org/10.1016/j.bcmd.2013.07.015.

Sukharev, S. I., Blount, P., Martinac, B., Blattner, F. R., & Kung, C. (1994). A large-conductance mechanosensitive channel in *E. coli* encoded by MscL alone. *Nature, 368*(6468), 265–268. http://dx.doi.org/10.1038/368265a0.

Syeda, R., Xu, J., Dubin, A. E., Coste, B., Mathur, J., Huynh, T., … Patapoutian, A. (2015). Chemical activation of the mechanotransduction channel Piezo1. *eLife, 4*. http://dx.doi.org/10.7554/eLife.07369.

Tabarean, I. V., & Morris, C. E. (2002). Membrane stretch accelerates activation and slow inactivation in Shaker channels with S3–S4 linker deletions. *Biophysical Journal, 82*(6), 2982–2994. http://dx.doi.org/10.1016/S0006-3495(02)75639-7.

Vogel, V. (2006). Mechanotransduction involving multimodular proteins: converting force into biochemical signals. *Annual Review of Biophysics and Biomolecular Structure, 35*, 459–488. http://dx.doi.org/10.1146/annurev.biophys.35.040405.102013.

Woo, S. H., Ranade, S., Weyer, A. D., Dubin, A. E., Baba, Y., Qiu, Z., … Patapoutian, A. (2014). Piezo2 is required for Merkel-cell mechanotransduction. *Nature, 509*(7502), 622–626. http://dx.doi.org/10.1038/nature13251.

Yan, Z., Bai, X. C., Yan, C., Wu, J., Li, Z., Xie, T., … Yan, N. (2015). Structure of the rabbit ryanodine receptor RyR1 at near-atomic resolution. *Nature, 517*(7532), 50–55. http://dx.doi.org/10.1038/nature14063.

Yang, X. N., Lu, Y. P., Liu, J. J., Huang, J. K., Liu, Y. P., Xiao, C. X., … Guleng, B. (2014). Piezo1 is as a novel trefoil factor family 1 binding protein that promotes gastric cancer cell mobility in vitro. *Digestive Diseases and Sciences, 59*(7), 1428–1435. http://dx.doi.org/10.1007/s10620-014-3044-3.

Zalk, R., Clarke, O. B., des Georges, A., Grassucci, R. A., Reiken, S., Mancia, F., … Marks, A. R. (2015). Structure of a mammalian ryanodine receptor. *Nature, 517*(7532), 44–49. http://dx.doi.org/10.1038/nature13950.

Zarychanski, R., Schulz, V. P., Houston, B. L., Maksimova, Y., Houston, D. S., Smith, B., … Gallagher, P. G. (2012). Mutations in the mechanotransduction protein PIEZO1 are associated with hereditary xerocytosis. *Blood, 120*(9), 1908–1915. http://dx.doi.org/10.1182/blood-2012-04-422253.

Zhang, X., Ren, W., DeCaen, P., Yan, C., Tao, X., Tang, L., … Yan, N. (2012). Crystal structure of an orthologue of the NaChBac voltage-gated sodium channel. *Nature, 486*(7401), 130–134. http://dx.doi.org/10.1038/nature11054.

Zhang, X., Wang, J., Feng, Y., Ge, J., Li, W., Sun, W., … Yang, M. (2012). Structure and molecular mechanism of an anion-selective mechanosensitive channel of small conductance. *Proceedings of the National Academy of Sciences of the United States of America, 109*(44), 18180–18185. http://dx.doi.org/10.1073/pnas.1207977109.

Zhao, Q., Wu, K., Geng, J., Chi, S., Wang, Y., Zhi, P., … Xiao, B. (2016). Ion permeation and mechanotransduction mechanisms of mechanosensitive piezo channels. *Neuron, 89*(6), 1248–1263. http://dx.doi.org/10.1016/j.neuron.2016.01.046.

CHAPTER SIX

In Touch With the Mechanosensitive Piezo Channels: Structure, Ion Permeation, and Mechanotransduction

J. Geng[a], Q. Zhao[a], T. Zhang[a] and B. Xiao[1]
Tsinghua University, Beijing, China
[1]Corresponding author: E-mail: xbailong@biomed.tsinghua.edu.cn

Contents

[a] These authors contributed equally to this work.

Current Topics in Membranes, Volume 79
ISSN 1063-5823
http://dx.doi.org/10.1016/bs.ctm.2016.11.006

Abstract

Mechanotransduction, the conversion of mechanical forces into biological signals, plays critical roles in various physiological and pathophysiological processes in mammals, such as conscious sensing of touch, pain, and sound, as well as unconscious sensing of blood flow-associated shear stress, urine flow, and bladder distention. Among the various molecules involved in mechanotransduction, mechanosensitive (MS) cation channels have long been postulated to represent one critical class of mechanotransducers that directly and rapidly converts mechanical force into electrochemical signals. Despite the awareness of their functional significance, the molecular identities of MS cation channels in mammals had remained elusive for decades till the groundbreaking finding that the Piezo family of genes, including Piezo1 and Piezo2, constitutes their essential components. Since their identification about 6 years ago, tremendous progress has been made in understanding their physiological and pathophysiological importance in mechanotransduction and their structure–function relationships of being the prototypic class of mammalian MS cation channels. On the one hand, Piezo proteins have been demonstrated to serve as physiologically and pathophysiologically important mechanotransducers for most, if not all, mechanotransduction processes. On the other hand, they have been shown to form a remarkable three-bladed, propeller-shaped homotrimeric channel complex comprising a separable ion-conducting pore module and mechanotransduction modules. In this chapter, we review the major advancements, with a particular focus on the structural and biophysical features that enable Piezo proteins to serve as sophisticated MS cation channels for force sensing, transduction, and ion conduction.

1. INTRODUCTION

Mechanosensitive (MS) channels represent a class of ion channels that gate upon mechanical force stimulation, which was originally proposed by Corey and Hudspeth (1979) when studying the auditory receptor cells. This was followed by recordings of the first MS current in chicken skeletal muscle cells by Sachs et al. in 1984 (Guharay & Sachs, 1984). Since then, there has been a long-lasting interest in characterizing their physiological and pathophysiological functions, cloning the encoding genes, and investigating their structure, gating, and ion permeation mechanisms. For instance, Kung et al. cloned the first prokaryotic MS channels, MscL, in 1994, enabling mechanistic understanding of the principles that allow MS channels to respond to mechanical force and conduct ions. They have proposed the "force-from-lipids" hypothesis for explaining the mechanogating mechanisms of certain MS channels (Sukharev, Blount, Martinac, Blattner, & Kung, 1994). In eukaryotes, the family of degenerin (DEG)/Epithelium Sodium Channel (ENaC)/Acid Sensing Ion Channel (ASIC) and some members of the Transient Receptor Potential (TRP) ion channels have been shown to play mechanotransduction roles in model organisms such as *Caenorhabditis elegans* and *Drosophila* (Brown, Liao, & Goodman, 2008; Cheng, Song, Looger, Jan, & Jan, 2010; Kang, Gao, Schafer, Xie, & Xu, 2010; Li, Feng, Sternberg, & Xu, 2006; O'Hagan, Chalfie, & Goodman, 2005; Sidi, Friedrich, & Nicolson, 2003; Walker, Willingham, & Zuker, 2000; Zhong, Hwang, & Tracey, 2010). However, to date, only few of them, for example, the TRPN channel, have been shown to sufficiently mediate mechanically activated currents when overexpressed in heterologous cell lines (Yan et al., 2013; Zhang et al., 2015). Furthermore, these MS candidates identified in invertebrates either have no homologues (e.g., TRPN) or no functional conservation (e.g., DEG/ENaC/ASIC) in mammals (Delmas, Hao, & Rodat-Despoix, 2011; Lumpkin, Marshall, & Nelson, 2010).

Mammalian MS cationic currents have been electrophysiologically recorded in many cell types, such as dorsal root ganglion (DRG) sensory neurons (McCarter, Reichling, & Levine, 1999), inner-ear hair cells (Brownell, Bader, Bertrand, & de Ribaupierre, 1985; Cunningham, Wachtel, & Abboud, 1995), nodose ganglions of the putative aortic baroreceptor neurons (Cunningham et al., 1995), smooth muscle cells (Davis, Donovitz, & Hood, 1992), skeletal muscle cells (Franco-Obregon

& Lansman, 1994), and endothelia cells (Li et al., 2014; Ranade, Qiu, et al., 2014), and proposed to play critical roles in various physiological processes, including embryonic development, touch, pain, hearing, proprioception, sensing flow-associated shear stress in blood vessels, regulation of urine flow, and bladder distention, as well as cell volume regulation, proliferation, migration, and fate determination. Some members of the K2P potassium channel family have been identified to be mechanosensitive (Patel et al., 1998), and the breakthrough in resolving their crystal structures allows mechanistic understanding of how they are gated by mechanical force (Brohawn, del Marmol, & MacKinnon, 2012; Dong et al., 2015). By contrast, the molecular identities of MS cation channels in mammals have remained unknown for decades, significantly hampering our knowledge of the mechanotransduction processes in vivo and the molecular mechanisms underlying the mechanogating and ion permeation properties.

Combining microarray analysis and RNAi-mediated knockdown of a total of 73 candidate genes, Patapoutian and colleagues have remarkably identified a single gene, *Piezo1* (also known as *Fam38A*), that is required for the mechanically activated cationic currents in the mouse Neuro2A neuroblastoma cell line. Further sequence homology analysis led to the identification of the homologous gene *Piezo2* in mammals (Coste et al., 2010). Remarkably, heterologous expression of Piezo1 or Piezo2 proteins in various mammalian cell lines is sufficient to mediate MS cationic currents, marking them the first family of genes identified in mammals that is both necessary and sufficient for generating MS cationic currents (Coste et al., 2010). Interestingly, other essential components involved in the hearing mechanotransduction channels have started to emerge as well. For example, the transmembrane channel-like (TMC) family of proteins, including TMC1 and TMC2, appears to be essential for the mechanotransduction current of mouse hair cells (Pan et al., 2013). Pan et al. (2013) have shown that mutations in TMC1 can alter calcium permeability, leading them to propose that TMCs could be the pore-forming subunits of the hearing mechanotransduction channel. However, it has not been demonstrated that TMC1/2 are sufficient to mediate MS in a heterologous expression system.

The groundbreaking discovery of the Piezo genes has heralded a new era of research in the field of mechanotransduction, and tremendous progress has been made toward characterizing the physiological significance, biophysical properties, and structural features of the Piezo proteins. In this chapter, we will summarize the major advancement in exploring this prototypic class of MS cation channels identified in mammals, with a particular emphasis on the recently resolved cryo-electron microscopy (cryo-EM) structure of Piezo1 and the structure-guided characterization of the molecular bases that underlie its ion permeation and mechanotransduction properties.

2. THE PIEZO FAMILY OF PROTEINS REPRESENTS A PHYSIOLOGICALLY AND PATHOPHYSIOLOGICALLY IMPORTANT CLASS OF MS CATION CHANNELS

The identification of the Piezo proteins as mammalian MS cation channels represents a critical first step toward understanding of the various mechanotransduction processes in vivo. Thus an intensive research effort has been oriented toward characterizing the physiological relevance of Piezo proteins in mechanotransduction. The Piezo family of proteins is evolutionally conserved in animals, plants, and protozoa, but not in yeast and bacteria, indicating their functional importance (Coste et al., 2010). In mouse, Piezo1 and Piezo2 are expressed in various MS tissues. For instance, Piezo1 is abundantly expressed in skin, bladder, lung, kidney, and colon, whereas Piezo2 has high expression in Dorsal Root Ganglia (DRG) primary sensory neurons, as well as in bladder, lung, and colon (Coste et al., 2010). Constitutive knockout of either Piezo1 or Piezo2 genes in mouse resulted in embryonic lethality, indicating their overall functional significance (Ranade, Woo, et al., 2014). Indeed, Piezo proteins have been shown to play critical roles in most, if not all, of the mechanotransduction processes, including sensing touch, pain, proprioception, blood flow-associated shear stress, urine flow and bladder distention, bone remodeling, volume regulation, cell migration, and fate determination (Faucherre, Nargeot, Mangoni, & Jopling, 2013; Martins et al., 2016; McHugh, Murdoch, Haslett, & Sethi, 2012; Ranade, Qiu, et al., 2014; Schrenk-Siemens et al., 2015; Woo, Lukacs, et al., 2015; Yang et al., 2014). Here we will only summarize those studies involving the characterization of Piezo knockout (KO) mice. For a comprehensive review in this research front, please refer to some other excellent review papers (Bagriantsev, Gracheva, & Gallagher, 2014; Honore, Martins, Penton, Patel, & Demolombe, 2015; Ranade, Syeda, & Patapoutian, 2015).

2.1 Piezo1 serves as an in vivo mechanotransducer in various biological processes

By characterizing Piezo1 KO mice, Piezo1 has been demonstrated to play critical roles in controlling proper vascular architecture. Two research groups have independently found that the constitutive Piezo1 KO mice were embryonic lethal at midgestation with defects in vascular formation (Li et al., 2014; Ranade, Qiu, et al., 2014). Li et al. have further shown specific deletion of Piezo1 in endothelia cells recapitulated the phenotypes observed in the constitutive KO mice, indicating that Piezo1 plays a critical role in endothelium-controlled vascular development. In endothelial cells, Piezo1 is responsible for flow-induced cationic currents and calcium influx, and promotes stress fiber organization and alignment in the direction of fluid flow. These studies suggest that Piezo1 might function as a flow sensor for monitoring blood flow-associated shear stress (Li et al., 2014; Ranade, Qiu, et al., 2014). Furthermore, Piezo1 is involved in endothelial cell migration and angiogenesis by promoting vascular endothelial growth factor—evoked phosphorylation of endothelial nitric oxide synthase (eNOS), resulting in the activation of eNOS (Li et al., 2014).

Using a Piezo1 LacZ reporter mouse line, Piezo1 was identified in smooth muscle cells as well (Retailleau et al., 2015). Interestingly, Piezo1 appears to have a trophic effect on the diameter and wall thickness of arterial wall of small arteries under hypertension or when its channel activity is enhanced at normal blood pressure. An increase of the Piezo1-mediated cytosolic calcium may lead to activation of the transglutaminases in smooth muscle cells, consequently contributing to arterial remodeling (Retailleau et al., 2015).

Piezo1 also plays a critical role in regulating volume homeostasis of red blood cells (RBCs) (Albuisson et al., 2013; Bae, Gnanasambandam, Nicolai, Sachs, & Gottlieb, 2013; Cahalan et al., 2015; Faucherre, Kissa, Nargeot, Mangoni, & Jopling, 2014; Shmukler et al., 2014; Zarychanski et al., 2012). Using Piezo1 conditional KO mice and a chemical activator of Piezo1, Cahalan et al. has shown that Piezo1 mediates MA currents and Ca^{2+} influx in RBCs. Blood-cell-specific deletion of Piezo1 in mice leads to overhydration and increased fragility of the RBCs. On the other hand, activation of Piezo1 by Yoda1, a chemical activator, induces calcium influx and subsequent dehydration of RBCs via downstream activation of the KCa3.2 Gardos channel. In line with this mechanism, gain-of-function mutations in Piezo1 are associated with dehydrated hereditary stomatocytosis in human, a disease characterized by severe dehydration and shrinkage of RBCs. Furthermore, knockdown of Piezo1 in zebrafish caused swelling and lysis of RBCs, thereby impairing their survival (Faucherre et al., 2014). Together, these data suggest that Piezo1 plays a conserved role in RBC volume homeostasis.

Piezo1 plays important roles in other examined cell types or tissues, including serving as the mechanotransduction channel in renal tubular epithelial cells (Peyronnet et al., 2013), regulating osmolarity sensing and ATP release in urothelial cells (Martins et al., 2016; Miyamoto et al., 2014), determining MS lineage choice in neural stem cells via monitoring matrix stiffness (Pathak et al., 2014).

2.2 Piezo2 serves as an in vivo mechanotransducer for somatosensation of mechanical force

In line with its high expression in DRG neurons, Piezo2 has been demonstrated to play a dominant role in somatosensation of touch and proprioception. Piezo2 mRNA was detected in about 20% of mouse DRG neurons comprising both nonnociceptive and nociceptive neurons. Small interfering RNA—mediated knockdown of Piezo2 in DRG neurons specifically diminished the rapidly adapting class of MA currents. Importantly, a series of elegant studies from Patapoutian and other groups have convincingly demonstrated that Piezo2 expressed in the Merkel cell—neurite complex mediates the mechanotransduction currents and gentle touch sensation behaviors in rodents (Ikeda et al., 2014). These studies settle the long-lasting controversy over whether Merkel cells, a specialized type of touch receptors residing in skin, are indeed mechanosensory cells for gentle touch, which was proposed about 150 years ago, and provide evidence for a two-receptor-site model for Merkel cell-neurite complex—mediated gentle touch sensation (Woo, Lumpkin, & Patapoutian, 2015). Interestingly, the role of Piezo2 in mediating gentle touch sensation appears to be well conserved in vertebrates. In zebrafish, *piezo2b*, a Piezo2 variant, mediates light touch response (Faucherre, Nargeot, Mangoni, & Jopling, 2013). Tactile-foraging waterfowls, whose feeding in murky waters almost exclusively relies on their acute sense of touch, appear to utilize Piezo2 for such tactile task (Schneider et al., 2014). Remarkably, Piezo2 has also been shown to be required for mechanotransduction in human stem cell-derived touch receptors, indicating their potential role as human touch receptor as well (Schrenk-Siemens et al., 2015).

Whether Piezo2 is required for mechanical nociception in mammals currently remains elusive. In mouse DRG neurons, Piezo2-mediated currents were increased by the bradykinin-PKA/PKC (protein kinase A/ protein kinase C) or cAMP-Epac1 signaling pathways, indicating the potential involvement of Piezo2 in mechanical hyperalgesia and allodynia (Dubin et al., 2012; Eijkelkamp et al., 2013). Piezo2 knockdown via intrathecal injection of Piezo2 antisense oligonucleotides significantly attenuated allodynia in two models of chronic and neuropathic pain and complete Freund's adjuvant (CFA)-induced mechanical hyperalgesia (Eijkelkamp et al., 2013; Singhmar et al., 2016). Knockdown of Piezo2 in whisker hair follicles presensitized with capsaicin injection diminished pain responses (Ikeda et al., 2014). These knockdown studies suggest that Piezo2 is involved in mechanical nociception. By contrast, characterization of DRG-specific Piezo2 KO mice did not reveal any obvious deficits in mechanical nociception (Ranade, Woo, et al., 2014). However, deleting the single Piezo gene in fly larvae resulted in specific deficit in responding to noxious mechanical stimuli, yet they still retain normal responsiveness to touch and thermal nociception (Kim, Coste, Chadha, Cook, & Patapoutian, 2012). Clearly, the involvement of Piezo2 in mechanical nociception requires more investigation.

Remarkably, Piezo2 may also function as the major mechanotransducer of mammalian proprioceptors, specialized somatosensory neurons that convey information about the stretch and tension experienced by muscles, tendons, skin, and joints. Piezo2 is expressed in sensory endings of proprioceptors that innervate muscle spindles and Golgi tendon organs in mice, and mediates stretch-induced firings of proprioceptors in muscle—nerve recordings. Importantly, proprioceptive neuron-specific knockout of Piezo2 resulted in mutant mice with severe uncoordinated body movements and abnormal limb positions. These studies provide strong evidence that Piezo2 may function as the long-sought mechanotransduction channel for proprioception, the elusive "sixth" sense (Woo, Lukacs, et al., 2015).

2.3 Piezo1 and Piezo2 are linked to human genetic diseases

The critical role of Piezo proteins in human physiology and pathophysiology has been validated by the identification of disease-causing mutations in Piezo1 and Piezo2 genes. Mutations in Piezo1 are associated with dehydrated hereditary stomatocytosis, lymphatic dysplasia, hereditary high phosphatidylcholine hemolytic anemia, and hemochromatosis-induced diabetes mellitus (Albuisson et al., 2013; Andolfo et al., 2015; Bae et al., 2013; Beneteau et al., 2014; Fotiou et al., 2015; Imashuku et al., 2016;

Lukacs et al., 2015; Sandberg, Nybo, Birgens, & Frederiksen, 2014; Szklarczyk et al., 2013; Zarychanski et al., 2012). On the other hand, mutations in Piezo2 are linked to Gordon syndrome, Marden–Walker syndrome, and distal arthrogryposis type 5 (McMillin et al., 2014). Intriguingly, nearly all of the disease-causing mutations examined so far slow the inactivation rate of Piezo1- or Piezo2-mediated currents, resulting in increased net cationic currents. Piezo2 mutants also show faster recovery from inactivation. Thus Piezo-associated etiology is most likely due to gain of function of the channel activity. However, the exact molecular mechanisms underlying the causal roles of these Piezo mutant proteins in the various disease conditions remain to be clarified. Together, all these studies derived either from animal models or human diseases highlight the physiological and pathophysiological importance of the Piezo proteins, as well as their potential as valid drug targets for developing novel therapeutics. However, key unaddressed questions include whether the in vivo roles of Piezo proteins are dependent on their mechanosensitivity, and how the ion permeation properties shape their in vivo roles. To answer these questions, one requires the understanding of how Piezo proteins function at the molecular level.

3. PIEZO PROTEINS FORM A DISTINCT CLASS OF MS CATION CHANNELS

Mammalian Piezos are large transmembrane (TM) proteins that are composed of about 2500–2800 amino acids and are predicted to encompass between 26 and 40 predicted transmembrane segments (TMs). In fact, human Piezo1 and Piezo2 are predicted to possess the highest number of TMs among all the human proteins. Adding a further layer of complexity to the Piezo proteins is their lack of apparent sequence homology with any known proteins including ion channels (Coste et al., 2010; Kamajaya, Kaiser, Lee, Reid, & Rees, 2014). For instance, unlike the 24-TM-containing voltage-gated Na^+ channel and Ca^{2+} channel, which are composed of four 6-TM repetitive domains, Piezo proteins do not have such repetitive domains. Thus it was unknown whether Piezo proteins are the pore-forming subunits of the MS cation channels, or they are nonconducting subunits of yet unidentified ion channels required for proper expression or for modulating channel properties. To address this fundamental question, Patapoutian and colleagues obtained several lines of evidence to strongly support that Piezo proteins are the pore-forming subunits (Coste et al., 2012).

3.1 Piezo-mediated MA currents display species-dependent pore properties

Coste et al. cloned the *Drosophila* Piezo (dPiezo) and characterized its biophysical properties (Coste et al., 2012). Despite the low amino acid sequence similarity, overexpression of dPiezo or mouse Piezo1/2 (mPiezo1/2) full-length cDNAs in various cell lines, including mammalian Neuro2A, C2C12, F11, NIH/3T3, HEK293T and Hela cells, is sufficient to mediate large MA cationic currents. The pressure for half-maximal activation (P_{50}) of dPiezo is about −30 mm Hg, similar to that of mPiezo1/2, indicating similar mechanosensitivity. The inactivation constant τ measured at −80 mV for dPiezo is ~6 ms, similar to that of mouse Piezo2 (~6 ms), but faster than that of mouse Piezo1 (~15 ms).

Importantly, dPiezo and mPiezo1/2 display distinct pore-related channel properties. First, dPiezo- and mPiezo1/2-mediated currents display dramatically different pharmacological sensitivity to the pore blocker, ruthenium red (RR) (Coste et al., 2012). RR has been shown to block the inward, but not outward, currents of mPiezo1/2 only when applied from the extracellular side, indicating that RR acts as a pore blocker of mPiezo1/2. In contrast, dPiezo is insensitive to RR blockage. Second, the unitary conductance of mechanically activated currents from mPiezo1 and dPiezo are drastically different. Under similar recording conditions, the slope conductance for mPiezo1 and dPiezo is about 30 and 3 pS, respectively. Therefore, dPiezo-dependent channels are a remarkable ninefold less conductive than mPiezo1-dependent channels.

Collectively, these data suggest that Piezos from distinct species are unlikely to function as accessary subunits to confer mechanosensitivity to other unknown pore-forming subunits that otherwise have to be ubiquitously expressed in all the tested mammalian cell lines and differentially regulated by Piezos in a species-dependent manner. Rather, these data indicate that Piezo proteins form the ion channel by themselves.

3.2 mPiezo1 proteins exist as oligomers

The ion-conducting pores of the vast majority of ion channels are formed by the assembly of TM domains from distinct subunits (e.g., voltage-gated K^+ channels and ligand-gated ion channels) or structurally repetitive domains within a protein (e.g., voltage-gated Na^+ and Ca^{2+} channels). Since Piezos lack repetitive TM motifs, they are more likely to assemble as oligomers if they indeed form ion channels. Coste et al. (2012) have employed three independent approaches to test this hypothesis.

First, they determined the number of subunits in Piezo complexes by expressing genetically engineered green fluorescent protein (GFP)-mPiezo1 fusion proteins in *Xenopus* oocytes, imaging individual spots with total internal reflection microscopy, and counting discrete photobleaching steps. The result of such single molecule photobleaching experiment is consistent with the idea that the GFP-mPiezo1 proteins exist as oligomers.

Second, they expressed and purified a glutathione S-transferase (GST)-tagged C-terminal mPiezo1 fusion protein (mPiezo1-GST), which was subjected to stoichiometry analysis using both native gel and cross-linking. The calculated molecular weight of the mPiezo1-GST fusion protein is ~318 kDa. When assayed with native gel, a major Piezo1-specific band migrated at the position near the 1236 KDa marker and a weak band near the 1048 kDa marker were detected (Fig. 1A and B). The higher band indicates a tetrameric complex, whereas the lower band likely represents a trimer. When assayed with formaldehyde cross-linking experiment, four Piezo1-specific bands were detected, although the molecular weight of the top three bands could not be accurately assigned given the largest molecular marker of only 460 kDa (Fig. 1C). Notably, the topmost band is always weak compared with the other three bands. These data suggest that mPiezo1-GST fusion proteins form oligomers likely as tetramers.

However, our subsequent biochemical and structural studies have clarified that the mPiezo1-GST protein band near the position of the 1236 kDa marker consists of a dimer of trimeric mPiezo1-GST complex instead of tetrameric complexes, as a result of the dimerized GST tags (Fig. 1F–L). We examined the mPiezo1-GST fusion proteins by negative staining EM, and observed apparent butterfly-like particles containing two identical halves that are organized in a back-to-back manner (Fig. 1G–I). By contrast, negative staining of the GST-tag-cleaved mPiezo1 proteins did not show the butterfly-like particles (Fig. 1J–L). Rather, particles with a threefold symmetry were clearly detected. These EM studies suggest that the dimerized GST tag could mediate the dimerization of mPiezo1. Native gel analysis also supports that the GST-tag-free mPiezo1 proteins form complexes with a molecular weight of ~900 kDa. Furthermore, FLAG-tagged mPiezo1 displayed a major band at ~900 kDa on native gel as well (Fig. 1F). Thus these data suggest that the major oligomeric state of the purified mPiezo1 is trimeric, with a minor population forming a dimer

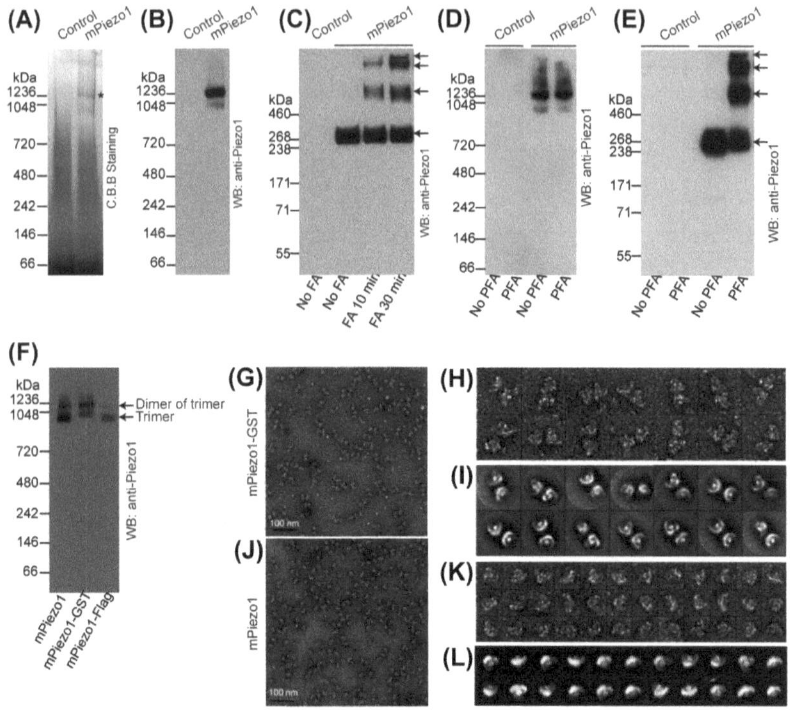

Figure 1 ***Determining the oligomeric status of mPiezo1.*** (A and B) The indicated samples were purified and separated on native gels and visualized by Coomassie Brilliant Blue (CBB) staining (A) or western blotting (WB) (B). *Asterisk* in (A) indicates a protein band specifically present in the Piezo1 sample. (C) Purified Piezo1—glutathione *S*-transferase (GST) proteins treated with or without formaldehyde (FA) with the indicated time period, separated on a denaturing gel and detected with the anti-Piezo1 antibody. Sample purified from cells without transfection served as a negative control. (D and E) Piezo1-GST-transfected HEK293T cells or untransfected cells treated with or without 0.25% paraformaldehyde (PFA) for 10 min. The cross-linked Piezo1-GST proteins were then purified and separated on native gel (D) or denaturing gels (E), followed by western blotting. (F) Native gel and western blotting analysis of GST-cleaved Piezo1 (mPiezo1), Piezo1-GST, and Piezo1-Flag samples with the anti-Piezo1 antibody. (G) Representative micrograph of the negatively stained Piezo1-GST proteins. (H) Raw particles of Piezo1-GST proteins. (I) 2D class averages of Piezo1-GST particles. (J) Representative micrograph of the negatively stained mPiezo1. (K) Raw particles of mPiezo1. (L) 2D class averages of mPiezo1 particles. *Adapted from Coste, B., Xiao, B., Santos, J.S., Syeda, R., Grandl, J., Spencer, K.S., … Patapoutian A. (2012). Piezo proteins are pore-forming subunits of mechanically activated channels.* Nature, 483*(7388), 176—181 and Ge, J., Li, W., Zhao, Q., Li, N., Chen, M., Zhi, P., … Yang, M. (2015). Architecture of the mammalian mechanosensitive Piezo1 channel.* Nature, 527*(7576), 64—69.*

of trimers on native gels. The majority of the mPiezo1-GST fusion proteins form dimers of trimers, as a result of the dimerized GST tag. Collectively, these data demonstrate that mPiezo1 forms a homotrimer (Ge et al., 2015). Intriguingly, unlike the family of voltage-gated channels with most of its members displaying fourfold symmetry, MS channels possess a variety of organizational symmetry spanning from two- to sevenfold (Bass, Strop, Barclay, & Rees, 2002; Brohawn et al., 2012; Chang, Spencer, Lee, Barclay, & Rees, 1998; Liu, Gandhi, & Rees, 2009), demonstrating their structural diversity.

It remains unknown what drives the unusual migration of the 1.8-mega-dalton mPiezo1-GST dimer of trimers near the 1.2 mega-dalton marker. The large native size of the fusion protein together with its numerous TMs might have resulted in its unusual mobility on native gels. Nevertheless, we could not completely exclude the possibility that mPiezo1 exists in other oligomeric states on the membrane or under different conditions in vitro, a scenario observed in previous studies of other ion channels, such as Orai channels (Hou, Pedi, Diver, & Long, 2012; Penna et al., 2008). However, when the mPiezo1-GST fusion proteins were cross-linked on living cells by paraformaldehyde prior to the purification procedure, the distribution pattern of the Piezo1-specific bands is similar to the protein sample subjected to cross-linking after purification. The topmost band, which may represent the weakly cross-linked dimer of trimer, appeared much weaker than the other three bands (Fig. 1D and E). These data suggest that mPiezo1 may exist as trimmers in living cells, which requires further validation.

3.3 Reconstituted mPiezo1-GST fusion proteins in lipid bilayers mediate RR-sensitive cationic currents

Coste et al. (2012) reconstituted the purified mPiezo1-GST fusion proteins into lipid bilayers in two distinct configurations, including droplet interface lipid bilayers (DIBs) assembled from two monolayers and proteoliposomes. In the first configuration, mPiezo1 was reconstituted into asymmetric bilayers aiming to mimic the cellular environment, in which the extracellular facing lipid monolayer is predominantly neutral in charge, whereas the intracellular facing leaflet is negatively charged. In contrast, the lipid composition of the bilayer in the second configuration is uniform. Remarkably, in the DIB configuration, brief, discrete, and spontaneous single channel openings were clearly detected in the mPiezo1-GST-reconstituted membranes, but not in controls. Importantly, RR efficiently blocked the current only when applied to the extracellular mimicking side. The asymmetric accessibility of RR block of reconstituted channels is in agreement with the data obtained from mPiezo1-overexpressing HEK293T cells. The single channel conductance was ~120 and 60 pS in symmetric 0.5 and 0. 2 M KCl, respectively. The reconstituted proteins also conducted sodium, consistent with Piezo1 proteins being nonselective cation channels. A similar pattern of activity was obtained from mPiezo1 reconstituted in asolectin liposomes. Collectively, these reconstitution experiments provide strong evidence that Piezo1 proteins are the pore-forming subunits. However, whether the reconstituted mPiezo1-GST proteins can respond to mechanical force stimulation, such as negative suction-elicited membrane stretch, was undetermined. Thus it remains unclear if Piezo1 proteins are intrinsically gated by membrane stretch, as shown for bacteria MscL and MscS and mammalian MS K2P channels (Brohawn, Su, & MacKinnon, 2014; Maingret, Patel, Lesage, Lazdunski, & Honore, 2000; Martinac, Adler, & Kung, 1990).

Collectively, based on all the evidence described previously, Patapoutian and colleagues conclude that Piezo proteins are the pore-forming subunits of MS cation channels, establishing them as a distinct novel class of ion channels. However, given that the ion-conducting pore was not identified, Nilius et al. has argued that it was still premature to claim that Piezos are indeed pore-forming subunits of the MS cation channels. The advancement in resolving the cryo-EM structure of the full-length mPiezo1 protein and functionally unveiling its ion-conducting pore provides definitive evidence to establish Piezo proteins as the prototypic class of MS cation channels in mammals.

4. THE THREE-DIMENSIONAL CRYO-EM STRUCTURE OF mPIEZO1

4.1 mPiezo1 protein purification and cryo-EM structure determination

Taking advantage of the technical breakthrough of application of cryomicroscopy (cryo-EM) for determining high-resolution three-dimensional (3D) structure of membrane proteins (Cao, Liao, Cheng, & Julius, 2013; Liao, Cao, Julius, & Cheng, 2013, 2014; Paulsen, Armache, Gao, Cheng, & Julius, 2015), in combination with our expertise in membrane protein engineering and expression and purification in mammalian expression systems, we have spearheaded the effort to resolve the 3D structure of Piezo proteins. The initial effort was focused on obtaining sufficient amount of and relatively homogenous Piezo protein samples by using the previously established heterologous expression and purification system (Coste et al., 2012). Piezo cDNAs either in full length, or truncated forms from different species including human, mouse, and *Drosophila*, were ligated to a sequence encoding a C-terminal GST tag with a precision protease cleavage site in between. Many constructs were tested for their expression by transient transfection in HEK293T cells. Meanwhile, a large number of different classes of detergents were screened for their compatibility with Piezo protein extraction and purification. Finally, a combination of the full-length mouse Piezo1 protein with C12E10 detergent was selected for protein purification and structural determination using the state-of-the-art Titan Krios microscope operated at a voltage of 300 kV with a K2 Summit direct electron detector. These collaborative efforts have eventually led to the determination of the 3D cryo-EM structure of the full-length (2547 amino acids) mPiezo1 at a medium resolution (Ge et al., 2015).

4.2 mPiezo1 forms a three-bladed, propeller-like homotrimeric architecture

Strikingly, mPiezo1 is organized into a unique three-bladed, propeller-shaped architecture with distinct domains resembling the typical structural components of a propeller, including three distal "Blades" and a central "Cap" in the extracellular side, and three distinct "Beams," which are ~80 Å in length and positioned in a 30-degree angle relative to the membrane plane in the intracellular side (Figs. 2 and 3). The

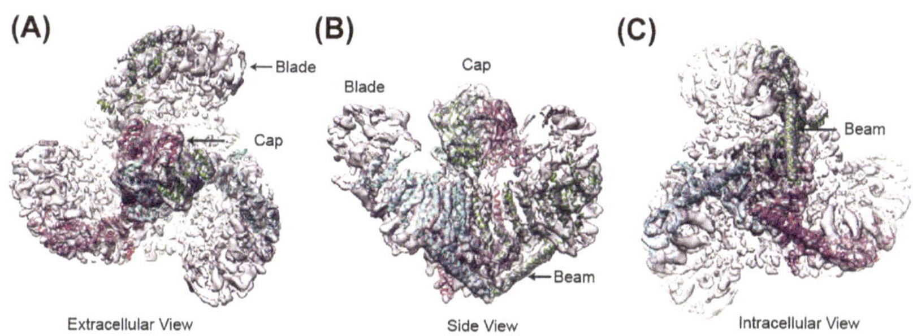

Figure 2 ***The cryo-electron microscopy structure of mPiezo1.*** Extracellular (A), side (B), and intracellular (C) views of the 3D cryo-electron microscopy density map superimposed with a polyalanine model and the crystal structure of the trimeric C-terminal extracellular domain. The three protomers are colored cyan, purple, and green, respectively. *Adapted from Ge, J., Li, W., Zhao, Q., Li, N., Chen, M., Zhi, P., … Yang, M. (2015). Architecture of the mammalian mechanosensitive Piezo1 channel.* Nature, *527(7576), 64–69.*

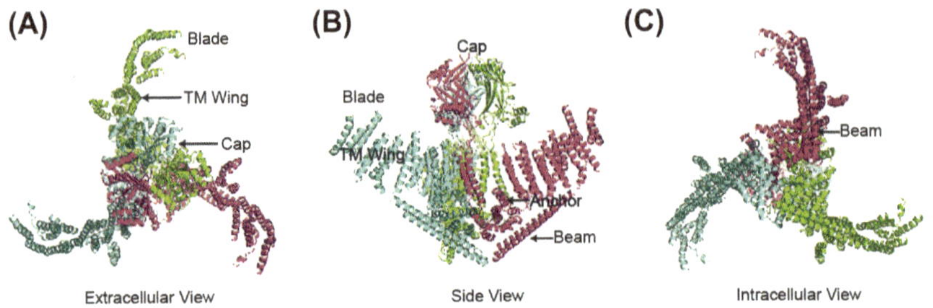

Figure 3 ***The transmembrane skeleton of mPiezo1.*** Extracellular (A), side (B), and intracellular (C) views of the separately colored polyalanine models of each subunit. *TM*, transmembrane. *Adapted from Ge, J., Li, W., Zhao, Q., Li, N., Chen, M., Zhi, P., … Yang, M. (2015). Architecture of the mammalian mechanosensitive Piezo1 channel.* Nature, *527(7576), 64–69.*

Beams appear to support the whole structure and mechanically connect the distal end of the TM portion and the Blade regions to the lower center of the trimeric complex. The TM region lies in between the extracellular and intracellular surface-exposing domains, serving as the skeleton backbone of the whole propeller-shaped structure. The diameter from the top view and the axial height of the structure are ~200 and 155 Å, respectively.

The striking similarity of the mPiezo1 structure with a three-bladed propeller apparently raises the possibility that mechanically activated Piezo channels might mechanistically function similar to a propeller that senses and transduces force. Thus from a functional point of view, the overall architecture of Piezo1 can be divided into a central pore module and three extended peripheral wings. The central pore module consists of the six core TMs that are converged at the center, the extracellular Cap, and a highly packed intracellular region located at the center (Fig. 3). This pore module displays a remarkable architectural similarity to other trimeric channels such as acid-sensing ion channel 1 (ASIC1) and ATP-gated P2X4 receptors, despite their lack of sequence homology (Gonzales, Kawate, & Gouaux, 2009; Kawate, Michel, Birdsong, & Gouaux, 2009). By contrast, the three peripheral wings, which contain the distinct extracellular Blades, peripheral TM helix (PH) skeleton, "Anchors," and intracellular Beams, are absent in other trimeric ion channels, thus defining the unique structural architecture of Piezo1 (Fig. 3).

5. PIEZO1 TOPOLOGY

5.1 The C-terminal extracellular domain trimerizes to form the cap structure

The apparent detachment of the Cap from the TM core indicates that it is likely a soluble region. A topological model predicts that the residues from 2210 to 2457 between the last two TMs constitute a large extracellular loop, termed C-terminal extracellular domain (CED). To test whether this region constitutes the apparent Cap structure, a deletion mutant of mPiezo1Δ2218−2453 was constructed, purified, and observed by negative-staining EM. Remarkably, two-dimensional (2D) classification of mPiezo1Δ2218−2453 particles revealed that the central Cap is absent in 2D class averages, confirming that this region indeed accounts for the Cap.

The soluble fragment protein (residues from 2218 to 2457) was then purified, crystalized, and resolved to 1.45-Å resolution. Interestingly, the CED forms a cradle-like structure with two α-helix bundles located at the bottom (α1 and α2) and on top (α3 and α4) (Ge et al., 2015). Between the two α-helix bundles, there are two layers of β-sheets, with one layer containing four long β-sheets (β1/β2/β5/β6) and the other consisting of two symmetrical motifs, each of which are formed by three β-sheets (β3/β4/β12 and β7/β10/β11). Two additional short β-sheets (β8 and β9) run antiparallel between the two β-sheet layers. The C-terminus of the CED turns back to the N-terminus of this domain, consistent with the prediction that the CED starts from the C-terminus of the second to last TM and ends at the N-terminus of the last TM.

Rees and colleagues have previously obtained the crystal structure of the corresponding region (residue S2162 to S2344) of *C. elegans* Piezo (Kamajaya et al., 2014). Although the sequence identity is only 22.6%, the 3D structures of these two fragment proteins are highly conserved.

Intriguingly, the mouse CED proteins form a trimeric complex in both solution and in the crystal lattice (Ge et al., 2015), a phenomenon not previously reported for the corresponding protein derived from *C. elegans* (Kamajaya et al., 2014). Remarkably, the crystallographic trimer fits well into the cryo-EM density map of the cap. These results demonstrate that the CED trimerizes to form the cap, further supporting that mPiezo1 forms a trimeric complex.

5.2 The cap and blades are extracellularly located, whereas the beams and C-terminus are at the intracellular side

To confirm the topological location of the CED and the C-terminus of mPiezo1, we utilized live labeling of the HEK293T cells expressing the mPiezo1 with a FLAG tag inserted either at a flexible region of the CED (after the residue 2419) based on its crystal structure or at the C-terminus of mPiezo1 (Ge et al., 2015). Using confocal microscopy, we found that the FLAG tag, only when inserted at the CED, but not at the C-terminus, was labeled on the plasma membrane surface. Consistently, Coste et al. (2015) detected a live-labeling signal of the Myc tag inserted at position 2336. These data demonstrate that the CED is located extracellularly, and the C-terminal tail is intracellular (termed the intracellular C-terminal domain, CTD) because it follows CED with only one TM and was not detected via live labeling. Consequently, both the central Cap and the three Blades are located extracellularly, whereas the Beams and CTDs are at the intracellular side. The N-terminal Myc tag was detected only after permeabilizing the expressing cells, indicating an intracellular localization. However, the exact location of the N-terminus in the 3D structure of mPiezo1 remains to be identified.

5.3 The anchor domain and subunit swapping

Interestingly, a unique hairpin structure, termed Anchor domain, exists at the interface of two adjacent subunits, which contains four helices $\alpha1-\alpha4$ and penetrates into the inner leaflet of the membrane, with a long helix ($\alpha4$) roughly parallel to the membrane (Fig. 3). This Anchor domain was assigned to connect the Outer Helix (OH) and Peripheral Helix 1 (PH1) with confidence. Given that the OH is formed by the second to last TM (around residues 2101−2213) (see later discussion), the Anchor domain is composed of amino acid residues from 2100 to 2190 where the most evolutionally conserved sequence motif PF(X2)E(X6)W (2129−2140) among Piezo homologues (Prole & Taylor, 2013) and the disease-causing mutation T2142 (T2127 of human Piezo1) have been identified (Andolfo et al., 2013), indicating its structural and functional importance. Indeed, Coste et al. have found that mutating E2133 affects unitary conductance, ion selectivity, and RR blockade (only when mutated to lysine but not to alanine). In the absence of the Piezo1 structural information, these results would have led to the suggestion that E2133 serves as a key pore determinant (Coste et al., 2015). Now, given its close proximity to the pore module, it seems more plausible that this residue most likely affects the essential pore properties via allosteric effect.

Intriguingly, the Anchor domain results in swapping the OH-CED–Inner Helix (IH)-containing region of one subunit into the neighboring subunit in a clockwise direction (viewed from the Cap), leading to a clear separation of the three subunits on the 3D structure (Figs. 2 and 3).

5.4 The TM topology of Piezo1 is complex and remains elusive

One of the most unique features of Piezo proteins is their unprecedentedly large number of predicted TMs (26–40 times). The relatively better density associated with the TM regions allowed us to build a de novo alanine model with 502 amino acids for the apparent TMs, the Beams, and the CTD. Together with the 227 amino acids of the CED, 719 of 2547 amino acids of each protomer were built into the density map (Fig. 3). The whole TM skeleton displays a three-winged arrangement, with each extended wing being slightly twisted (Fig. 3). Unexpectedly, each wing has only 14 apparently resolved TMs, much lower than the predicted number. However, given the low resolution at the peripheral regions, it is totally possible that additional TMs might exist, but this was not resolved in the current Piezo1 structure. Nevertheless, the trimeric complex comprises at least 42 TMs with each subunit having 14 apparently resolved TMs (Ge et al., 2015), marking Piezo proteins as ion channels with the highest number of TMs among all ion channels known so far.

Given the better resolution and more reliable connectivity among those TMs close to the center of the trimeric complex, we termed the two innermost TMs that are converged at the center as Inner Helix (IH) and Outer Helix (OH), respectively, and those TMs located in the extended wings as Peripheral Helix (PH1–12). The OH and IH could be tentatively traced to the N- and C-terminus of the CED, respectively, although the lack of reliable density at the junction points made this assignment ambiguous (Fig. 2). However, our subsequent functional characterizations have validated that the pore-lining IH is formed by the last TM, whereas OH is composed of the second to last TM (Zhao et al., 2016). The OH is connected to the PH1 via the four alpha helices–constituted anchor domain. Continuous connection from PH1 to PH4 appears to be reliable, whereas tracing the connectivity of the rest of the PHs becomes challenging.

Coste et al. (2015) have carried out extensive efforts to experimentally determine the TM topologic model. They generated a total of 48 constructs with the insertion of Myc tag into predicted loops and then used live labeling to locate those extracellularly facing tag positions. These studies led to the identification of nine extracellular loops. Together with bioinformatics analysis and identification of intracellular phosphorylation sites, they have proposed possible topologic models of 38, 18 or 10 TMs (Coste et al., 2015). A topology module of about 18 TMs appears to be consistent with the observed 14 evident TMs in each subunit. Nevertheless, the TM topology remains elusive and clearly requires high-resolution structure for validation.

6. THE CENTRAL ION-CONDUCTING PORE

Despite the complexity of the TM topology and the large size of the protein, it is striking to find that the ion-conducting pore of mPiezo1 channels is encoded by the last ~360 residues (residues 2189–2547), which comprise the last two putative TMs, CED (a large extracellular loop region containing ~240 residues), and the intracellular CTD of ~70 residues (Fig. 4A). Remarkably, although Piezo proteins lack sequence homology with known proteins including ion channels, the Piezo1 pore module shares unexpected structural similarity with other types of trimeric channels, including ASICs (~500 residues) and P2X receptors (~400 residues) (Gonzales et al., 2009; Kawate et al., 2009). We review the major studies listed later that lead to this important finding, which provides a valuable structural framework for in-depth characterization of the ion-conducting and gating mechanisms of Piezo1 channels.

6.1 The mPiezo1 structure reveals a continuous central channel

As pore-forming subunits, Piezo proteins are supposed to encode ion-conducting pores. Indeed, the central slice of the rotationally averaged density map of mPiezo1 revealed a continuous central channel along the Z-axis, including an extracellular vestibule (EV) within the cap region, a TM vestibule (MV) enclosed by the three IHs, and an intracellular vestibule (IV), which is formed by the CTD (Ge et al., 2015). Thus this central region, including the IH–OH pairs, CED, and CTD, comprises the pore module of Piezo1 channels (Fig. 4A). This pore module displays a remarkable architectural similarity to trimeric ASIC1 and ATP-gated P2X4 receptors that

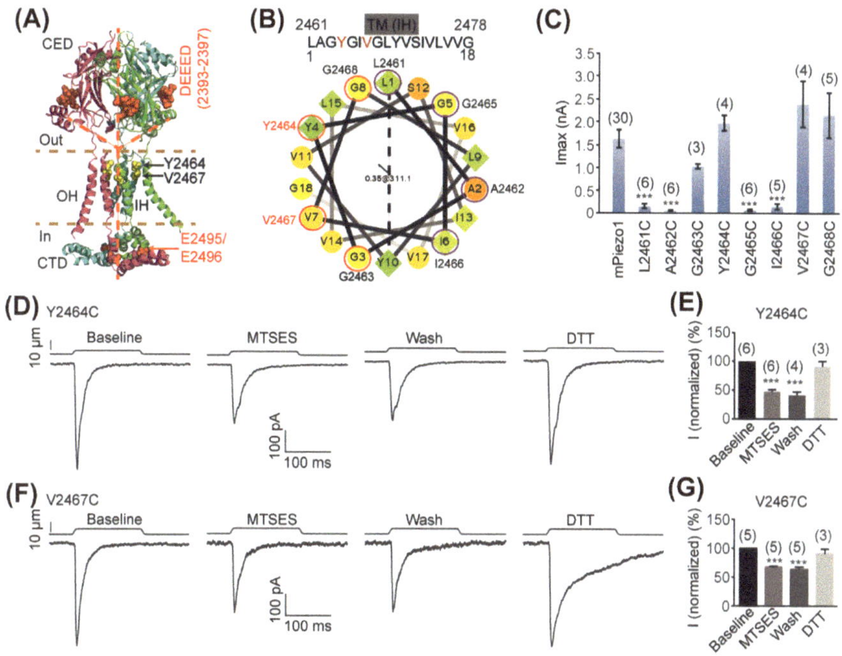

Figure 4 ***The pore module and the pore-lining Inner Helix (IH).*** (A) Putative locations of Y2464 and V2467 on the IH on the 3D structure of the pore module. Other important residues DEEED (2393–2397) and E2495/E2496 are also labeled. (B) Helical wheel prediction of residues residing in the last putative transmembrane (TM). The hydrophilic residues are represented by *circles* and the hydrophobic residues by *diamonds*. The most hydrophobic residue is green, and the amount of green decreases proportionally to the hydrophobicity, with zero hydrophobicity coded as yellow. Cysteine mutations of the residues with *red circles* were subjected to SCAM analysis. Y2464 and V2467 highlighted in red can be covalently modified by 2-sulfonatoethyl methanethiosulfonate sodium salt (MTSES). Cysteine mutations of the residues with *purple circles* had drastically reduced mechanically activated (MA) currents. (C) Average of maximal MA currents. One-way analysis of variance (ANOVA) with Dunn's comparison to mPiezo1. (D and F) Representative traces of MA currents of cells transfected with the indicated constructs at baseline, during 1 mM MTSES application, upon washing and application of 0.5 mM DTT. (E and G) Average of currents normalized to baseline in cells transfected with the indicated constructs recorded under the indicated condition. One-way ANOVA with Dunn's comparison to mPiezo1. All *bar panels* represent mean ± SEM, and the number of cells tested is shown above the bars. ***$p < .001$, **$p < .01$. *Adapted from Zhao, Q., Wu, K., Geng, J., Chi, S., Wang, Y., Zhi, P., … Xiao B. (2016). Ion permeation and mechanotransduction mechanisms of mechanosensitive piezo channels.* Neuron, *89(6), 1248–1263.*

essentially contain two TMs and a large extracellular domain in each promoter (Gonzales et al., 2009; Kawate et al., 2009), but distinct from known potassium- (Doyle et al., 1998) or sodium-selective channels (Payandeh, Scheuer, Zheng, & Catterall, 2011; Zhang et al., 2012), or nonselective cation channels like TRPV1 (Liao et al., 2013) and TRPA1 (Paulsen et al., 2015).

6.2 Primary sequence assignment of the central pore module

The structurally revealed OH-CED-IH-CED-constituted pore module has been tentatively assigned to the most distal C-terminal region (residues from 2189 to 2547) of mPiezo1 that contains the last two putative TMs, CED, and CTD (Ge et al., 2015). However, the lack of side-chain information and reliable density connections between CEDs, which alone have atomic-resolution structures, and the flanking OH–IH TM pairs introduce important uncertainties about the assignment of primary sequence into the pore region (Ge et al., 2015).

In line with this ambiguity, the assignment of the structurally indicated pore region of mPiezo1 appears to contradict a previous study showing that human Piezo1 lacking the entire C-terminus starting from residue 2218 (the corresponding residue 2234 in mPiezo1) retains stretch-activated channel activities (Bae et al., 2013), indicating that this region is not essential for ion conduction. Furthermore, Coste et al. have reported that mutations of E2133 affect unitary conductance and ion selectivity of the channel (Coste et al., 2015). However, neutralizing E2133 to alanine did not abolish the blocking effect of RR, a pore blocker of Piezo1 channels. Therefore it was not clear whether this residue resides in the pore or not. On the basis of the mPiezo1 structure, E2133 is putatively located in the Anchor domain instead of the structurally indicated pore region (Ge et al., 2015). Thus the assignment of the structurally revealed pore module clearly requires further functional validation.

6.3 Residues 2189–2547 of mPiezo1 encode the central pore module

To functionally validate the pore-forming region of Piezo proteins, we took advantage of the distinct pore properties between mPiezo1 and dPiezo as described earlier. We generated a series of chimeras between mPiezo1 and dPiezo by swapping the proposed pore region (2189–2547 of mPiezo1) that has an overall 31% sequence homology between the two species, focused on RR sensitivity of mechanically activated (MA) whole-cell currents and unitary conductance of stretch-activated currents recorded in the cell-attached configuration (Zhao et al., 2016). Remarkably, replacing residues 2189–2547 of mPiezo1 with the corresponding region of dPiezo resulted in a chimeric channel (mP1/dP-Pore) that was insensitive to RR blockage and had dPiezo-like single-channel properties. These data suggest that residues 2189–2547 of mPiezo1 determine the fundamental pore properties, including unitary conductance and pore blockage. Incredibly, the portion of the pore module of mPiezo1 can be replaced with mASIC1 to generate a hybrid channel that can respond to mechanical force and acid, further supporting that residues 2189–2547 of mPiezo1 are able to form a separable pore module (Zhao et al., 2016).

6.4 The last putative TM forms the pore-lining IH

The lack of reliable density at the junction points of CED and the IH and OH made the assignment of these two helices into the pore module ambiguous. However, the observation that the last two TM-containing region determine the essential pore properties suggests that the two TMs constitute the IH and OH. To test whether the last TM forms the pore-lining IH of the Piezo1 channel, we employed the substituted-cysteine accessibility method (SCAM) that has been widely used for probing channel pore positions (Akabas, Stauffer, Xu, & Karlin, 1992; Li, Yang, & Canessa, 2011). Eight consecutive residues (2461–2468) that likely span two turns of alpha helix in the extracellular side of the last TM were substituted into cysteine, and assayed for their covalent modification by the thiol-reactive reagent 2-sulfonatoethyl methanethiosulfonate sodium salt (MTSES) (Fig. 4B). Among the mutants, only Y2464C and V2467C were covalently modified by MTSES, demonstrating periodic accessibility of these pore-facing residues in the last TM (Fig. 4D–G). Notably, all mutants located in the opposite side of the helix (L2461, A2462, G2465, I2466) had drastically reduced MA currents (Fig. 4C), indicating that proper interaction of these residues with either the lipid membrane or residues in OH is required for maintaining the pore integrity, in agreement with the observation that swapping the IH or OH of mPiezo1 and dPiezo resulted in chimeric channels with drastically reduced MA currents as well (Zhao et al., 2016). Furthermore, despite the lack of charge-selecting residues in the putative pore helix, it appears that the helical side with residues accessible by MTSES is more hydrophilic than the other side (Fig. 4B). Collectively, these studies strongly suggest that the last TM forms the pore-lining IH, consistent with the structural assignment.

7. MOLECULAR BASIS FOR PIEZO ION PERMEATION, ION SELECTIVITY, AND PORE BLOCKADE

mPiezo1 is a nonselective cation channel, conducting Ca^{2+}, Na^+, K^+, and Cs^+ with a slight preference for Ca^{2+}, and has a low permeability for Cl^- (Coste et al., 2010, 2015). The establishment of the pore-forming domain allowed a detailed characterization of the molecular basis underlying the ion permeation, ion selection, and pore blockade of Piezo channels. We have characterized various parts along the ion-conducting pathway (Zhao et al., 2016).

7.1 CED mediates efficient ion conduction and cation selectivity

Based on the trimeric crystal structure of the CED, we have hypothesized that the three fenestration openings located in the extracellular Cap structure may allow cations to enter or exit the TM pore (Fig. 4A). Using chimeras, deletion, and site-directed mutagenesis, we made several important findings (Zhao et al., 2016). First, CED determines the distinct unitary conductance and RR sensitivity between mPiezo1 and dPiezo. Second, deleting the CED or mutating a large patch of negatively charged residues (DEEED at the position from 2393 to 2397) at the fenestration sites drastically reduced the MA currents. Thirdly, the DEEED (2393–2397)-AAAAA mutant had significantly higher permeability to chloride ($P_{Cl}/P_{Na} = \sim 27\%$) than mPiezo1 ($P_{Cl}/P_{Na} = \sim 8\%$), but normal Ca^{2+} permeability over monovalent Cs^+. Based on these data, we have proposed that Piezo1 may employ the patch of negatively charged residues DEEED (2393–2397) located at the fenestration sites of the extracellular Cap structure for selecting cation over anion entry to the pore and ensuring efficient ion conduction by enriching cations via electrostatic effect (Fig. 4A).

7.2 E2495 and E2496 are the key pore determinants

Negatively charged residues along the ion-conducting pathway of cation channels are generally involved in determining pore properties including unitary conductance, ion selectivity, and pore blockage (Owsianik, Talavera, Voets, & Nilius, 2006). The last putative TM of Piezo1 that lines the pore does not contain acidic residues, but is followed by several ones in the intracellular CTD. Combining site-directed mutagenesis and characterization of the unitary conductance, ion selectivity, and pore blockage, we have identified that two critical residues, E2495 and E2496, potentially residing in the CTD-constituted IV (Fig. 4A), play critical roles in discriminating divalent Ca^{2+} from monovalent Cs^{+}, as well as determining unitary conductance and RR sensitivity (Zhao et al., 2016). Mutating the corresponding residues, E2769 and E2770, in mouse Piezo2 to alanine also resulted in altered ion selectivity and loss of RR blockade (Zhao et al., 2016), indicating similar roles of these residues in determining Piezo2 channel pore properties.

Negative charges at this position may analogously act as a selectivity filter to control ion permeation properties and pore blockage, reminiscent of other Ca^{2+}-selective channels such as voltage-gated Ca^{2+} channels (VGCCs) (Yang, Ellinor, Sather, Zhang, & Tsien, 1993) or Orai channels (Prakriya et al., 2006; Yeromin et al., 2006). However, the intracellular localization of these residues might explain the relatively low preference of Ca^{2+} over monovalent cations for Piezo1 channels when compared with VGCCs (Yang et al., 1993) or Orai channels (Hou et al., 2012), which have a ring of acidic residues at the extracellular side of their ion-conducting pores. In contrast to other cation channels, no acidic residues are located in the pore-lining helix of Piezo1 channels.

7.3 The molecular basis of Piezo pore blockade

Pharmacological tool compounds are valuable for exploring the structure–function relationships of ion channels. Mammalian Piezo1 and Piezo2 are blocked by various nonspecific blockers, including RR, gadolinium (Gd^{3+}) (Coste et al., 2010), FM1-43 (Eijkelkamp et al., 2013), and the peptide toxin GsMTx-4 (Bae, Sachs, & Gottlieb, 2011). RR is a polycationic pore blocker for mPiezo1 and mPiezo2, but not for dPiezo (Coste et al., 2012). The species-specific blocking effect of RR enables a chimeric approach to locate the action site of RR, which is indicative of the location of the ion-conducting pore. Coste et al. have found that replacing the long C-terminal half of dPiezo with mPiezo1 resulted in chimeras, including $dP^{1-1,316}/mP1^{1,316-2,547}$ and $dP^{1-1,752}/mP1^{1,715-2,547}$, obtaining RR sensitivity. Consistently, a reverse chimera, $mP1^{1-1,973}/dP^{1,930-2,548}$, became RR insensitive (Coste et al., 2015). These results suggest that the RR-determining region is located in the C-terminus of mPiezo1 containing residues 1974–2547. They further identified the glutamate residue at 2133 (E2133) is involved in RR sensitivity. When mutated to lysine, the E2133K mutant became RR insensitive. However, neutralizing this residue (E2133A) did not affect the RR sensitivity

(Coste et al., 2015). Given that a direct binding of the polycationic RR is mediated through ionic interaction, these data suggest that E2133 is unlikely the binding site for RR. Indeed, E2133 is putatively localized to the Anchor region instead of the pore region (Ge et al., 2015). However, given the proximity between the Anchor region and the pore module, it is very likely that E2133 may allosterically affect the binding of RR to the pore.

Based on the putatively assigned pore module into the last two TM-containing C-terminal region from residues 2189 to 2547, we generated a series of chimeras within this region and found that the species difference of RR sensitivity is mediated by the extracellular Cap-constituting CED (residues 2213–2458) (Zhao et al., 2016). Intriguingly, both mPiezo1-ΔCED and mPiezo1-DEEED (2393–2397)-AAAAA remained to be effectively blocked by RR, indicating that CED and these acidic residues are not essential for RR sensitivity of mPiezo1, despite the critical role of CED in determining the distinct responses of mPiezo1 and dPiezo to RR. Unexpectedly, E2495 and E2496, two residues likely located in the intracellular side, were found to be critical for RR responsiveness (Zhao et al., 2016). The E2495A and E2496A mutations in mPiezo1 or E2769A and E2770A in mPiezo2 impaired the blocking effect of RR. Interestingly, the corresponding residues in dPiezo are E2466 and D2467. Thus the RR-binding site might also exist in dPiezo but is likely prevented from being accessed by its distinct CED, consistent with the data that mP1/dP-IH/CTD retained RR sensitivity (Zhao et al., 2016). This possibility, together with the key residues in CED that determine the distinct RR response of mPiezo1 and dPiezo, requires further investigation.

Intriguingly, RR can only block Piezo1 channels when applied from the extracellular side (Coste et al., 2012), suggesting that E2495 and E2496 might not be accessed by RR from the intracellular side. Instead, as a large positively charged molecule, RR can permeate through the pore along the EV and MV to reach these residues located in the IV. These data suggest that Piezo1 channels might possess a wide ion-conducting pore. Indeed, organic cations, such as tetramethyl ammonium and tetraethyl ammonium, can permeate Piezo1 channels, indicating a pore diameter larger than 8 Å (Gnanasambandam, Bae, Gottlieb, & Sachs, 2015). Furthermore, the lack of charge-selecting residues in the putative pore-lining TM may also account for the ability of Piezo1 to conduct cations but to much less extent as well as anions, and to allow large ions such as RR to access the pore. Thus characterization of the pore blocker such as RR may reveal not only its own working mechanism but also the pore properties.

7.4 Piezo proteins are bona fide pore-forming subunits of MS cation channels

It is generally believed that mutations of residues that alter unitary conductance, ion selectivity, and pore blockage provide ultimately functional proof that the tested protein is indeed a pore-forming subunit of an ion channel (Nilius & Honore, 2012; Owsianik et al., 2006). Together with the observation that a continuous central pore exists in the mPiezo1 structure and reconstituted mPiezo1 proteins are sufficient to give rise to RR-sensitive cationic currents, the results described from Sections 7.1–7.3 provide definitive evidence that Piezos are *bona fide* pore-forming proteins, establishing them as the first prototypic class of mammalian MS cation channels.

8. THE MOLECULAR BASIS FOR PIEZO MECHANOTRANSDUCTION

8.1 The mechanogating properties of Piezo channels

Piezo1 can respond to various forms of mechanical stimulation, including poking, suction, and flow-induced shear stress (Coste et al., 2010; Ranade, Qiu, et al., 2014; Yang et al., 2014), and appear to be activated by membrane tension with exquisite sensitivity (Cox et al., 2016; Lewis & Grandl, 2015). Lewis et al. have shown the mechanical sensitivity of Piezo1 can be tuned by cellular membrane tension. They revealed the effect of membrane curvature on Piezo1 channel activity, and determined the tension of half-maximal activation (T_{50}) is $\sim$2.7 mN/m, which is lower than that for MscS and MscL ($T_{50} = \sim 5$ and 10 mN/m, respectively) (Lewis & Grandl, 2015). By characterizing Piezo1 activity in membrane blebs that are largely free of cytoskeleton, Cox et al. (2016) have found that Piezo1 can be activated by membrane tension, leading them to propose that Piezo1 senses force directly transmitted through the bilayer. However, to obtain definitive evidence for this hypothesis, it needs to be demonstrated that Piezo proteins reconstituted into lipid bilayers can directly respond to change in membrane tension, which has not been reported yet.

Nevertheless, given their sophisticated mechanosensitivity, Piezo channels might possess delicate force-sensing and mechanotransduction mechanisms. The striking similarity of the mPiezo1 structure to a three-bladed propeller apparently suggests that Piezo channels might mechanistically function in a similar manner to how a propeller senses and transduces force, which is to employ the peripheral wings as designated mechanotransduction modules to gate the central ion-conducting pore module. It is remarkable that Piezo channels appear to possess all the needed structural components to fulfill their mechanical sensing and transduction roles.

8.2 The flexible blade domain as force sensors?

The extracellular protruding blade regions appear to be ideal for shear stress or other forms of mechanical stimulation. In line with this idea, 3D classification revealed that these domains are highly mobile and occupy variable positions (Ge et al., 2015). The most significant one is that the rotational spacing between adjacent blades could vary between 100 and 140 degrees. Other identifiable conformational variations include the tilt of the Blade relatively to the plasma membrane plane and curvature changes on the helicoidal surface. Given the critical role of Piezo1 in sensing flow-associated shear stress (Ranade, Qiu, et al., 2014), the Blade domains could function as designated flow sensors.

8.3 The extended PH wing for sensing membrane tension?

The 12 apparently resolved PHs from the same protomer are roughly organized as six helical pairs extending from the central axis to the periphery of the complex (Fig. 3B), and connected to the extracellular Blades The PH wings appear not only twisted within the membrane but also tilted out of the membrane at the peripheral end. Cross-sectional area expansion with opening is believed to account for the membrane-tension-dependent energy difference between the open and closed conformations of MS channels. Thus the twisted and tilted PH wings could be involved in sensing membrane tensions by changing the exposure area in the membrane.

8.4 The beam domains for mechanotransduction?

It is remarkable that Piezo1 channels possess three beam-like structures exposed at the intracellular surface (Fig. 3). The Beam is about 80 Å in length and positioned in a 30-degree angle relative to the membrane plane. It originates peripherally at the intracellular side of the PH7-8 pair, and ends near the central axis of the channel complex, where it forms extensive interactions with the Anchor and the CTD (Fig. 3C). This unique organization suggests that the three Beam structures may not only support the whole TM skeleton, but also transmit conformational changes from peripheral TMs and the extracellular Blade regions to the central region where the ion-conduction pore is located. It is also reasonable to speculate that the Beams would be ideal interaction sites for Piezo interacting proteins to modulate their mechanosensitivity. For instance, STOML3 and TRPP2 proteins have been reported to interact with Piezo1 and modulate its function (Peyronnet et al., 2013; Qi et al., 2015).

8.5 The N-terminal non-pore-containing region of mPiezo1 might serves as a sufficient mechanotransduction module

Given that the respective amino acid compositions for the distinct Blade, PH wing, and Blade domains are unclear, it remains challenging to dissect out the proposed roles of these individual domains. To obtain a proof of concept that the peripheral region indeed can function as an integrated mechanotransduction module, we took advantage of the structural similarity of the Piezo1 pore module and ASIC1 channels and engineered a chimeric construct [Piezo1(1-2190)-ASIC1] by replacing the pore region of mPiezo1 with the mechanoinsensitive mouse ASIC1 channel (528 residues) (Zhao et al., 2016). Remarkably, the expressed chimeras became mechanically activated, albeit with reduced mechanosensitivity and maximum MA currents. The MA currents were partially blocked by amiloride at 100 μM concentration that effectively blocked ASIC1 but not Piezo1. Intriguingly, a portion of the Piezo1(1-2190)-ASIC1-transfected cells also responded to acid. Given that Piezo1 channels are instead inhibited by acid (Bae, Sachs, & Gottlieb, 2015), these data collectively suggest that the mechanical force- or acid-induced current in Piezo1(1-2190)-ASIC1-transfected cells is mediated by the ASIC channel pore of the hybrid protein (Zhao et al., 2016).

Thus we have for the first time demonstrated that the non-pore-containing region of mPiezo1 might function as an intrinsic mechanosensing and transducing module to sufficiently gate the trimeric mASIC1 (Zhao et al., 2016). These studies also support that the pore-encoding region of mPiezo1 can function as a separable and switchable pore module that can be replaced by other trimeric channel pores, and represent a remarkable example of engineering artificial ion channels with desired properties in a module-like manner. Interestingly, Zhang et al. have demonstrated that the N-terminal ankyrin repeats of the fly MS no mechanoreceptor potential C (NOMPC) channel are required for its mechanosensitivity (Zhang et al., 2015). Thus despite the lack of sequence or structural similarity between Piezo1 channels and NOMPC channels, both these types of eukaryotic MS cation channels employ separable mechanotransduction and ion-conducting modules to fulfill their specialized mission in mechanical force sensing, gating, and ion conduction.

9. A PROPOSED WORKING MODEL FOR MECHANOSENSITIVE PIEZO CHANNELS

Based on the revelation of the three-bladed, propeller-shaped structure and functional characterization of the separable ion-conducting pore and mechanotransduction modules, we propose that Piezo proteins consist of distinct and separable modules, which are responsible for ion conduction, mechanical force sensing, and transduction to coordinately fulfill their function as sophisticated MS channels (Fig. 5A and B). The extracellular protruding Blade domains and the extending PH wings could be utilized for sensing various mechanical stimulations such as shear stress, poking, and stretching. Motions of the peripheral force-sensing domains could be transmitted to the central ion-conduction pore via the Beam domains that display extensive interactions with the Anchor and CTD domains (Fig. 5C). The putative roles of the distinct domains remain to be determined.

10. CONCLUSION AND PERSPECTIVE

The identification of the MS Piezo family of ion channels no doubt has heralded a new era of research on mechanotransduction mechanisms in mammals. Powered with transgenic animal models, disease linkage studies, site-directed mutagenesis, electrophysiology, and cryo-EM and single particle reconstruction technique for determining complex membrane protein structure, the study of Piezo channels has advanced rapidly as evidenced by validating their physiological and pathophysiological roles, demonstrating their pore-forming and mechanogating properties, determining the 3D structure of mPiezo1, and revealing the molecular bases underlying their ion permeation and mechanotransduction properties.

Despite these remarkable advancements during the past 6 years, many fundamental questions still remain and are positioned to be addressed. For instance, how do the mechanosensitivity and ion permeation properties of Piezo channels determine their in vivo physiological and pathophysiological roles? How does the pore module control ion permeation and selection properties? How do the peripheral mechanotransduction modules gate the central pore module? Does Piezo2 share a similar structural architecture

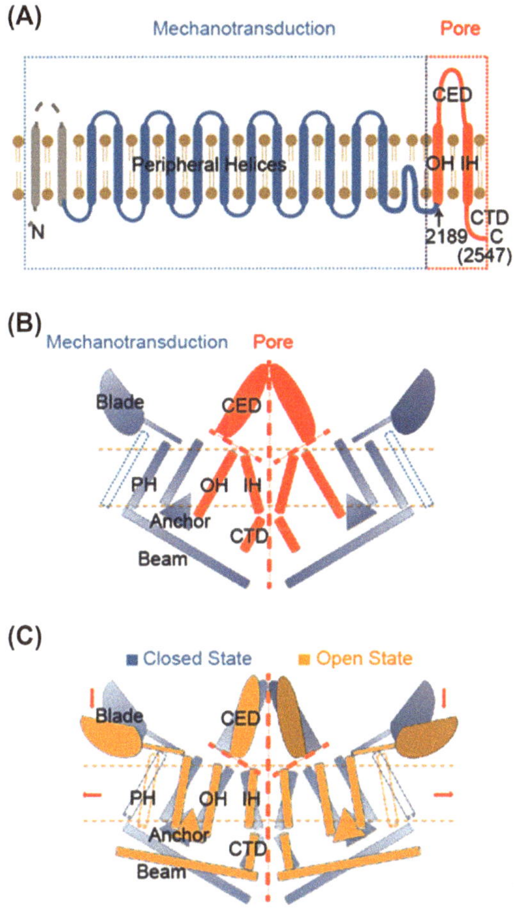

Figure 5 ***A proposed working model for mechanosensitive Piezo channels.*** Illustration of the separable mechanotransduction and pore modules in either a topology model (A) or a structure model (B) of the mPiezo1 channel. (C) A proposed model of the force-induced gating of Piezo channels. The blue and orange models represent the closed and open state channels, respectively. *Red dashed lines* indicate the possible ion-conduction pathways. Presumably, force-induced motion (*red arrows*) of the peripheral blade or peripheral helices (PHs) leads to conformational arrangement and gating of the channel. *CED*, C-terminal extracellular domain; *CTD*, C-terminal domain; *IH*, inner helix; *OH*, outer helix. *Adapted from Ge, J., Li, W., Zhao, Q., Li, N., Chen, M., Zhi, P., … Yang, M. (2015). Architecture of the mammalian mechanosensitive Piezo1 channel.* Nature, *527(7576), 64–69 and Zhao, Q., Wu, K., Geng, J., Chi, S., Wang, Y., Zhi, P., … Xiao B. (2016). Ion permeation and mechanotransduction mechanisms of mechanosensitive piezo channels.* Neuron, *89(6), 1248–1263.*

and working mechanism with Piezo1? Can we succeed in targeting Piezo channels for achieving novel therapeutics? Addressing these questions will definitively lead to a more thorough appreciation of the role of mechanotransduction in health and disease, and the working principles of mammalian MS ion channels.

ACKNOWLEDGMENTS

This work was supported by grants from the National Natural Science Foundation of China (31422027, 31371118, 31630090), the Ministry of Science and Technology (2015CB910102, 2016YFA0500402), and the Ministry of Education (Young Thousand Talent Program) to B.X.

REFERENCES

Akabas, M. H., Stauffer, D. A., Xu, M., & Karlin, A. (1992). Acetylcholine receptor channel structure probed in cysteine-substitution mutants. *Science, 258*(5080), 307–310.

Albuisson, J., Murthy, S. E., Bandell, M., Coste, B., Louis-Dit-Picard, H., Mathur, J., … Patapoutian, A. (2013). Dehydrated hereditary stomatocytosis linked to gain-of-function mutations in mechanically activated PIEZO1 ion channels. *Nature Communications, 4*, 1884.

Andolfo, I., 'Alper, S. L., De Franceschi, L., Auriemma, C., Russo, R., De Falco, L., … Iolascon, A. (2013). Multiple clinical forms of dehydrated hereditary stomatocytosis arise from mutations in PIEZO1. *Blood, 121*(19), 3925–3935. S3921–3912.

Andolfo, I., Russo, R., Manna, F., Shmukler, B. E., Gambale, A., Vitiello, G., … Iolascon, A. (2015). Novel Gardos channel mutations linked to dehydrated hereditary stomatocytosis (xerocytosis). *American Journal of Hematology, 90*(10), 921–926.

Bae, C., Gnanasambandam, R., Nicolai, C., Sachs, F., & Gottlieb, P. A. (2013). Xerocytosis is caused by mutations that alter the kinetics of the mechanosensitive channel PIEZO1. *Proceedings of the National Academy of Sciences of the United States of America, 110*(12), E1162–E1168.

Bae, C., Sachs, F., & Gottlieb, P. A. (2011). The mechanosensitive ion channel Piezo1 is inhibited by the peptide GsMTx4. *Biochemistry, 50*(29), 6295–6300.

Bae, C., Sachs, F., & Gottlieb, P. A. (2015). Protonation of the human PIEZO1 ion channel stabilizes inactivation. *Journal of Biological Chemistry, 290*(8), 5167–5173.

Bagriantsev, S. N., Gracheva, E. O., & Gallagher, P. G. (2014). Piezo proteins: Regulators of mechanosensation and other cellular processes. *Journal of Biological Chemistry, 289*(46), 31673–31681.

Bass, R. B., Strop, P., Barclay, M., & Rees, D. C. (2002). Crystal structure of *Escherichia coli* MscS, a voltage-modulated and mechanosensitive channel. *Science, 298*(5598), 1582–1587.

Beneteau, C., Thierry, G., Blesson, S., Le Vaillant, C., Picard, V., Bene, M. C., … Le Caignec, C. (2014). Recurrent mutation in the PIEZO1 gene in two families of hereditary xerocytosis with fetal hydrops. *Clinical Genetics, 85*(3), 293–295.

Brohawn, S. G., del Marmol, J., & MacKinnon, R. (2012). Crystal structure of the human K2P TRAAK, a lipid- and mechano-sensitive K+ ion channel. *Science, 335*(6067), 436–441.

Brohawn, S. G., Su, Z., & MacKinnon, R. (2014). Mechanosensitivity is mediated directly by the lipid membrane in TRAAK and TREK1 K+ channels. *Proceedings of the National Academy of Sciences of the United States of America, 111*(9), 3614–3619.

Brown, A. L., Liao, Z., & Goodman, M. B. (2008). MEC-2 and MEC-6 in the *Caenorhabditis elegans* sensory mechanotransduction complex: Auxiliary subunits that enable channel activity. *The Journal of General Physiology, 131*(6), 605–616.

Brownell, W. E., Bader, C. R., Bertrand, D., & de Ribaupierre, Y. (1985). Evoked mechanical responses of isolated cochlear outer hair cells. *Science, 227*(4683), 194–196.

Cahalan, S. M., Lukacs, V., Ranade, S. S., Chien, S., Bandell, M., & Patapoutian, A. (2015). Piezo1 links mechanical forces to red blood cell volume. *Elife, 4*.

Cao, E., Liao, M., Cheng, Y., & Julius, D. (2013). TRPV1 structures in distinct conformations reveal activation mechanisms. *Nature, 504*(7478), 113–118.

Chang, G., Spencer, R. H., Lee, A. T., Barclay, M. T., & Rees, D. C. (1998). Structure of the MscL homolog from *Mycobacterium tuberculosis*: A gated mechanosensitive ion channel. *Science, 282*(5397), 2220–2226.

Cheng, L. E., Song, W., Looger, L. L., Jan, L. Y., & Jan, Y. N. (2010). The role of the TRP channel NompC in *Drosophila* larval and adult locomotion. *Neuron, 67*(3), 373–380.

Corey, D. P., & Hudspeth, A. J. (1979). Ionic basis of the receptor potential in a vertebrate hair cell. *Nature, 281*(5733), 675–677.

Coste, B., Mathur, J., Schmidt, M., Earley, T. J., Ranade, S., Petrus, M. J., … Patapoutian, A. (2010). Piezo1 and Piezo2 are essential components of distinct mechanically activated cation channels. *Science, 330*(6000), 55–60.

Coste, B., Murthy, S. E., Mathur, J., Schmidt, M., Mechioukhi, Y., Delmas, P., & Patapoutian, A. (2015). Piezo1 ion channel pore properties are dictated by C-terminal region. *Nature Communications, 6*, 7223.

Coste, B., Xiao, B., Santos, J. S., Syeda, R., Grandl, J., Spencer, K. S., … Patapoutian, A. (2012). Piezo proteins are pore-forming subunits of mechanically activated channels. *Nature, 483*(7388), 176–181.

Cox, C. D., Bae, C., Ziegler, L., Hartley, S., Nikolova-Krstevski, V., Rohde, P. R., … Martinac, B. (2016). Removal of the mechanoprotective influence of the cytoskeleton reveals PIEZO1 is gated by bilayer tension. *Nature Communications, 7*, 10366.

Cunningham, J. T., Wachtel, R. E., & Abboud, F. M. (1995). Mechanosensitive currents in putative aortic baroreceptor neurons in vitro. *Journal of Neurophysiology, 73*(5), 2094–2098.

Davis, M. J., Donovitz, J. A., & Hood, J. D. (1992). Stretch-activated single-channel and whole cell currents in vascular smooth muscle cells. *American Journal of Physiology, 262*(4 Pt 1), C1083–C1088.

Delmas, P., Hao, J., & Rodat-Despoix, L. (2011). Molecular mechanisms of mechanotransduction in mammalian sensory neurons. *Nature Reviews. Neuroscience, 12*(3), 139–153.

Dong, Y. Y., Pike, A. C., Mackenzie, A., McClenaghan, C., Aryal, P., Dong, L., … Carpenter, E. P. (2015). K2P channel gating mechanisms revealed by structures of TREK-2 and a complex with Prozac. *Science, 347*(6227), 1256–1259.

Doyle, D. A., Morais Cabral, J., Pfuetzner, R. A., Kuo, A., Gulbis, J. M., Cohen, S. L., … MacKinnon, R. (1998). The structure of the potassium channel: Molecular basis of K+ conduction and selectivity. *Science, 280*(5360), 69–77.

Dubin, A. E., Schmidt, M., Mathur, J., Petrus, M. J., Xiao, B., Coste, B., & Patapoutian, A. (2012). Inflammatory signals enhance piezo2-mediated mechanosensitive currents. *Cell Reports, 2*(3), 511–517 [Electronic Resource].

Eijkelkamp, N., Linley, J. E., Torres, J. M., Bee, L., Dickenson, A. H., Gringhuis, M., … Wood, J. N. (2013). A role for Piezo2 in EPAC1-dependent mechanical allodynia. *Nature Communications, 4*, 1682.

Faucherre, A., Kissa, K., Nargeot, J., Mangoni, M. E., & Jopling, C. (2014). Piezo1 plays a role in erythrocyte volume homeostasis. *Haematologica, 99*(1), 70–75.

Faucherre, A., Nargeot, J., Mangoni, M. E., & Jopling, C. (2013). piezo2b regulates vertebrate light touch response. *Journal of Neuroscience, 33*(43), 17089–17094.

Fotiou, E., Martin-Almedina, S., Simpson, M. A., Lin, S., Gordon, K., Brice, G., … Ostergaard, P. (2015). Novel mutations in PIEZO1 cause an autosomal recessive generalized lymphatic dysplasia with non-immune hydrops fetalis. *Nature Communications, 6*, 8085.

Franco-Obregon, A., Jr., & Lansman, J. B. (1994). Mechanosensitive ion channels in skeletal muscle from normal and dystrophic mice. *Journal of Physiology, 481*(Pt 2), 299–309.

Ge, J., Li, W., Zhao, Q., Li, N., Chen, M., Zhi, P., … Yang, M. (2015). Architecture of the mammalian mechanosensitive Piezo1 channel. *Nature, 527*(7576), 64–69.

Gnanasambandam, R., Bae, C., Gottlieb, P. A., & Sachs, F. (2015). Ionic selectivity and permeation properties of human PIEZO1 channels. *PLoS One, 10*(5), e0125503.

Gonzales, E. B., Kawate, T., & Gouaux, E. (2009). Pore architecture and ion sites in acid-sensing ion channels and P2X receptors. *Nature, 460*(7255), 599–604.

Guharay, F., & Sachs, F. (1984). Stretch-activated single ion channel currents in tissue-cultured embryonic chick skeletal muscle. *Journal of Physiology, 352*, 685–701.

Honore, E., Martins, J. R., Penton, D., Patel, A., & Demolombe, S. (2015). The piezo mechanosensitive ion channels: May the force Be with You! *Reviews of Physiology, Biochemistry and Pharmacology, 169*, 25–41.

Hou, X., Pedi, L., Diver, M. M., & Long, S. B. (2012). Crystal structure of the calcium release-activated calcium channel Orai. *Science, 338*(6112), 1308–1313.

Ikeda, R., Cha, M., Ling, J., Jia, Z., Coyle, D., & Gu, J. G. (2014). Merkel cells transduce and encode tactile stimuli to drive Abeta-afferent impulses. *Cell, 157*(3), 664–675.

Imashuku, S., Muramatsu, H., Sugihara, T., Okuno, Y., Wang, X., Yoshida, K., … Kanno, H. (2016). PIEZO1 gene mutation in a Japanese family with hereditary high phosphatidylcholine hemolytic anemia and hemochromatosis-induced diabetes mellitus. *International Journal of Hematology, 104*(1), 125–129.

Kamajaya, A., Kaiser, J. T., Lee, J., Reid, M., & Rees, D. C. (2014). The structure of a conserved piezo channel domain reveals a topologically distinct beta sandwich fold. *Structure, 22*(10), 1520–1527.

Kang, L., Gao, J., Schafer, W. R., Xie, Z., & Xu, X. Z. (2010). *C. elegans* TRP family protein TRP-4 is a pore-forming subunit of a native mechanotransduction channel. *Neuron, 67*(3), 381–391.

Kawate, T., Michel, J. C., Birdsong, W. T., & Gouaux, E. (2009). Crystal structure of the ATP-gated P2X(4) ion channel in the closed state. *Nature, 460*(7255), 592–598.

Kim, S. E., Coste, B., Chadha, A., Cook, B., & Patapoutian, A. (2012). The role of *Drosophila* Piezo in mechanical nociception. *Nature, 483*(7388), 209–212.

Lewis, A. H., & Grandl, J. (2015). Mechanical sensitivity of Piezo1 ion channels can be tuned by cellular membrane tension. *Elife, 4*.

Li, J., Hou, B., Tumova, S., Muraki, K., Bruns, A., Ludlow, M. J., … Beech, D. J. (2014). Piezo1 integration of vascular architecture with physiological force. *Nature, 515*(7526), 279–282.

Li, T., Yang, Y., & Canessa, C. M. (2011). Outlines of the pore in open and closed conformations describe the gating mechanism of ASIC1. *Nature Communications, 2*, 399.

Li, W., Feng, Z., Sternberg, P. W., & Xu, X. Z. (2006). A *C. elegans* stretch receptor neuron revealed by a mechanosensitive TRP channel homologue. *Nature, 440*(7084), 684–687.

Liao, M., Cao, E., Julius, D., & Cheng, Y. (2013). Structure of the TRPV1 ion channel determined by electron cryo-microscopy. *Nature, 504*(7478), 107–112.

Liao, M., Cao, E., Julius, D., & Cheng, Y. (2014). Single particle electron cryo-microscopy of a mammalian ion channel. *Current Opinion in Structural Biology, 27*, 1–7.

Liu, Z., Gandhi, C. S., & Rees, D. C. (2009). Structure of a tetrameric MscL in an expanded intermediate state. *Nature, 461*(7260), 120–124.

Lukacs, V., Mathur, J., Mao, R., Bayrak-Toydemir, P., Procter, M., Cahalan, S. M., ... Krock, B. L. (2015). Impaired PIEZO1 function in patients with a novel autosomal recessive congenital lymphatic dysplasia. *Nature Communications, 6*, 8329.

Lumpkin, E. A., Marshall, K. L., & Nelson, A. M. (2010). The cell biology of touch. *Journal of Cell Biology, 191*(2), 237–248.

Maingret, F., Patel, A. J., Lesage, F., Lazdunski, M., & Honore, E. (2000). Lysophospholipids open the two-pore domain mechano-gated K(+) channels TREK-1 and TRAAK. *Journal of Biological Chemistry, 275*(14), 10128–10133.

Martinac, B., Adler, J., & Kung, C. (1990). Mechanosensitive ion channels of *E. coli* activated by amphipaths. *Nature, 348*(6298), 261–263.

Martins, J. R., Penton, D., Peyronnet, R., Arhatte, M., Moro, C., Picard, N., ... Demolombe, S. (2016). Piezo1-dependent regulation of urinary osmolarity. *Pflugers Arch, 468*(7), 1197–1206.

McCarter, G. C., Reichling, D. B., & Levine, J. D. (1999). Mechanical transduction by rat dorsal root ganglion neurons in vitro. *Neuroscience Letters, 273*(3), 179–182.

McHugh, B. J., Murdoch, A., Haslett, C., & Sethi, T. (2012). Loss of the integrin-activating transmembrane protein Fam38A (Piezo1) promotes a switch to a reduced integrin-dependent mode of cell migration. *PLoS One, 7*(7), e40346.

McMillin, M. J., Beck, A. E., Chong, J. X., Shively, K. M., Buckingham, K. J., Gildersleeve, H. I., ... Bamshad, M. J. (2014). Mutations in PIEZO2 cause Gordon syndrome, Marden-Walker syndrome, and distal arthrogryposis type 5. *American Journal of Human Genetics, 94*(5), 734–744.

Miyamoto, T., Mochizuki, T., Nakagomi, H., Kira, S., Watanabe, M., Takayama, Y., ... Tominaga, M. (2014). Functional role for Piezo1 in stretch-evoked Ca(2)(+) influx and ATP release in urothelial cell cultures. *Journal of Biological Chemistry, 289*(23), 16565–16575.

Nilius, B., & Honore, E. (2012). Sensing pressure with ion channels. *Trends in Neurosciences, 35*(8), 477–486.

O'Hagan, R., Chalfie, M., & Goodman, M. B. (2005). The MEC-4 DEG/ENaC channel of *Caenorhabditis elegans* touch receptor neurons transduces mechanical signals. *Nature Neuroscience, 8*(1), 43–50.

Owsianik, G., Talavera, K., Voets, T., & Nilius, B. (2006). Permeation and selectivity of TRP channels. *Annual Review of Physiology, 68*, 685–717.

Pan, B., Geleoc, G. S., Asai, Y., Horwitz, G. C., Kurima, K., Ishikawa, K., ... Holt, J. R. (2013). TMC1 and TMC2 are components of the mechanotransduction channel in hair cells of the mammalian inner ear. *Neuron, 79*(3), 504–515.

Patel, A. J., Honore, E., Maingret, F., Lesage, F., Fink, M., Duprat, F., & Lazdunski, M. (1998). A mammalian two pore domain mechano-gated S-like K+ channel. *EMBO Journal, 17*(15), 4283–4290.

Pathak, M. M., Nourse, J. L., Tran, T., Hwe, J., Arulmoli, J., Le, D. T., ... Tombola, F. (2014). Stretch-activated ion channel Piezo1 directs lineage choice in human neural stem cells. *Proceedings of the National Academy of Sciences of the United States of America, 111*(45), 16148–16153.

Paulsen, C. E., Armache, J. P., Gao, Y., Cheng, Y., & Julius, D. (2015). Structure of the TRPA1 ion channel suggests regulatory mechanisms. *Nature, 520*(7548), 511–517.

Payandeh, J., Scheuer, T., Zheng, N., & Catterall, W. A. (2011). The crystal structure of a voltage-gated sodium channel. *Nature, 475*(7356), 353–358.

Penna, A., Demuro, A., Yeromin, A. V., Zhang, S. L., Safrina, O., Parker, I., & Cahalan, M. D. (2008). The CRAC channel consists of a tetramer formed by Stim-induced dimerization of Orai dimers. *Nature, 456*(7218), 116–120.

Peyronnet, R., Martins, J. R., Duprat, F., Demolombe, S., Arhatte, M., Jodar, M., ... Patel, A. (2013). Piezo1-dependent stretch-activated channels are inhibited by Polycystin-2 in renal tubular epithelial cells. *EMBO Reports, 14*(12), 1143–1148.

Prakriya, M., Feske, S., Gwack, Y., Srikanth, S., Rao, A., & Hogan, P. G. (2006). Orai1 is an essential pore subunit of the CRAC channel. *Nature, 443*(7108), 230–233.

Prole, D. L., & Taylor, C. W. (2013). Identification and analysis of putative homologues of mechanosensitive channels in pathogenic protozoa. *PLoS One, 8*(6), e66068.

Qi, Y., Andolfi, L., Frattini, F., Mayer, F., Lazzarino, M., & Hu, J. (2015). Membrane stiffening by STOML3 facilitates mechanosensation in sensory neurons. *Nature Communications, 6*, 8512.

Ranade, S. S., Qiu, Z., Woo, S. H., Hur, S. S., Murthy, S. E., Cahalan, S. M., ... Patapoutian, A. (2014). Piezo1, a mechanically activated ion channel, is required for vascular development in mice. *Proceedings of the National Academy of Sciences of the United States of America, 111*(28), 10347–10352.

Ranade, S. S., Syeda, R., & Patapoutian, A. (2015). Mechanically activated ion channels. *Neuron, 87*(6), 1162–1179.

Ranade, S. S., Woo, S. H., Dubin, A. E., Moshourab, R. A., Wetzel, C., Petrus, M., ... Patapoutian, A. (2014). Piezo2 is the major transducer of mechanical forces for touch sensation in mice. *Nature, 516*(7529), 121–125.

Retailleau, K., Duprat, F., Arhatte, M., Ranade, S. S., Peyronnet, R., Martins, J. R., ... Honoré, E. (2015). Piezo1 in smooth muscle cells is involved in hypertension-dependent arterial remodeling. *Cell Reports, 13*(6), 1161–1171 [Electronic Resource].

Sandberg, M. B., Nybo, M., Birgens, H., & Frederiksen, H. (2014). Hereditary xerocytosis and familial haemolysis due to mutation in the PIEZO1 gene: A simple diagnostic approach. *International Journal of Laboratory Hematology, 36*(4), e62–65.

Schneider, E. R., Mastrotto, M., Laursen, W. J., Schulz, V. P., Goodman, J. B., Funk, O. H., ... Bagriantsev, S. N. (2014). Neuronal mechanism for acute mechanosensitivity in tactile-foraging waterfowl. *Proceedings of the National Academy of Sciences of the United States of America, 111*(41), 14941–14946.

Schrenk-Siemens, K., Wende, H., Prato, V., Song, K., Rostock, C., Loewer, A., ... Siemens, J. (2015). PIEZO2 is required for mechanotransduction in human stem cell-derived touch receptors. *Nature Neuroscience, 18*(1), 10–16.

Shmukler, B. E., Vandorpe, D. H., Rivera, A., Auerbach, M., Brugnara, C., & Alper, S. L. (2014). Dehydrated stomatocytic anemia due to the heterozygous mutation R2456H in the mechanosensitive cation channel PIEZO1: A case report. *Blood Cells, Molecules and Diseases, 52*(1), 53–54.

Sidi, S., Friedrich, R. W., & Nicolson, T. (2003). NompC TRP channel required for vertebrate sensory hair cell mechanotransduction. *Science, 301*(5629), 96–99.

Singhmar, P., Huo, X., Eijkelkamp, N., Berciano, S. R., Baameur, F., Mei, F. C., ... Kavelaars, A. (2016). Critical role for Epac1 in inflammatory pain controlled by GRK2-mediated phosphorylation of Epac1. *Proceedings of the National Academy of Sciences of the United States of America, 113*(11), 3036–3041.

Sukharev, S. I., Blount, P., Martinac, B., Blattner, F. R., & Kung, C. (1994). A large-conductance mechanosensitive channel in *E. coli* encoded by mscL alone. *Nature, 368*(6468), 265–268.

Szklarczyk, R., Wanschers, B. F., Nijtmans, L. G., Rodenburg, R. J., Zschocke, J., Dikow, N., ... van den Heuvel, L. P. (2013). A mutation in the FAM36A gene, the human ortholog of COX20, impairs cytochrome c oxidase assembly and is associated with ataxia and muscle hypotonia. *Human Molecular Genetics, 22*(4), 656–667.

Walker, R. G., Willingham, A. T., & Zuker, C. S. (2000). A *Drosophila* mechanosensory transduction channel. *Science, 287*(5461), 2229–2234.

Woo, S. H., Lukacs, V., de Nooij, J. C., Zaytseva, D., Criddle, C. R., Francisco, A., … Patapoutian, A. (2015). Piezo2 is the principal mechanotransduction channel for proprioception. *Nature Neuroscience, 18*(12), 1756–1762.

Woo, S. H., Lumpkin, E. A., & Patapoutian, A. (2015). Merkel cells and neurons keep in touch. *Trends in Cell Biology, 25*(2), 74–81.

Yan, Z., Zhang, W., He, Y., Gorczyca, D., Xiang, Y., Cheng, L. E., … Jan, Y. N. (2013). *Drosophila* NOMPC is a mechanotransduction channel subunit for gentle-touch sensation. *Nature, 493*(7431), 221–225.

Yang, J., Ellinor, P. T., Sather, W. A., Zhang, J. F., & Tsien, R. W. (1993). Molecular determinants of Ca2+ selectivity and ion permeation in L-type Ca2+ channels. *Nature, 366*(6451), 158–161.

Yang, X. N., Lu, Y. P., Liu, J. J., Huang, J. K., Liu, Y. P., Xiao, C. X., … Guleng, B. (2014). Piezo1 is as a novel trefoil factor family 1 binding protein that promotes gastric cancer cell mobility in vitro. *Digestive Diseases and Sciences, 59*(7), 1428–1435.

Yeromin, A. V., Zhang, S. L., Jiang, W., Yu, Y., Safrina, O., & Cahalan, M. D. (2006). Molecular identification of the CRAC channel by altered ion selectivity in a mutant of Orai. *Nature, 443*(7108), 226–229.

Zarychanski, R., Schulz, V. P., Houston, B. L., Maksimova, Y., Houston, D. S., Smith, B., … Gallagher, P. G. (2012). Mutations in the mechanotransduction protein PIEZO1 are associated with hereditary xerocytosis. *Blood, 120*(9), 1908–1915.

Zhang, W., Cheng, L. E., Kittelmann, M., Li, J., Petkovic, M., Cheng, T., … Jan, Y. N. (2015). Ankyrin repeats convey force to gate the NOMPC mechanotransduction channel. *Cell, 162*(6), 1391–1403.

Zhang, X., Ren, W., DeCaen, P., Yan, C., Tao, X., Tang, L., … Yan, N. (2012). Crystal structure of an orthologue of the NaChBac voltage-gated sodium channel. *Nature, 486*(7401), 130–134.

Zhao, Q., Wu, K., Geng, J., Chi, S., Wang, Y., Zhi, P., … Xiao, B. (2016). Ion permeation and mechanotransduction mechanisms of mechanosensitive piezo channels. *Neuron, 89*(6), 1248–1263.

Zhong, L., Hwang, R. Y., & Tracey, W. D. (2010). Pickpocket is a DEG/ENaC protein required for mechanical nociception in *Drosophila* larvae. *Current Biology: CB, 20*(5), 429–434.

CHAPTER SEVEN

Piezo2 in Cutaneous and Proprioceptive Mechanotransduction in Vertebrates[a]

E.O. Anderson, E.R. Schneider and S.N. Bagriantsev[1]
Yale University, New Haven, CT, United States
[1]Corresponding author: E-mail: slav.bagriantsev@yale.edu

Contents

Abstract

Mechanosensitivity is a fundamental physiological capacity, which pertains to all life forms. Progress has been made with regard to understanding mechanosensitivity in bacteria, flies, and worms. In vertebrates, however, the molecular identity of mechanotransducers in

[a] This work was supported by grants from National Science Foundation (1453167), National Institutes of Health (1R01NS097547-01A1) and American Heart Association (14SDG17880015) to S.N.B. E.R.S. was partially supported by a training grant from National Institutes of Health T32HD007094 and a postdoctoral fellowship from the Arnold and Mabel Beckman Foundation. E.O.A. is a fellow of The Gruber Foundation and an Edward L. Tatum Fellow. Correspondence should be addressed to S.N.B (slav.bagriantsev@yale.edu).

Current Topics in Membranes, Volume 79
ISSN 1063-5823
http://dx.doi.org/10.1016/bs.ctm.2016.11.002

somatic and neuronal cells has only started to appear. The Piezo family of mechanogated ion channels marks a pivotal milestone in understanding mechanosensitivity. Piezo1 and Piezo2 have now been shown to participate in a number of processes, ranging from arterial modeling to sensing muscle stretch. In this review, we focus on Piezo2 and its role in mediating mechanosensation and proprioception in vertebrates.

1. INTRODUCTION

Piezo2 was identified by homology with Piezo1, a mechanogated ion channel cloned by the Ardem Patapoutian group from the neuroblastoma N2A cells (Coste et al., 2010). The two molecules are the only known members of an evolutionary conserved family of proteins and are among the largest ion channels cloned to date. Mouse Piezo1 and Piezo2 are ~2500 and 2800 amino acids long nonselective cation channels with ~42% identity. Though structural information on Piezo2 is not available, predictions based on functional analysis and the cryo-EM structure of Piezo1 suggest that functional Piezo2 is formed by three subunits with 16–18 transmembrane helixes (Fig. 1) (Coste et al., 2015; Ge et al., 2015; Zhao et al., 2016).

Piezo homologs are found in plants, unicellular eukaryotes (but not yeast), and invertebrates, but the distinction between Piezo1 and Piezo2 begins in vertebrates, where the two proteins carry out a number of mechanosensory functions in various cells and tissues. Known Piezo1 functions pertain to

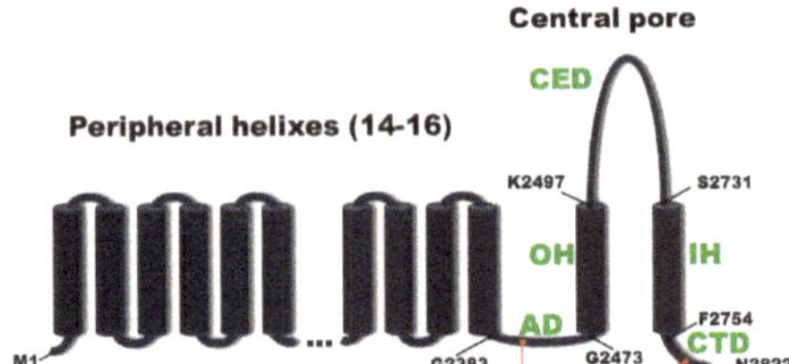

Figure 1 ***Hypothetical topology of Piezo2.*** Shown is a hypothetical transmembrane topology diagram of mouse Piezo2 based on the cryo-EM (Ge et al., 2015) and functional studies of mouse Piezo1 (Coste et al., 2015; Zhao et al., 2016). *AD,* anchor domain; *CED,* C-terminal extracellular domain; *CTD,* C-terminal domain; *IH,* inner helix; *OH,* outer helix. E2416 in the anchor domain alters pore properties of mouse Piezo2 (Coste et al., 2015). E2797 in the C-terminal domain is homologous to E2727 in human Piezo2. The deletion of E2727 in hPiezo2 prolongs kinetics of MA current inactivation in vitro (Dubin et al., 2012).

mechanotransduction in different types of somatic cells and has been extensively reviewed elsewhere (Bagriantsev, Gracheva, & Gallagher, 2014; Glokowska & Gallagher, 2015; Honore, Martins, Penton, Patel, & Demolombe, 2015; Ranade, Syeda, & Patapoutian, 2015; Volkers, Mechioukhi, & Coste, 2015; Wu, Lewis, & Grandl, 2016). Similar to Piezo1, Piezo2 is widely expressed in somatic cells: chondrocytes (Lee et al., 2014), odontoblasts (Khatibi Shahidi et al., 2015), endothelial cells (Ferrari, Bogen, Green, & Levine, 2015), and astrocytes (Choi, Sun, & Jakobs, 2015). In contrast to Piezo1, Piezo2 is also expressed in Merkel cells (Ikeda et al., 2014; Maksimovic et al., 2014; Woo et al., 2014), outer hair cells (Wu, Grillet, et al., 2016), endothelial cells in the brain (Wu, Lewis, et al., 2016), enterochromaffin cells of the gut (Wang et al., 2016), and in the neurons of the somatosensory ganglia, where it plays a key role in mechanosensation and proprioception.

Somatosensory ganglia house a very diverse array of somatosensory neurons, which project to the skin of the face (trigeminal ganglia, TG) or body (dorsal root ganglia, DRG) to detect the principal components of cutaneous senses: temperature, chemical irritants, painful touch, and light touch (Le Pichon & Chesler, 2014). In addition, the DRG houses proprioceptive neurons innervating muscle. Proprioceptors are absent from TG and reside in the mesencephalic ganglion in the brainstem (Lazarov, 2007). In most vertebrates, the vast majority of TG and DRG neurons are nonmyelinated (C-type) nociceptors and thermoreceptors, which detect temperature and painful chemical and mechanical stimuli. Light touch is detected by low-threshold mechanoreceptors (LTMRs). Most LTMRs are myelinated Aβ-fibers, but there are also thinly myelinated Aδ- and nonmyelinated C-fibers (Zimmerman, Bai, & Ginty, 2014). Aβ-LTMRs can be further segregated by the firing pattern they produce in response to mechanical stimulation: rapidly adapting mechanoreceptors (RA-LTMRs), which only fire with onset and offset of a stimulus; and slowly adapting mechanoreceptors (SA-LTMRs), which fire for the duration of a mechanical stimulus (Fleming & Luo, 2013). Mechanosensitive channels in the nerve terminals of LTMRs trigger action potential generation in these cells (Ranade et al., 2015; Sharif-Naeini, 2015). Recent studies determined that Piezo2 is a key mechanotransducer in a subset of LTMRs and in proprioceptors. Here, we review the role of Piezo2 in cutaneous and proprioceptive mechanotransduction of vertebrates.

2. SOMATOSENSORY NEURONS

2.1 Piezo2 and fast mechanoactivated current in mouse dorsal root ganglia neurons

Understanding mechanosensation in the different types of somatosensory end organs (Merkel cell–neurite complexes, Meissner and Pacinian corpuscles and others; see Zimmerman et al., 2014) requires the identification of the molecular mechanism that translates tissue deformation into action potential firing in the corresponding LTMR. This necessitates knowledge of the molecules that mediate mechanoactivated current, modulate kinetics of inactivation, generate action potentials, and shape the pattern of firing adaptation. In addition, this requires understanding the contribution from the somatic components of the end organs, such as Merkel cells in Merkel cell–neurite complexes or lamellar cells in Pacinian corpuscles. Recently, a great breakthrough was made with regard to the role of Piezo2 in Merkel cell–neurite complexes (Ikeda et al., 2014; Maksimovic et al., 2014; Woo et al., 2014), but by and large the molecular basis of peripheral neuronal mechanosensitivity remains poorly understood.

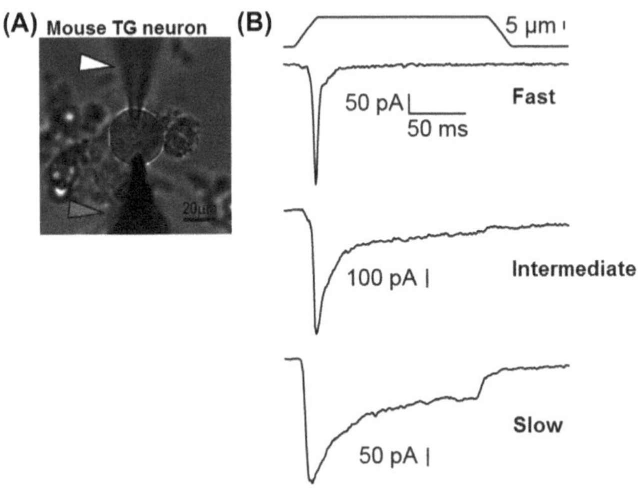

Figure 2 ***Mechanoactivated currents in mouse trigeminal neurons.*** (A) An image of a mouse trigeminal (TG) neuron with an electrode (*white arrowhead*) and blunt probe (*blue arrowhead*) in the working position to record MA current. (B) Representative whole cell currents from mouse trigeminal neurons showing exemplar fast, intermediate, and slow MA currents based on the rate of exponential current decay (τ). Currents were obtained from a holding potential of −60 mV, using a probe moving at 800 µm/s velocity. (For experimental details, see Schneider et al., 2014).

Somatosensory neurons are innately mechanosensitive, i.e., they can convert mechanical stimuli into excitatory current and mechanoactivated action potentials even in the absence of other tissue components. In culture, mechanosensitivity of dissociated TG and DRG neurons is often studied by patch clamp recording in the voltage clamp mode (McCarter, Reichling, & Levine, 1999). Stimulation with a blunt glass probe evokes mechanoactivated excitatory current (MA current, Fig. 2), a necessary prerequisite for the generation of action potentials.

Different neurons produce specific types of MA currents, which are thought to be mediated by more than one mechanogated ion channel, judging from ion selectivity, pharmacological and kinetic characteristics (Coste, Crest, & Delmas, 2007; Drew et al., 2007; Drew & Wood, 2007; Hu & Lewin, 2006; McCarter & Levine, 2006; Poole, Herget, Lapatsina, Ngo, & Lewin, 2014; Qi et al., 2015). A salient feature of all MA currents is the time constant of near-exponential decay (τ), which occurs immediately after initial channel opening. A detailed analysis of the decay kinetics reveals a complex underlying mechanism, which involves adaptation and inactivation (Hao & Delmas, 2010; Rugiero, Drew, & Wood, 2010). Traditionally, most studies classify neuronal MA currents as fast (or rapid, $\tau < 10$ ms), intermediate (10 ms $< \tau < 30$ ms), and slow ($\tau > 30$ ms) adapting (Fig. 2B). Here, we will use the term "inactivation" with regard to MA current (fast, intermediately, or slowly inactivating) and "adaptation" to refer to the pattern of action potential firing (rapidly or slowly adapting).

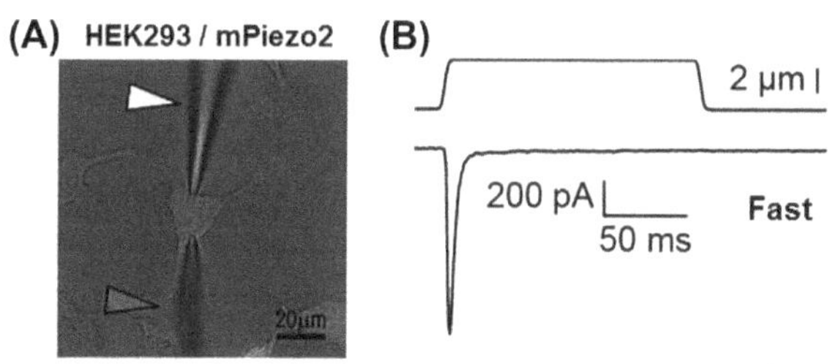

Figure 3 ***Mouse Piezo2 generates fast inactivating MA current in HEK293 cells.*** (A) An image of an HEK293 cell expressing mouse Piezo2 during voltage clamp recording to obtain MA current in response to mechanical stimulation with a glass probe. *White arrowhead*: electrode; *blue arrowhead*: mechanical probe. (B) Representative fast MA current ($\tau < 10$ ms) trace from mPiezo2 in HEK293 cell obtained in the whole cell configuration from a holding potential of −60 mV. Currents were obtained from a holding potential of −80 mV, using a probe moving at 800 μm/s velocity. Solutions used (mM): bath, 140 NaCl, 5 KCl, 2.5 $CaCl_2$, 1 $MgCl_2$, 10 glucose, 10 HEPES/NaOH pH 7.4; pipette: 133CsCl, 5 EGTA, 1 $CaCl_2$, 1 $MgCl_2$, 4 MgATP, 0.4 Na_2GTP, 10 HEPES/CsOH pH 7.3.

Piezo2 was the first ion channel firmly linked to an MA current in somatosensory neurons. Piezo2 is expressed in 20–50% of mammalian DRG and TG neurons, including LTMRs and nociceptors (Alamri, Bron, Brock, & Ivanusic, 2015; Bron, Wood, Brock, & Ivanusic, 2014; Coste et al., 2010; Lou, Duan, Vong, Lowell, & Ma, 2013; Ranade et al., 2014). In one neuronal subset, C-type LTMRs, Piezo2 expression was shown to rely on the transcriptional factor Runx1 (Lou et al., 2013). Fast MA current generated by a subset of mouse DRG neurons is remarkably similar to the MA current produced by Piezo2 in HEK293T cells (Fig. 3), both in terms of kinetics and pharmacology.

Small interfering RNA-mediated knockdown of Piezo2 in cultured DRG neurons leads to a specific reduction of most of fast MA current, demonstrating that it is Piezo2 dependent (Coste et al., 2010; Lou et al., 2013). These data are corroborated through DRG-specific Piezo2 knockouts (Ranade et al., 2014; Woo et al., 2015). Thus, Piezo2 mediates most, if not all, fast MA current in mouse DRG neurons (Coste et al., 2010; Ranade et al., 2014; Woo et al., 2015), but the molecular identity of intermediate and slow MA current is yet to be established.

2.2 Possible role of Piezo2 in slow mechanoactivated current

Functional studies suggested that fast MA current is largely (but not exclusively) present in mechanoreceptors, while slow MA current is present in nociceptors (Coste et al., 2007; Drew et al., 2007; Lechner, Frenzel, Wang, & Lewin, 2009; Poole et al., 2014). Overall, Aδ- and Aβ-type LTMRs are rare cells in rodent DRG, where 60–70% of neurons are C-type nociceptors and thermoreceptors (Kobayashi et al., 2005; Le Pichon & Chesler, 2014). Some LTMR subtypes account for only a few percent of the total neuronal population, which may obscure their identification in vitro by electrophysiological analyses. For example, RA-LTMRs account for mere 6% of all mouse thoracic DRG neurons (Li et al., 2011), yet they innervate a host of mechanosensitive end organs, including lanceolate endings in the hairy skin, Meissner and Pacinian corpuscles (Zimmerman et al., 2014). It is therefore plausible that such rare neurons will avoid a definitive classification in an en masse electrophysiological analysis of dissociated DRG neurons, leaving the possibility that some types of LTMRs express mechanotransducers with intermediate or slow kinetics of inactivation. In support of this, studies of cat mesenteric

Pacinian corpuscles showed that mechanical stimulation of the corresponding LTMRs that have had the somatic components of the end organ manually removed evokes a slowly decaying receptor potential, indicating the presence of a mechanotransducer with slow MA current (Loewenstein & Mendelson, 1965; Mendelson & Lowenstein, 1964). In duck TG, sensory ganglia with an unusually high population of LTMRs innervating Pacinian-like corpuscles in the bill, most neurons express Piezo2 and exhibit slow MA current (Schneider et al., 2014). Thus, it is possible that RA-LTMRs innervating Pacinian corpuscles are subserved by mechanotransducers with slow MA current, possibly mediated by Piezo2. In support of this, a recent study showed that people carrying non-functional Piezo2 alleles display general losses in vibration detection in the frequency range perceived by Pacinian corpuscles (Chesler et al., 2016). So far, however, Piezo2 has been associated only with fast MA current, though it is expressed in various LTMRs, including RA-LTMRs innervating Meissner corpuscles (Ranade et al., 2014). Whether RA-LTMRs innervating Pacinian and Meissner corpuscles are subserved by different mechanotransducer channels or by the same channel with modified inactivation kinetics is unknown. These considerations provide strong rationale to identify novel slow-type mechanotransducers in nociceptors and LTMRs, as well as and molecules and pathways that prolong inactivation kinetics of Piezo2.

2.3 Functional regulation of Piezo2 in mouse dorsal root ganglia neurons

While the quest for slow mechanotransducer(s) is ongoing, recent studies have revealed mechanisms shaping Piezo2 function. In vitro, the deletion of E2727 in the distal C-terminal region of human Piezo2 (E2797 in mouse Piezo2, Fig. 1) causes a twofold prolongation of inactivation (Coste et al., 2013). Interestingly, this and other Piezo2 mutations were found to be linked to distal arthrogryposis type 5, suggesting that Piezo2 channelopathy could lead to severe developmental malformations (Coste et al., 2013; McMillin et al., 2014; Okubo et al., 2015). Piezo2 inactivation can also be reversibly prolonged by osmotic swelling and pharmacological activation of the protein kinase A (PKA)/protein kinase C (PKC) pathway in neurons and heterologous cells (Dubin et al., 2012; Jia, Ikeda, Ling, Viatchenko-Karpinski, & Gu, 2016), establishing that Piezo2 kinetics are modulated by intracellular factors. Most likely, these and other mechanisms are involved in sustained prolongation of Piezo2 kinetics in neuronal mechanoreceptor subtypes as discussed earlier, warranting further research.

The mechanism of Piezo2 fine-tuning in neurons has only started to emerge. Piezo2 activity is regulated by G-protein signaling. The inclusion of GTP in the recording pipette leads to a slow run-up of Piezo2 current in neurons and HEK293 cells (Jia, Ikeda, Ling, & Gu, 2013). Part of the mechanism probably includes the activation the PKA/PKC pathway, which can be engaged directly by GTP, through the bradykinin receptor β2 (Dubin et al., 2012), or cAMP sensor Epac1, leading to potentiation of Piezo2 current (Eijkelkamp et al., 2013; Singhmar et al., 2016). Given that the PKA/PKC pathway can be engaged through a host of receptors, Piezo2 activity is expected to respond to a number of signaling molecules. These may converge onto Piezo2 directly, e.g., via phosphorylation, or indirectly, by changing the molecular composition and physical properties of the plasma membrane. Indeed, Piezo2 was shown to interact with STOML3 (Poole et al., 2014), an important regulator of mechanosensitivity (Wetzel et al., 2007). STOML3 expression potentiates Piezo2 activity via a mechanism that leads to cholesterol-dependent stiffening of the plasma membrane (Qi et al., 2015). In support of this, it was shown that an increase in membrane tension caused by intracellular hypotonicity-induced cell swelling reversibly activates Piezo2 (Jia et al., 2016). Conversely, phosphoinositide depletion from the plasma membrane leads to suppression of Piezo2 (Borbiro, Badheka, & Rohacs, 2015). Thus, Piezo2 activity is tightly regulated in neurons by various cellular factors and pathways, including components of the plasma membrane, most of which have yet to be elucidated (Narayanan et al., 2016).

3. LIGHT TOUCH

3.1 Insights from mice

A transgenic mouse expressing *Piezo2–GFP* showed that not only is Piezo2 protein present in the cell bodies of DRG neurons but it is also found in the nerve terminal in the skin (Ranade et al., 2014). In particular, Piezo2 was identified in LTMRs innervating the hair follicles, Merkel cells, Meissner corpuscles (Ranade et al., 2014), in proprioceptive muscle spindles, and Golgi tendon organs (Woo et al., 2015). Conditional knockout of Piezo2 in DRG neurons yields a reduction in the total number of mechanosensitive myelinated Aβ-fibers as tested through an ex vivo skin nerve preparation (Ranade et al., 2014). The remaining mechanosensitive Aβ-fibers required more force to elicit firing and had reduced firing frequency compared to control. SA-LTMRs exhibit the most striking spiking deficits in the absence of Piezo2, with a dramatic reduction in spikes on both the onset (dynamic phase) and hold (static phase) of the mechanical stimulus. Both RA- and SA-LTMRs displayed deficits in velocity detection in the absence of Piezo2. This deficit was not seen in C-fibers that detect noxious heat, cold, and mechanical stimulation (Ranade et al., 2014). Mice lacking Piezo2 in sensory neurons demonstrate light touch deficits compared to WT, as assayed by stimulation by von Frey filaments, response to a piece of tape affixed to the back, and a two-plate vibration preference test (Ranade et al., 2014). Both ambient and noxious temperature sensation, in addition to noxious mechanosensation, were not impaired in these knockout mice (Ranade et al., 2014), indicating that Piezo2 is not a key component of nociceptor function.

Accumulated evidence shows that Piezo2 is a major but not the only mechanotransducer in murine DRG. Indeed, some fast MA current remains in the neurons even after knockout (Ranade et al., 2014; Woo et al., 2015). This could either be explained by an incomplete Cre-mediated excision or the presence of other mechanotransducer(s). The deletion of Piezo2 does not appear to affect the percentage of neurons with intermediate and slow MA currents. While most of the remaining MA currents probably arise from nociceptors, electrophysiological recordings in dissociated DRG as well other considerations discussed earlier suggest that some LTMR subtypes could also express slow MA current (Coste et al., 2007; Drew et al., 2007; Lechner et al., 2009; Poole et al., 2014). Further, there is also a possibility that some ion channels might not be expressed in neuronal soma in dissociated system, and thus mechanotransducers observed in DRG neurons in culture may not fully reflect the multitude of MA current in LTMR endings in vivo. Indeed, while the deletion of Piezo2 profoundly suppresses both LTMR spiking and behavioral responses to light touch, it does not abrogate either process, warranting further research in this area.

3.2 Insight from other vertebrates

Extensive studies of the role of Piezo2 in mouse have established that the channel is a key component in light touch detection. Other studies have aimed to understand if the role of Piezo2 is conserved in other vertebrates. In zebrafish, a homolog of Piezo2 called *piezo2b*, likely the result of genome duplication, was identified in the somatosensory Rohan–Beard cells (Faucherre, Nargeot, Mangoni, & Jopling, 2013). Morpholino oligonucleotide knockdown of *piezo2b* in zebrafish embryos significantly reduced light touch responses in these embryos compared to wild type, but had no effect on noxious mechanical or chemical responses (Faucherre et al., 2013). A test of the mechanosensitivity of *piezo2b* via overexpression in cell culture followed by mechanical stimulation in voltage clamp would determine if this conserved phenotype is due to Piezo2 mechanosensitivity or some other mechanism.

Tactile specialist organisms provide a unique perspective to studying mechanosensation (Schneider, Gracheva, & Bagriantsev, 2016). Some species of ducks, including the domesticated Pekin duck, are tactile-guided foragers (Zweers, 1977). The glabrous skin covering both dorsal and ventral surfaces of the duck bill is rich in Grandry (Meissner) and Herbst (Pacinian) corpuscles—the principal detectors of the lightest forms of touch and vibration (Berkhoudt, 1980; Saxod, 1996). The corpuscles are innervated by trigeminal RA-LTMRs (Arends & Dubbeldam, 1984; Dubbeldam, Brauch, & Don, 1981; Gottschaldt, 1974; Gregory, 1973; Leitner & Roumy, 1974). Histological analysis showed that TG of several species of tactile-foraging duck are rich in large-diameter neurons expressing Piezo2, where 84% of TG neurons stain positive for Piezo2 mRNA. This is a far greater proportion of Piezo2-expressing neurons that the typical 20–30% of Piezo2 mRNA-expressing neurons found using the same method in duck DRG, mouse TG or DRG, or TG of visually foraging birds (Fig. 4) (Schneider et al., 2014). Together with the exceptionally high density (>100 mm^2) of Meissner- and Pacinian-like corpuscles in the duck bill

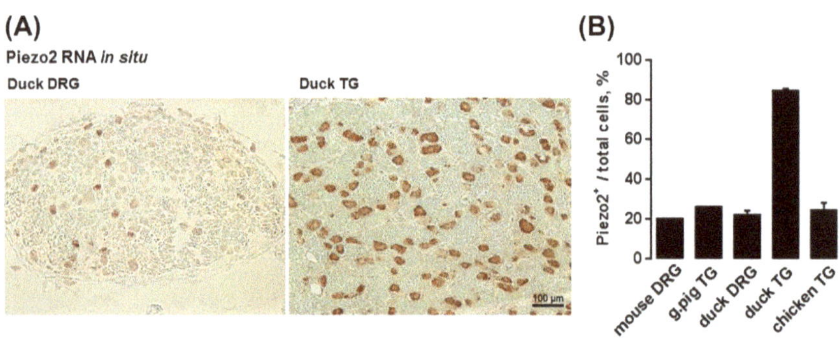

Figure 4 ***Piezo2 expression in trigeminal ganglia (TG) and dorsal root ganglia (DRG) of rodents and birds.*** (A) Representative RNA in situ hybridization images from duck TG and DRG using Piezo2 antisense probe. (B) Quantification of Piezo2 mRNA-expressing neurons from TG and DRG of mice (Coste et al., 2010), guinea pig (Bron et al., 2014) and birds (Schneider et al., 2014).

(Berkhoudt, 1980), these data suggest that duck TG contains high proportion of Piezo2$^+$ RA-LTMRs, providing a molecular basis for the tactile-driven foraging behavior.

Electrophysiological analysis showed that 80% of duck TG neurons are mechanosensitive in vitro. Unexpectedly, even though >80% of duck TG neurons express Piezo2, all three types of MA current (fast, intermediate and slow) are present in, respectively, 20%, 20%, and 60% of the neurons. Possibly, some neurons express factors that modulate Piezo2 inactivation kinetics, or Piezo2 coexists with unidentified mechanotransducers with slow MA current (Schneider et al., 2014). The duck TG is a particularly interesting system for studying mechanosensitivity because of its rich population of RA-LTMRs innervating Meissner- and Pacinian-like corpuscles in the glabrous skin of the duck bill. Further study of the role of Piezo2 and other proteins in this system could provide insight on the basis of mechanosensitivity of Meissner and Pacinian corpuscles, which are innervated by RA-LTMRs that are very rare in mouse DRG.

4. PROPRIOCEPTION

A subset of mechanosensitive DRG neurons are proprioceptors marked by the expression of parvalbumin (Pvalb) (Arber, Ladle, Lin, Frank, & Jessell, 2000). Proprioceptors are SA-LTMRs that innervate muscle and the Golgi tendon organ and provide information on body and limb position (Proske & Gandevia, 2012). A recent study showed that Piezo2 is the main mechanotransducer in mouse proprioceptors (Woo et al., 2015). Woo et al. used two conditional Piezo2 knockout mouse strains obtained by crossing a Piezo2$^{fl/fl}$ strain to *Pvalb-Cre*, resulting in Piezo2 knockout in parvalbumin-expressing proprioceptors and some cutaneous mechanoreceptors; and *HoxB8-Cre* to target Piezo2 in a broader population of DRG neurons, inclusive of proprioceptors, in the caudal (lower thoracic, lumbar) body segments (Woo et al., 2015).

Piezo2^{cKO} mice retained the wild-type number of proprioceptors in DRG but exhibited profoundly impaired coordination in both pairs of limbs in *Pvalb-Cre; Piezo2^{cKO}* and in hind limbs only in *HoxB8-Cre*; *Piezo2^{cKO}*, consistent with the caudal bias of HoxB8-Cre activity. Comparison of muscle spindle firing properties between WT and Piezo2^{cKO} mice in an ex vivo muscle stretch preparation revealed a profound impairment of baseline and stretch-activated spiking in muscle-innervating afferents in Piezo2^{cKO} mice, demonstrating that Piezo2 is vital for stretch-induced proprioceptor activity (Woo et al., 2015).

Dissociation of DRG neurons from *Pvalb-Cre; TdTomato* allowed for identification of a proprioceptor-rich population in vitro. Mechanical stimulation of TdTomato-positive neurons predominately yields fast MA currents in 89.5% of the cells, similar to the kinetics of Piezo2 in HEK293 (Fig. 3B). In contrast, only 8.3% of fluorescent DRG neurons from *Pvalb-Cre; TdTomato; Piezo2*^{cKO} mice exhibited fast MA current. Interestingly, a small subset of these neurons has intermediate MA current, and the proportion of these neurons increases with Piezo2 conditional knockout (Woo et al., 2015). This suggests that another mechanotransducer ion channel could exist in this subset of neurons, although the fact that some *Pvalb*$^{+}$ neurons are cutaneous mechanoreceptors means that neurons with intermediate MA current are not necessarily proprioceptors. Nevertheless, considering the severe behavioral phenotypes of proprioceptor-specific Piezo2 knockout, the channel appears to be the principal mechanotransducer in proprioceptors. Recently, an elegant study by the Ana Gomis group corroborated this conclusion using proprioceptive neurons from the mesencephalic ganglion, which contains a rather homogeneous population of proprioceptors (Florez-Paz, Bali, Kuner, & Gomis, 2016).

Interestingly, even though proprioceptors express Piezo2-dependent MA current with fast kinetics of inactivation, they exhibit a slowly adapting pattern of afferent discharge (Proske & Gandevia, 2012; Woo et al., 2015). The discontinuity between MA current inactivation and firing adaptation rates is puzzling. It is possible that the fast MA current measured in a dissociated soma does not fully reflect the situation in the afferent ending, where the kinetics of inactivation could be prolonged by unknown factors. Alternatively, the geometry of the muscle spindle could lead to a sequential engagement and disengagement of Piezo2 molecules, leading to repetitive generation of MA currents. It is also possible that slow adaptation requires a contribution from the somatic components surrounding the afferent ending in the muscle spindle, which are yet to be identified. The latter scenario has been shown to exist in the Merkel cell—neurite complexes.

5. MERKEL CELLS

Static mechanical stimuli, rough surface textures, as well as guard hair and whisker deflections are detected by Merkel cell—neurite complexes. The complex is composed of a modified mechanosensitive epithelial cell (Merkel cell) and a slowly adapting type I LTMR (SAI-LTMR). Recent works demonstrated that both the somatic and neural components of the complex are mechanosensitive and depend on the expression of Piezo2 (Ikeda et al., 2014; Maksimovic et al., 2014; Woo et al., 2014).

Voltage clamp recording from Merkel cells in situ in the mouse whisker follicles revealed that the cells are mechanosensitive, exhibiting fast MA current (Ikeda et al., 2014). A similar current was detected in vitro in Merkel cells isolated from cutaneous touch domes (Maksimovic et al., 2014; Woo et al., 2014). The MA current closely resembles Piezo2 activity in cultured cell lines both in terms of fast kinetics of inactivation and inhibition by Gd^{3+} and ruthenium red (Coste et al., 2010). The injection of an antibody or shRNA against Piezo2 into the whisker follicle suppressed MA current in Merkel cells in situ (Ikeda et al., 2014), whereas skin-specific knockout of Piezo2 in mice (*Krt14-Cre;Piezo2*$^{fl/fl}$*;Atoh1*^{GFP}) abolished MA current of isolated Merkel cells in vitro (Woo et al., 2014).

Interestingly, these studies also revealed that mechanical stimulation of mouse Merkel cells in current clamp mode yielded sustained depolarization and in some cases, Ca^{2+}-dependent action potentials (Ikeda et al., 2014; Woo et al., 2014). This mechanically evoked depolarization is absent in *Krt14-Cre* and *Piezo2*$^{fl/fl}$ mice (Woo et al., 2014). Recordings from SAI-LTMRs ex vivo showed that conditional knockout of Piezo2 in Merkel cells greatly reduced the number of spikes fired in the static phase of the stimulation, while firing in the dynamic phase stayed mostly intact (Ikeda et al., 2014; Maksimovic et al., 2014; Woo et al., 2014). Rather than firing for the duration of the stimulus, Piezo2 CKO fibers only fire for part of the static phase, thus displaying intermediate adaptation to mechanical stimulation (Maksimovic et al., 2014; Woo et al., 2014). This suggests that Piezo2-mediated activation of Merkel cells is not necessary for SAI-LTMR firing, but contributes to sustained nerve firing in the static phase of stimulation. To better probe the role of Merkel cell excitation in SAI-LTMR firing, Maksimovic et al. made a mouse that expresses channelrhodopsin in Merkel cells. Light-induced activation of channelrhodopsin-expressing Merkel cells ex vivo gave rise to sustained afferent firing with an impaired dynamic phase (Maksimovic et al., 2014). This suggests that afferent stimulation is important for dynamic firing, while Merkel cells are most important for sustained firing during the static phase. As discussed earlier, Piezo2 is present in SAI terminals (Ranade et al., 2014), so it is likely that Piezo2 underlies SAI-LTMR mechanosensitivity during the dynamic phase.

Skin-specific Merkel cell Piezo2 deficiency also causes behavioral light touch deficits. Piezo2 CKO mice exhibit decreased sensitivity to von Frey filament stimulation of 1.5 g and lower (Woo et al., 2014). Mice injected with capsaicin in the facial region demonstrate nocifensive responses to low-speed whisker stimulation, while follicular injection of shRNA against Piezo2 significantly reduces this nocifensive behavior (Ikeda et al., 2014). This impairment is strikingly similar to that caused by the injection of the voltage-gated calcium channel blockers Cd^{2+}, felodipine, and ω-conotoxin MVIIC (Ikeda et al., 2014), indicating that inhibition of Merkel cell activation can lead to behavioral light touch deficits. Together, these data support the idea that Piezo2 is a principal mechanotransducer in Merkel cells that helps the detection of light touch and whisker deflection in vivo.

6. NOCICEPTION

RNA in situ hybridization revealed that Piezo2 expression is not restricted to A- and C-type LTMRs. The channel is found in 60% of peripherin-expressing neurons, a group encompassing both peptidergic and nonpeptidergic C-fibers (Goldstein, House, & Gainer, 1991), and in 24% of $TRPV1^+$ polymodal peptidergic nociceptors (Coste et al., 2010). This colocalization with TRPV1 is also observed on the functional level. In dissociated mouse DRG culture, small diameter neurons can be identified that display both fast, Piezo2-like MA currents and currents induced by capsaicin, a specific agonist of TRPV1 (Borbiro et al., 2015; Caterina et al., 2000). In such neurons, activation of TRPV1 by capsaicin almost completely eliminates fast MA current, an effect that can be reproduced in HEK293 cells coexpressing TRPV1 and Piezo2 (Borbiro et al., 2015). The mechanism of this inhibition is based on Ca^{2+}-dependent depletion of phospholipids in the plasma membrane. Capsaicin is known to cause local analgesia, including numbing of mechanosensory perception, which follows the initial sensation of burning. The inhibition of Piezo2-mediated MA current in response to TRPV1 activation in nociceptors provides a molecular explanation for the analgesic effect of capsaicin (Borbiro et al., 2015).

Despite the expression of Piezo2 in murine nociceptors, it does not seem to be necessary for noxious mechanical sensation under normal conditions, as Piezo2 CKO in DRG neurons does not significantly change behavioral responses to pain (Ranade et al., 2014). However, Piezo2 appears to play a role in mechanical allodynia, a specific facet of pain sensitization whereby normally innocuous stimuli are perceived as painful as a result of tissue damage or inflammation. Nerve dissection leads to the production of cAMP, which contributes to mechanical allodynia via a mechanism involving activation of the cAMP sensor Epac1. Mechanical allodynia can be directly induced by the injection of a cAMP analog and Epac1 agonist 8-pCPT (Eijkelkamp et al., 2013). In dissociated large diameter DRG neurons, 8-pCPT is capable of potentiating fast MA current, a trend that holds true in HEK293T cells cotransfected with human Piezo2 and Epac1. Consistently, intrathecal injection of oligodeoxynucleotides against Piezo2 in mice partially rescues Epac1-mediated reduction in threshold to mechanical stimulation by von Frey filaments, both in 8-pCPT and injury models (Eijkelkamp et al., 2013).

The mechanism of the Epac1—Piezo2 cross talk is probably indirect and involves the activation of the Ras-like GTPase Rap1 via GDP to GTP exchange. Rap1, in turn, activates a host of signaling cascades, including PKC (Breckler et al., 2011), which was shown to facilitate Piezo2 current in neurons (Dubin et al., 2012). These results are also in line with the reported positive effect of GTP on Piezo2 current (Jia et al., 2013). A recent study determined that the Epac2—Piezo2 pathway is modulated by the phosphokinase GRK2, which phosphorylates Epac1, inhibiting the ensuing GDP—GTP exchange in Rap1 (Singhmar et al., 2016).

Interestingly, the Epac1-mediated allodynia persists in animals with a genomic ablation of $Na_V1.8$, a voltage-gated ion channel critical for the generation of action potential in the majority of nociceptors (Akopian, Sivilotti, & Wood, 1996; Renganathan, Cummins, & Waxman, 2001; Sangameswaran et al., 1996; Shields et al., 2012). Thus the phenomenon appears to occur in a very specific subset of neurons, which await identification. A recent study detected Piezo2 expression in vascular endothelial cells. The injection of oligodeoxynucleotides against Piezo2 in the skin protects against endothelin- and oxaliplatin-induced pain, supporting a role for the channel in hyperalgesia (Ferrari et al., 2015). Overall, the involvement of Piezo2 in nociception has only started to emerge and requires further study.

While recent studies have revealed a likely role of Piezo2 in various types of painful responses, the channel is not the principal mechanotransducer of high-threshold mechanoreceptors (HTMRs). First, the deletion of Piezo2 in DRG neurons does not significantly alter pain responses (Ranade et al., 2014). Second, HTMRs are probably subserved by mechanotransducers with a Piezo2-independent slowly inactivating MA current (Hao & Delmas, 2010; Lechner et al., 2009; Rugiero et al., 2010), which exhibits properties different from the fast MA current mediated by Piezo2 (Coste et al., 2010; Hu & Lewin, 2006; Ranade et al., 2014). The identity of this molecule remains to be determined.

7. CONCLUSIONS AND PERSPECTIVES

Cutaneous and proprioceptive mechanosensation is dependent on mechanogated ion channels present in the neural and somatic components of mechanosensory end-organs. Evidence show that Piezo2 is expressed in SAI- and RAI-LTMRs, nociceptors, and in somatic Merkel cells. Additionally, knockout studies in mice show that Piezo2 is critical for many aspects of light touch and proprioception. In agreement with the mouse data, humans carrying non-functional Piezo2 display profound mechanosensory and proprioceptive deficits (Chesler et al., 2016; Delle Vedove et al., 2016; Mahmud et al., 2016). Knockdown experiments also indicate that it may play a role in the development of hyperalgesia. However, electrophysiological and behavioral studies demonstrate that Piezo2 is not the sole somatosensory mechanotransducer. It is firmly established that Piezo2 mediates fast inactivating MA current in neuronal mechanoreceptors, but the transducers of intermediate and slow inactivating MA current (Hong et al., 2016), and their physiological specialization in innocuous and noxious mechanosensation require further investigation.

The identification of neuronal mechanotransducers will help understand the relationship between the kinetics of MA current inactivation in the afferent ending and pattern of firing adaptation. This probably involves cross talk between mechanotransducers, voltage-gated ion channels that generate the action potential, and somatic components of the mechanoreceptive end organs, such as Merkel cells. Understanding the molecular identity and functional relationship between these components will reveal the mechanism of cutaneous receptor fine-tuning, which allows the end organs to detect mechanical stimuli of specific force, duration, and frequency. Further research using standard laboratory rodents and tactile specialists is needed to understand mechanosensation, which remains the least well-understood sense at the cellular and molecular level.

REFERENCES

Akopian, A. N., Sivilotti, L., & Wood, J. N. (1996). A tetrodotoxin-resistant voltage-gated sodium channel expressed by sensory neurons. *Nature, 379*, 257–262.

Alamri, A., Bron, R., Brock, J. A., & Ivanusic, J. J. (2015). Transient receptor potential cation channel subfamily V member 1 expressing corneal sensory neurons can be subdivided into at least three subpopulations. *Frontiers in Neuroanatomy, 9*, 71.

Arber, S., Ladle, D. R., Lin, J. H., Frank, E., & Jessell, T. M. (2000). ETS gene Er81 controls the formation of functional connections between group Ia sensory afferents and motor neurons. *Cell, 101*, 485–498.

Arends, J. J., & Dubbeldam, J. L. (1984). The subnuclei and primary afferents of the descending trigeminal system in the mallard (*Anas platyrhynchos* L.). *Neuroscience, 13*, 781–795.

Bagriantsev, S. N., Gracheva, E. O., & Gallagher, P. G. (2014). Piezo proteins: Regulators of mechanosensation and other cellular processes. *Journal of Biological Chemistry, 289*, 31673–31681.

Berkhoudt, H. (1980). The morphology and distribution of cutaneous mechanoreceptors (Herbst and Grandry corpuscles) in bill and tongue of the mallard (*Anas platyrhynchos* L.). *Netherlands Journal of Zoology, 30*, 1–34.

Borbiro, I., Badheka, D., & Rohacs, T. (2015). Activation of TRPV1 channels inhibits mechanosensitive Piezo channel activity by depleting membrane phosphoinositides. *Science Signaling, 8*, ra15.

Breckler, M., Berthouze, M., Laurent, A. C., Crozatier, B., Morel, E., & Lezoualc'h, F. (2011). Rap-linked cAMP signaling Epac proteins: Compartmentation, functioning and disease implications. *Cellular Signalling, 23*, 1257–1266.

Bron, R., Wood, R. J., Brock, J. A., & Ivanusic, J. J. (2014). Piezo2 expression in corneal afferent neurons. *Journal of Comparative Neurology, 522*.

Caterina, M. J., Leffler, A., Malmberg, A. B., Martin, W. J., Trafton, J., Petersen-Zeitz, K. R., … Julius, D. (2000). Impaired nociception and pain sensation in mice lacking the capsaicin receptor. *Science, 288*, 306–313.

Chesler, A. T., Szczot, M., Bharucha-Goebel, D., Ceko, M., Donkervoort, S., Laubacher, C., … Bönnemann, C. G. (2016). The role of PIEZO2 in human mechanosensation. *The New England Journal of Medicine, 375*, 1355–1364.

Choi, H. J., Sun, D., & Jakobs, T. C. (2015). Astrocytes in the optic nerve head express putative mechanosensitive channels. *Molecular Vision, 21*, 749–766.

Coste, B., Crest, M., & Delmas, P. (2007). Pharmacological dissection and distribution of NaN/Nav1.9, T-type Ca^{2+} currents, and mechanically activated cation currents in different populations of DRG neurons. *Journal of General Physiology, 129*, 57–77.

Coste, B., Houge, G., Murray, M. F., Stitziel, N., Bandell, M., Giovanni, M. A., … Patapoutian, A. (2013). Gain-of-function mutations in the mechanically activated ion channel Piezo2 cause a subtype of distal arthrogryposis. *Proceedings of the National Academy of Sciences of the United States of America, 110*, 4667–4672.

Coste, B., Mathur, J., Schmidt, M., Earley, T. J., Ranade, S., Petrus, M. J., … Patapoutian, A. (2010). Piezo1 and Piezo2 are essential components of distinct mechanically activated cation channels. *Science, 330*, 55–60.

Coste, B., Murthy, S. E., Mathur, J., Schmidt, M., Mechioukhi, Y., Delmas, P., & Patapoutian, A. (2015). Piezo1 ion channel pore properties are dictated by C-terminal region. *Nature Communications, 6*, 7223.

Delle Vedove, A., Storbeck, M., Heller, R., Hölker, I., Hebbar, M., Shukla, A., … Wirth, B. (2016). Biallelic loss of proprioception-related PIEZO2 causes muscular atrophy with perinatal respiratory distress, arthrogryposis, and scoliosis. *The American Journal of Human Genetics, 99*, 1206–1216.

Drew, L. J., Rugiero, F., Cesare, P., Gale, J. E., Abrahamsen, B., Bowden, S., … Wood, J. N. (2007). High-threshold mechanosensitive ion channels blocked by a novel conopeptide mediate pressure-evoked pain. *PLoS One, 2*, e515.

Drew, L. J., & Wood, J. N. (2007). FM1-43 is a permeant blocker of mechanosensitive ion channels in sensory neurons and inhibits behavioural responses to mechanical stimuli. *Molecular Pain, 3*, 1.

Dubbeldam, J. L., Brauch, C. S., & Don, A. (1981). Studies on the somatotopy of the trigeminal system in the mallard, *Anas platyrhynchos* L. III. Afferents and organization of the nucleus basalis. *Journal of Comparative Neurology, 196*, 391–405.

Dubin, A. E., Schmidt, M., Mathur, J., Petrus, M. J., Xiao, B., Coste, B., & Patapoutian, A. (2012). Inflammatory signals enhance piezo2-mediated mechanosensitive currents. *Cell Reports, 2*, 511–517.

Eijkelkamp, N., Linley, J. E., Torres, J. M., Bee, L., Dickenson, A. H., Gringhuis, M., … Wood, J. N. (2013). A role for Piezo2 in EPAC1-dependent mechanical allodynia. *Nature Communications, 4*, 1682.

Faucherre, A., Nargeot, J., Mangoni, M. E., & Jopling, C. (2013). piezo2b regulates vertebrate light touch response. *Journal of Neuroscience, 33*, 17089–17094.

Ferrari, L. F., Bogen, O., Green, P., & Levine, J. D. (2015). Contribution of Piezo2 to endothelium-dependent pain. *Molecular Pain, 11*, 65.

Fleming, M. S., & Luo, W. (2013). The anatomy, function, and development of mammalian Abeta low-threshold mechanoreceptors. *Frontiers in Biology, 8*.

Florez-Paz, D., Bali, K. K., Kuner, R., & Gomis, A. (2016). A critical role for Piezo2 channels in the mechanotransduction of mouse proprioceptive neurons. *Scientific Reports, 6*, 25923.

Ge, J., Li, W., Zhao, Q., Li, N., Chen, M., Zhi, P., … Yang, M. (2015). Architecture of the mammalian mechanosensitive Piezo1 channel. *Nature, 527*.

Glokowska, E., & Gallagher, P. G. (2015). Disorders of erythrocyte volume homeostasis. *International Journal of Laboratory Hematology, 37*(Suppl. 1), 85–91.

Goldstein, M. E., House, S. B., & Gainer, H. (1991). NF-L and peripherin immunoreactivities define distinct classes of rat sensory ganglion cells. *Journal of Neuroscience Research, 30*, 92–104.

Gottschaldt, K. M. (1974). The physiological basis of tactile sensibility in the beak of geese. *Journal of Comparative Physiology, 95*, 29–47.

Gregory, J. E. (1973). An electrophysiological investigation of the receptor apparatus of the duck's bill. *Journal of Physiology, 229*, 151–164.

Hao, J., & Delmas, P. (2010). Multiple desensitization mechanisms of mechanotransducer channels shape firing of mechanosensory neurons. *Journal of Neuroscience, 30*, 13384–13395.

Hong, G.-S., Lee, B., Wee, J., Chun, H., Kim, H., Jung, J., … Oh, U. (2016). Tentonin 3/TMEM150c confers distinct mechanosensitive currents in dorsal-root ganglion neurons with proprioceptive function. *Neuron, 91*, 107–118.

Honore, E., Martins, J. R., Penton, D., Patel, A., & Demolombe, S. (2015). The piezo mechanosensitive ion channels: May the force be with you! *Reviews of Physiology, Biochemistry and Pharmacology, 169*, 25–41.

Hu, J., & Lewin, G. R. (2006). Mechanosensitive currents in the neurites of cultured mouse sensory neurones. *Journal of Physiology, 577*, 815–828.

Ikeda, R., Cha, M., Ling, J., Jia, Z., Coyle, D., & Gu, J. G. (2014). Merkel cells transduce and encode tactile stimuli to drive abeta-afferent impulses. *Cell, 157*, 664–675.

Jia, Z., Ikeda, R., Ling, J., & Gu, J. G. (2013). GTP-dependent run-up of Piezo2-type mechanically activated currents in rat dorsal root ganglion neurons. *Molecular Brain, 6*, 57.

Jia, Z., Ikeda, R., Ling, J., Viatchenko-Karpinski, V., & Gu, J. G. (2016). Regulation of Piezo2 mechanotransduction by static plasma membrane tension in primary afferent neurons. *Journal of Biological Chemistry, 291*, 9087–9104.

Khatibi Shahidi, M., Krivanek, J., Kaukua, N., Ernfors, P., Hladik, L., Kostal, V., … Fried, K. (2015). Three-dimensional imaging reveals new compartments and structural adaptations in odontoblasts. *Journal of Dental Research, 94*.

Kobayashi, K., Fukuoka, T., Obata, K., Yamanaka, H., Dai, Y., Tokunaga, A., & Noguchi, K. (2005). Distinct expression of TRPM8, TRPA1, and TRPV1 mRNAs in rat primary afferent neurons with adelta/c-fibers and colocalization with trk receptors. *Journal of Comparative Neurology, 493*, 596–606.

Lazarov, N. E. (2007). Neurobiology of orofacial proprioception. *Brain Research Reviews, 56*, 362–383.
Le Pichon, C. E., & Chesler, A. T. (2014). The functional and anatomical dissection of somatosensory subpopulations using mouse genetics. *Frontiers in Neuroanatomy, 8*, 21.
Lechner, S. G., Frenzel, H., Wang, R., & Lewin, G. R. (2009). Developmental waves of mechanosensitivity acquisition in sensory neuron subtypes during embryonic development. *EMBO Journal, 28*, 1479–1491.
Lee, W., Leddy, H. A., Chen, Y., Lee, S. H., Zelenski, N. A., McNulty, A. L., … Liedtke, W. B. (2014). Synergy between Piezo1 and Piezo2 channels confers high-strain mechanosensitivity to articular cartilage. *Proceedings of the National Academy of Sciences of the United States of America, 111*, E5114–E5122.
Leitner, L. M., & Roumy, M. (1974). Mechanosensitive units in the upper bill and in the tongue of the domestic duck. *Pflügers Archiv, 346*, 141–150.
Li, L., Rutlin, M., Abraira, V. E., Cassidy, C., Kus, L., Gong, S., … Ginty, D. D. (2011). The functional organization of cutaneous low-threshold mechanosensory neurons. *Cell, 147*, 1615–1627.
Loewenstein, W. R., & Mendelson, M. (1965). Components of receptor adaptation in a Pacinian corpuscle. *Journal of Physiology, 177*, 377–397.
Lou, S., Duan, B., Vong, L., Lowell, B. B., & Ma, Q. (2013). Runx1 controls terminal morphology and mechanosensitivity of VGLUT3-expressing C-mechanoreceptors. *Journal of Neuroscience, 33*, 870–882.
Mahmud, A. A., Nahid, N. A., Nassif, C., Sayeed, M. S., Ahmed, M. U., Parveen, M., … Michaud, J. L. (2016). Loss of the proprioception and touch sensation channel PIEZO2 in siblings with a progressive form of contractures. *Clinical genetics* [Epub 09/09/2016].
Maksimovic, S., Nakatani, M., Baba, Y., Nelson, A. M., Marshall, K. L., Wellnitz, S. A., … Lumpkin, E. A. (2014). Epidermal Merkel cells are mechanosensory cells that tune mammalian touch receptors. *Nature, 509*, 617–621.
McCarter, G. C., & Levine, J. D. (2006). Ionic basis of a mechanotransduction current in adult rat dorsal root ganglion neurons. *Molecular Pain, 2*, 28.
McCarter, G. C., Reichling, D. B., & Levine, J. D. (1999). Mechanical transduction by rat dorsal root ganglion neurons in vitro. *Neuroscience Letters, 273*, 179–182.
McMillin, M. J., Beck, A. E., Chong, J. X., Shively, K. M., Buckingham, K. J., Gildersleeve, H. I., … Bamshad, M. J. (2014). Mutations in Piezo2 cause Gordon syndrome, Marden-Walker syndrome, and distal arthrogryposis type 5. *American Journal of Human Genetics, 94*, 734–744.
Mendelson, M., & Lowenstein, W. R. (1964). Mechanisms of receptor adaptation. *Science, 144*, 554–555.
Narayanan, P., Sondermann, J., Rouwette, T., Karaca, S., Urlaub, H., Mitkovski, M., … Schmidt, M. (2016). Native Piezo2 interactomics identifies Pericentrin as a novel regulator of Piezo2 in somatosensory neurons. *Journal of Proteome Research, 15*, 2676–2687.
Okubo, M., Fujita, A., Saito, Y., Komaki, H., Ishiyama, A., Takeshita, E., … Sasaki, M. (2015). A family of distal arthrogryposis type 5 due to a novel Piezo2 mutation. *American Journal of Medical Genetics A*, (5).
Poole, K., Herget, R., Lapatsina, L., Ngo, H. D., & Lewin, G. R. (2014). Tuning Piezo ion channels to detect molecular-scale movements relevant for fine touch. *Nature Communications, 5*, 3520.
Proske, U., & Gandevia, S. C. (2012). The proprioceptive senses: Their roles in signaling body shape, body position and movement, and muscle force. *Physiological Reviews, 92*, 1651–1697.

Qi, Y., Andolfi, L., Frattini, F., Mayer, F., Lazzarino, M., & Hu, J. (2015). Membrane stiffening by STOML3 facilitates mechanosensation in sensory neurons. *Nature Communications, 6*, 8512.

Ranade, S. S., Syeda, R., & Patapoutian, A. (2015). Mechanically activated ion channels. *Neuron, 87*, 1162–1179.

Ranade, S. S., Woo, S. H., Dubin, A. E., Moshourab, R. A., Wetzel, C., Petrus, M., … Patapoutian, A. (2014). Piezo2 is the major transducer of mechanical forces for touch sensation in mice. *Nature, 516*, 121–125.

Renganathan, M., Cummins, T. R., & Waxman, S. G. (2001). Contribution of Na(v)1.8 sodium channels to action potential electrogenesis in DRG neurons. *Journal of Neurophysiology, 86*, 629–640.

Rugiero, F., Drew, L. J., & Wood, J. N. (2010). Kinetic properties of mechanically activated currents in spinal sensory neurons. *Journal of Physiology, 588*, 301–314.

Sangameswaran, L., Delgado, S. G., Fish, L. M., Koch, B. D., Jakeman, L. B., Stewart, G. R., … Herman, R. C. (1996). Structure and function of a novel voltage-gated, tetrodotoxin-resistant sodium channel specific to sensory neurons. *Journal of Biological Chemistry, 271*, 5953–5956.

Saxod, R. (1996). Ontogeny of the cutaneous sensory organs. *Microscopy Research and Technique, 34*, 313–333.

Schneider, E. R., Gracheva, E. O., & Bagriantsev, S. N. (2016). Evolutionary specialization of tactile perception in vertebrates. *Physiology, 31*, 193–200.

Schneider, E. R., Mastrotto, M., Laursen, W. J., Schulz, V. P., Goodman, J. B., Funk, O. H., … Bagriantsev, S. N. (2014). Neuronal mechanism for acute mechanosensitivity in tactile-foraging waterfowl. *Proceedings of the National Academy of Sciences of the United States of America, 111*, 14941–14946.

Sharif-Naeini, R. (2015). Contribution of mechanosensitive ion channels to somatosensation. *Progress in Molecular Biology and Translational Science, 131*, 53–71.

Shields, S. D., Ahn, H. S., Yang, Y., Han, C., Seal, R. P., Wood, J. N., … Dib-Hajj, S. D. (2012). Nav1.8 expression is not restricted to nociceptors in mouse peripheral nervous system. *Pain, 153*, 2017–2030.

Singhmar, P., Huo, X., Eijkelkamp, N., Berciano, S. R., Baameur, F., Mei, F. C., … Kavelaars, A. (2016). Critical role for Epac1 in inflammatory pain controlled by GRK2-mediated phosphorylation of Epac1. *Proceedings of the National Academy of Sciences of the United States of America, 113*.

Volkers, L., Mechioukhi, Y., & Coste, B. (2015). Piezo channels: From structure to function. *Pflügers Archiv, 467*, 95–99.

Wang, F., Knutson, K., Alcaino, C., Linden, D. R., Gibbons, S. J., Kashyap, P., … Beyder, A. (2016). Mechanosensitive ion channel Piezo2 is important for enterochromaffin cell response to mechanical forces. *Journal of Physiology*.

Wetzel, C., Hu, J., Riethmacher, D., Benckendorff, A., Harder, L., Eilers, A., … Lewin, G. R. (2007). A stomatin-domain protein essential for touch sensation in the mouse. *Nature, 445*, 206–209.

Woo, S. H., Lukacs, V., de Nooij, J. C., Zaytseva, D., Criddle, C. R., Francisco, A., … Patapoutian, A. (2015). Piezo2 is the principal mechanotransduction channel for proprioception. *Nature Neuroscience, 18*.

Woo, S. H., Ranade, S., Weyer, A. D., Dubin, A. E., Baba, Y., Qiu, Z., … Patapoutian, A. (2014). Piezo2 is required for Merkel-cell mechanotransduction. *Nature, 509*, 622–626.

Wu, Z., Grillet, N., Zhao, B., Cunningham, C., Harkins-Perry, S., Coste, B., … Ulrich, M. (2016). Mechanosensory hair cells express two molecularly distinct mechanotransduction channels. *Nature Neuroscience*.

Wu, J., Lewis, A. H., & Grandl, J. (2016). Touch, tension, and transduction - The function and regulation of Piezo ion channels. *Trends in Biochemical Sciences* [Epub 10/17/2016].

Zhao, Q., Wu, K., Geng, J., Chi, S., Wang, Y., Zhi, P., ... Xiao, B. (2016). Ion permeation and mechanotransduction mechanisms of mechanosensitive piezo channels. *Neuron, 89*, 1248–1263.
Zimmerman, A., Bai, L., & Ginty, D. D. (2014). The gentle touch receptors of mammalian skin. *Science, 346*, 950–954.
Zweers, G. A. (1977). *Mechanics of the feeding of the mallard (*Anas Platyrhynchos, *L.; Aves, Anseriformes)* (1st ed.). S Karger Pub.

CHAPTER EIGHT

Mechanosensitive Piezo Channels in the Gastrointestinal Tract

C. Alcaino, G. Farrugia and A. Beyder
Mayo Clinic College of Medicine, Rochester, MN, United States

Contents

Abstract

Sensation of mechanical forces is critical for normal function of the gastrointestinal (GI) tract and abnormalities in mechanosensation are linked to GI pathologies. In the GI tract there are several mechanosensitive cell types—epithelial enterochromaffin cells, intrinsic and extrinsic enteric neurons, smooth muscle cells and interstitial cells of Cajal. These cells use mechanosensitive ion channels that respond to mechanical forces by altering transmembrane ionic currents in a process called mechanoelectrical coupling. Several mechanosensitive ionic conductances have been identified in the mechanosensory GI cells, ranging from mechanosensitive voltage-gated sodium and calcium

Current Topics in Membranes, Volume 79
ISSN 1063-5823
http://dx.doi.org/10.1016/bs.ctm.2016.11.003

channels to the mechanogated ion channels, such as the two-pore domain potassium channels K2P (TREK-1) and nonselective cation channels from the transient receptor potential family. The recently discovered Piezo channels are increasingly recognized as significant contributors to cellular mechanosensitivity. Piezo1 and Piezo2 are nonselective cationic ion channels that are directly activated by mechanical forces and have well-defined biophysical and pharmacologic properties. The role of Piezo channels in the GI epithelium is currently under investigation and their role in the smooth muscle syncytium and enteric neurons is still not known. In this review, we outline the current state of knowledge on mechanosensitive ion channels in the GI tract, with a focus on the known and potential functions of the Piezo channels.

1. THE GASTROINTESTINAL TRACT MECHANOSENSITIVITY

Electromechanical organs, such as the heart and GI tract, are electrically excitable tissues with a primary mechanical function. These organs generate, and are subject to, mechanical forces that need to be detected as physiologic stimuli and as feedback signals to maintain normal function. Mechanosensitivity is critical for normal GI function and abnormalities in mechanosensitivity lead to disease. For example, in the stomach, distention is a critical determinant not only of gastric motility but also of satiety. Consequently, abnormalities in mechanosensation are associated with diseases such as obesity (Acosta et al., 2015). In the colon, alterations in mechanosensitivity lead to disorders of defecation, such as constipation (Neshatian et al., 2015), and may also be involved in the pathogenesis of colon cancer (Eisenhoffer et al., 2012; Fernandez-Sanchez et al., 2015).

2. MECHANOSENSITIVE CELLS IN THE GASTROINTESTINAL TRACT

At a fundamental level, all cells are mechanosensitive because of the need to sense normal physiologic forces (Sachs & Morris, 1998), such as cell crowding in the epithelium (Eisenhoffer et al., 2012). Given the importance of mechanosensation and ubiquity of mechanical forces, cells have developed several mechanisms of mechanosensation, ranging from integrins for mechanical interaction with substrates to mechanosensitivity of the nuclear envelope to modulate gene expression (reviewed in Eyckmans, Boudou, Yu, & Chen, 2011).

In the mechanically active tissues such as the gut, bladder, and heart, there are tissue-specific forces that are important for normal function. In these

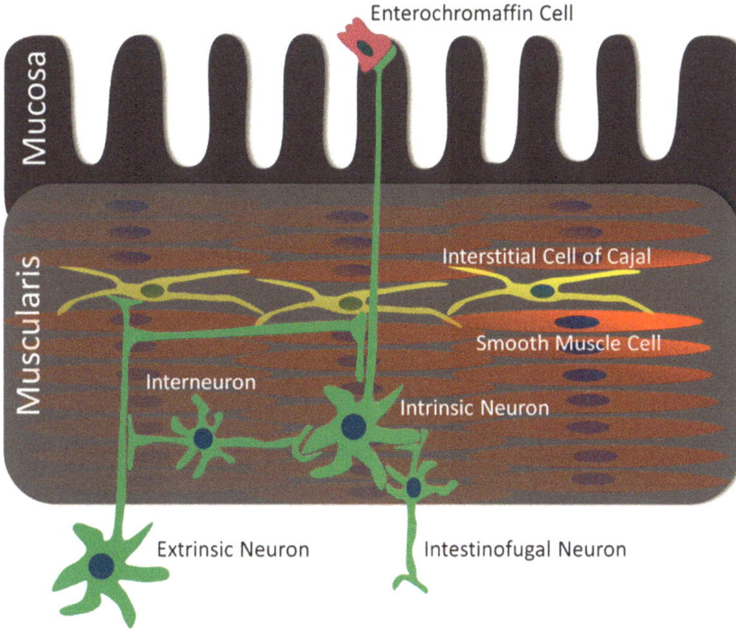

Figure 1 Mechanosensitive cells in the gastrointestinal (GI) tract. Mechanosensitive cells in the GI tract include epithelial enterochromaffin cells, smooth muscle cells (SMCs), interstitial cells of Cajal (ICCs), intrinsic neurons, including interneurons, intestinofugal neurons, and extrinsic neurons.

tissues, specific cells sense acute mechanical deformation. In the gastrointestinal (GI) tract several cell types are mechanosensitive (Fig. 1). The mechanosensitive cells in the GI tract are present within all layers of the wall: mucosa, submucosa, smooth muscle, and the submucosal and myenteric (between circular and longitudinal muscle) plexuses. In the epithelium, in response to mechanical forces, enterochromaffin (EC) cells release serotonin, which has well-documented effects on motility and secretion (reviewed by Mawe & Hoffman, 2013). In the mucosa, submucosa, and muscle layers, both intrinsic enteric and extrinsic nerves respond to mechanical stimuli. In the smooth muscle syncytium, both the smooth muscle pacemakers—the interstitial cells of Cajal (ICC) and smooth muscle cells (SMC) are mechanosensitive.

3. MECHANOSENSITIVE ION CHANNELS

Mechanosensory cells detect mechanical forces and transduce them into electrical signals using mechanosensitive ion channels (reviewed in

Hamill & Martinac, 2001). Mechanosensitive ion channels are transmembrane proteins that form ion conduction pores with gates that are strongly altered by mechanical forces (reviewed in Arnadottir & Chalfie, 2010; Hamill & Martinac, 2001; Ranade, Syeda, & Patapoutian, 2015). Previous work has identified mechanosensitive ionic currents in GI mechanosensory cells, and in some cases the molecular identities of the mechanosensitive ion channels are known. However, in majority of cases, the identities of the GI tract mechanosensitive ion channels are not known. The Piezo family of ion channels are an example of mechanosensitive ion channels whose molecular identity was only recently determined. Piezo1 and Piezo2 mechanosensitive ion channels are expressed throughout the length of the GI tract (Coste et al., 2010) and have distinct biophysical properties, such as nonselective cationic permeability, kinetics described by fast activation, and inactivation that is slow-to-medium (Piezo1) (Coste et al., 2010, 2012; Gottlieb & Sachs, 2012) or fast (Piezo2) (Coste et al., 2010, 2012). Pharmacologically, the Piezo channels are inhibited by the mechanosensitive channel blockers gadolinium (Gd^{3+}) and ruthenium red (RR), and specifically by the peptide blocker GsMTx-4 (Bae, Sachs, & Gottlieb, 2011; Coste et al., 2010, 2012; Gottlieb & Sachs, 2012). The Piezo channels have quickly found significance in several mechanosensitive systems, ranging from the sensation of light touch to preventing red blood cells from deformation-induced rupture.

In the following sections, we summarize the state of knowledge on mechanosensitive ion channels in the specific GI mechanosensitive cells, starting with the GI epithelium and working toward the extrinsic nerves that have their cell bodies outside the GI tract. Where possible, we will focus on the potential roles of the Piezo channels in GI mechanotransduction.

3.1 Gastrointestinal epithelium mechanosensitivity

GI epithelium is a sensor of both static and acute forces. Sensation of static forces is critical for cell density homeostasis, while sensation of acute forces is important for the processes of digestion and motility.

3.1.1 Static force detection by the gastrointestinal epithelium

Similar to other epithelia, such as skin and bladder, the GI epithelium is mechanosensitive. The GI epithelium serves as an important first point of interaction with the external environment (Fig. 1). In the intestinal epithelium, static forces are important for normal epithelial cell development and turnover. Studies show that mechanical forces are important for epithelial health, since abnormalities lead to diseases, including cancer

(Fernandez-Sanchez et al., 2015; Yang et al., 2014). For example, mechanical pressure caused by hyperproliferative adjacent crypts overexpressing active Notch or by exogenous force application was associated with increased Ret and β-catenin signaling, leading to aberrant crypt foci (Fernandez-Sanchez et al., 2015). The link between mechanical force and downstream biochemical pathway activation is not known. However, recent work shows that Piezo1 channels in the GI epithelium are important sensors of cell crowding (Eisenhoffer et al., 2012) and migration (Yang et al., 2014). During epithelial apoptosis, extrusion occurs when dying cells send signals to surrounding epithelial cells to contract and remove the dying cell (Slattum & Rosenblatt, 2014). In normal homeostasis, overcrowding due to proliferation and migration induces extrusion of live cells that helps control the number of cells in the epithelia, which occurs at sites where there is highest crowding. In vivo studies showed that disruption of Piezo1 channels by pharmacological agents (Gd^{3+}) or genetic knock down prevents extrusion and induces epithelial cell mass formation (Eisenhoffer et al., 2012). Formation of cell masses can promote tumorigenesis. Therefore, Piezo1 may provide the link between mechanical force and intracellular biochemical pathway activation, suggesting an important role of Piezo1 channels in the epithelium tumor-suppressive mechanisms (Slattum & Rosenblatt, 2014). These data are exciting, but further work is required to determine the molecular mechanism linking Piezo channels cellular phenotype and cancer formation.

3.1.2 Acute force detection by the gastrointestinal epithelium

The sensation of acute mechanical forces—such as intestinal contents deforming the epithelium is critical to normal day-to-day GI function. Mechanosensitive ion channels are important for acute force sensation in epithelia. In renal tubular epithelial cells, Piezo1 is responsible for a mechanosensitive ionic current (Peyronnet et al., 2013) and in urothelium, Piezo1 mediates mechanosensitive Ca^{2+} influx and ATP release (Miyamoto et al., 2014). In the gut epithelium, acute force sensation is performed by the specialized mechanosensory epithelial cells called EC cells (reviewed by Mawe & Hoffman, 2013) (Fig. 1). The EC cells have significant developmental and functional similarities to the Merkel cells in the skin that contribute to light touch sensation via Piezo2 channels (Ikeda & Gu, 2014; Ranade, Woo, et al., 2014; Woo et al., 2014). EC cells synthesize, store, and release a large amount of serotonin (5-HT) (Cote et al., 2003) in response to mechanical (Bulbring & Crema, 1959) and chemical (Racke

& Schworer, 1991; Racke, Reimann, Schworer, & Kilbinger, 1996; Schworer, Katsoulis, & Racke, 1992) stimuli. In turn, EC cell 5-HT is critical for normal GI secretion (Brown, 1996), motility (Bulbring & Lin, 1958), and sensation and is also an important hormone involved in cardiac function (Cote et al., 2003), metabolic health (Crane et al., 2015), and bone health (Yadav et al., 2010).

Mucosa forces, such as intraluminal pressure, produce stimulus-dependent 5-HT release from EC cells (Bertrand, 2006; Bertrand, Hu, Mach, & Bertrand, 2008; Bulbring & Crema, 1959; Patel, Bian, Quaiserova-Mocko, Galligan, & Swain, 2007), within milliseconds of mechanical stimulation (Bertrand, 2004). This mucosal 5-HT stimulates peristalsis (Bulbring & Crema, 1958) and secretion (Frieling, Wood, & Cooke, 1992) via several 5-HT receptors on mucosa-projecting neurons (Bertrand, Kunze, Furness, & Bornstein, 2000; Foxx-Orenstein, Kuemmerle, & Grider, 1996; Kadowaki, Wade, & Gershon, 1996; Neya, Mizutani, & Yamasato, 1993) and epithelium (Hoffman et al., 2012). Intriguingly similar to the Merkel cell, which forms Merkel cell—neurite complex (Haeberle et al., 2004), the enteroendocrine cells akin to EC cells may make direct contacts with submucosal nerves (Bohorquez et al., 2015).

While the concept of EC cell mechanosensitivity is well known, the molecular mechanisms of EC cell mechanosensitivity remain elusive mainly due to difficulties with identification, isolation, and culture of primary epithelial cells. Therefore, most of the current knowledge is derived from pharmacologic manipulation of GI tissues and immortalized cell lines. The pancreatic carcinoid tumor BON cell line produces 5-HT and releases it in response to rotational shaking. Serotonin release in response to this mechanical stimulus was associated with an intracellular Ca^{2+} increase (Christofi et al., 2004; Kim, Javed, Yu, Christofi, & Cooke, 2001) and depended on $G_{\alpha q}$ (Kim et al., 2001), adenosine (Chin et al., 2012; Christofi et al., 2004), and purine P2X and P2Y (Christofi et al., 2004; Linan-Rico et al., 2013) receptors. These studies suggest a central role for purines, and specifically ATP, in mechanically stimulated 5-HT release from the EC cells (reviewed in Cooke, Wunderlich, & Christofi, 2003). However, it is unknown whether these receptors serve as primary mechanosensors.

Several ion channels have been reported in EC cells, but none in relation to EC mechanosensitivity. Specifically, studies have described TRPA1 (Cho, Callaghan, Bron, Bravo, & Furness, 2014; Nozawa et al., 2009) and L-type calcium channels (Lomax, Gallego, Novalbos, Garcia, &

Warhurst, 1999; Raghupathi et al., 2013). It is accepted that cytoplasmic Ca^{2+} is important for 5-HT exocytosis in EC cells. However, whether mechanical activation of these ion channels is important for modulating 5-HT release is unknown.

In our recent work, we examined EC cell mechanosensitivity by leveraging the similarities between EC cells and Merkel cells in both function (Nakatani, Maksimovic, Baba, & Lumpkin, 2014; Raybould, Cooke, & Christofi, 2004) and development (Li, Ray, Singh, Johnston, & Leiter, 2011; Roach et al., 2013; Wright et al., 2015; Yang, Bermingham, Finegold, & Zoghbi, 2001). Merkel cells are the specialized skin epithelium light touch sensors, and recent studies have shown that Piezo2 is important for Merkel cell mechanotransduction (Ikeda et al., 2014; Maksimovic et al., 2014; Woo et al., 2014). We discovered that Piezo2 channels are important for EC cell mechanosensitivity (Wang et al., 2016) (Fig. 2). We found Piezo2 mRNA in both human and murine small bowel epithelia. Piezo2 immunolabeling showed that these channels were specifically distributed in 5-HT positive EC cells in human jejunum. We also showed that in a transgenic mouse model TPH1-CFP, where CFP expression is encoded in EC cells, Piezo2 label localizes specifically within the CFP positive cells in the epithelium (Fig. 2A–C). We then used an EC cell model and showed that single cell mechanical stimulation elicited nonselective rapidly inactivating currents consistent with Piezo2 (Fig. 2D–F). These currents were blocked by Gd^{3+}, GsMTx-4, and Piezo2 siRNA but not nontargeted siRNA (Fig. 2G and H). Pressure application resulted in 5-HT release, which was also prevented by treatment with pharmacological blockers (Gd^{3+} and D-GsMTx-4) and Piezo2 siRNA. Importantly, mucosal pressure increased 5-HT release and mucosal secretion in mouse small bowel and pressure-induced increase in secretion was significantly reduced by the Piezo2 blockers $Gd^{3+}$and RR and the specific Piezo blocker D-GsMtx-4 (Wang et al., 2016) (Fig. 2I).

Further work is required to determine the EC cell mechanotransduction mechanism linking Piezo2 activation by force to 5-HT release and downstream effects on tissue physiology and in vivo effects. Functional studies on primary EC cells will be critical to confirm and extend the findings that implicate Piezo2 in EC mechanosensation. Further, given the lethality of constitutive Piezo2 mouse knockouts, cell-specific Piezo2 knockout models will be required to unequivocally determine the Piezo2-dependent mechanism of mechanotransduction in EC cells.

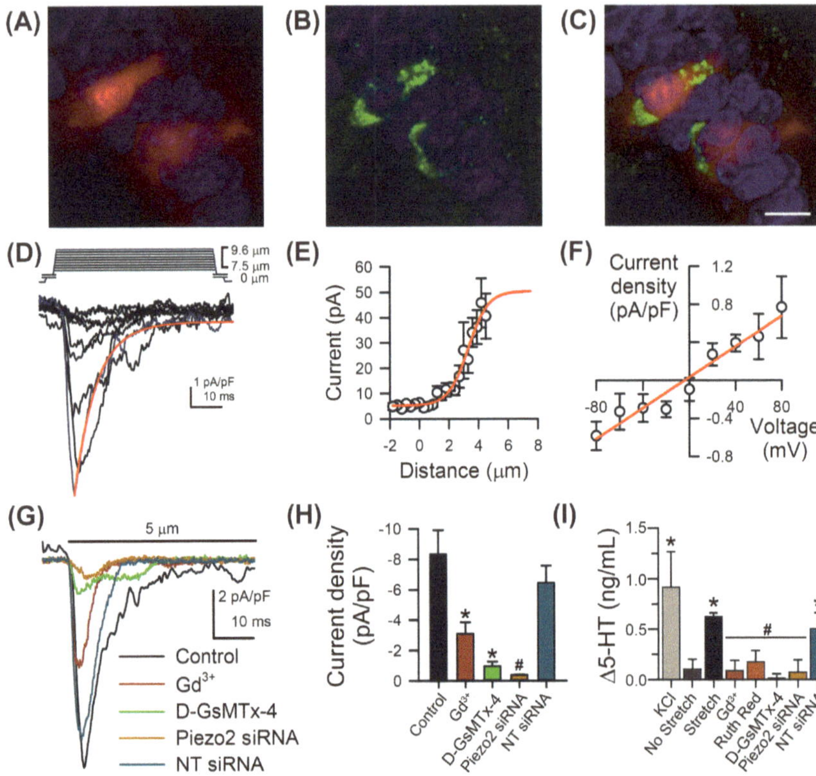

Figure 2 Piezo2 is important for enterochromaffin (EC) cell mechanotransduction. Immunohistochemistry showing that mouse (A) TPH1-CFP positive EC cells [red] label with (B) Piezo2 [green] (C). Nuclei labeled with DAPI, scale bar 10 μm. (D) Inward currents evoked by mechanical stimulation of EC cell model (QGP-1) are rapidly activating and inactivating [*red line*]. *Blue trace* represents peak current. (E) Current-deformation data are fit with two-state Boltzmann [*red line*] and (F) current–voltage relationship is linear [*red line*]. (G) The mechanically induced inward currents are blocked by Gd^{3+}, D-GsMTx-4, and Piezo2 siRNA. (H) Averaged peak current in control QGP-1 cells is inhibited by Gd^{3+}, D-GsMTx-4, and Piezo2 siRNA but not nontargeting (NT) siRNA. *$p < .05$ compared to no stretch. #$p < .05$ compared to NT siRNA. (I) Blockade of stretch-dependent 5-HT release from EC cell model QGP-1 by Piezo channel blockers Gd^{3+}, D-GsMTx-4, RR, and Piezo2 siRNA but not NT siRNA. *$p < .05$. #$p > .05$ compared to stretch. *Modified with permission from Wang, F., Knutson, K., Alcaino, C., Linden, D. R., Gibbons, S. J., Kashyap, P. K., … Beyder, A. (2016). Mechanosensitive ion channel Piezo2 is important for enterochromaffin cell response to mechanical forces.* The Journal of Physiology. *http://dx.doi.org/10.1113/JP272718.*

3.2 Gastrointestinal smooth muscle mechanosensitivity

Bayliss (1902) first described the active contractile responses to counteract increased transmural pressure in arterial wall smooth muscle more than a century ago. GI smooth muscle mechanosensitivity was suggested by

Bulbring (1955), when she measured muscle responses to stretch in the absence of neuronal input, the so-called myogenic contractions. The smooth muscle syncytium contains SMCs and their pacemakers—the ICC. Both cell types are mechanosensitive and little is currently known about how much each of these cell types contributes toward myogenic responses.

3.2.1 Gastrointestinal smooth muscle cell mechanosensitivity

The SMC is the workhorse of the smooth muscle organs like the gut (Fig. 1). SMC uses voltage-gated ion channels to convert electrical energy into contractions in a process called excitation–contraction coupling (Bolton, Prestwich, Zholos, & Gordienko, 1999). Therefore, voltage-gated ion channels are critical for normal SMC function. Studies show that both the mechanosensitive voltage-gated ion channels and mechanogated ion channels are expressed in SMCs.

3.2.2 Voltage-gated ion channel mechanosensitivity

Voltage-gated channels are important determinants of smooth muscle membrane potential and excitability. Both Ca_V and Na_V channels are expressed in SMC and have been found to respond to mechanical stimulation. The L-type voltage-gated calcium channels ($Ca_V1.2$) are sine *qua non* requirement for the contraction of SMC (Sanders, Koh, & Ward, 2006). A robust set of evidence indicates that $Ca_V1.2$ channels mediate SMC mechanosensitivity (Farrugia et al., 1999). GI L-type channel was sensitive to mechanical shear stress (Farrugia et al., 1999; Strege, Holm, et al., 2003) and osmotic stress (Kim, Rhee, & Kang, 2007). Mechanical stimulation increase peak current and speed up activation and inactivation, with no effect on voltage dependency (Farrugia et al., 1999; Lyford et al., 2002). $Ca_V1.2$ mechanosensitivity depended on the lipid bilayer (Kraichely, Strege, Sarr, Kendrick, & Farrugia, 2009) and not cytoskeleton (Strege, Holm, et al., 2003). In the gastric smooth muscle, while cytoskeletal disruption by cytochalasin significantly diminished L-type currents in isotonic solutions, cytoskeletal disruption did not alter L-type currents when cells were stimulated by hypoosmotic solutions (Kim et al., 2007). Similarly, a dihydropyridine-sensitive inward current increase has been found in rat myocytes in response to application of negative and positive pressure and after cell swelling. These currents were thought to be voltage-dependent Ca^{2+} currents (Langton, 1993). In the rat, uterine SMC stretch induced contractions due to Ca^{2+} influx, which modulates oxytocin-induced rhythmic contractions (Kasai, Tsutsumi, Taketani, Endo, & Iino, 1995).

A mechanosensitive voltage-gated sodium channel $Na_V1.5$ (encoded by *SCN5A*), as in ICC, is also found in SMC (Holm et al., 2002). Although multiple Na_V isoforms have been reported in GI SMC and ICC, the TTX-insensitive $Na_V1.5$ has been identified in human jejunum and colon and sodium currents have been shown to be activated by shear stress in SMC in a cytoskeleton-dependent manner, with membrane lipids also playing a role (Ou et al., 2002; Strege, Holm, et al., 2003). Pharmacological block of $Na_V1.5$ by the local anesthetic lidocaine reduced the rate of slow wave rise and increased slow wave duration, resulting in an overall decrease of the slow wave frequency (Strege, Ou, et al., 2003). There is also an important link to disease. *SCN5A* mutations from patients with irritable bowel syndrome (IBS) have been shown to generate $Na_V1.5$ currents of smaller density and reduced mechanosensitivity (Saito et al., 2009).

3.2.3 Potassium channel mechanosensitivity

Mechanogated K^+ channels TREK-1 were found in stomach (Ordway, Petrou, Kirber, Walsh, & Singer, 1995) and colonic SMC (Koh & Sanders, 2001). These channels are thought to contribute to membrane hyperpolarization and relaxation in certain regions of the GI tract, where relaxation is required upon mechanical stimulation, such as in gastric fundus and ascending colon that serve an important storage function (Sanders & Koh, 2006). The two-pore domain mechanosensitive K^+ channel TREK-1, known to be mechanosensitive (Berrier et al., 2013), is present and functionally relevant for mechanosensation in GI smooth muscle (Hwang et al., 2008; Sanders & Koh, 2006). Patch-clamp experiments in freshly dissociated SMC from toad stomach have shown an increase in channel activation after pressure, which was increased with the application of fatty acids (Ordway et al., 1995). Other potassium channels such as BK_{Ca} are also activated by mechanical stretch (Wang, Huang, et al., 2010), generating a large outward current that hyperpolarizes the cell membrane and reduces SMC excitability in the colon.

3.2.4 Nonselective cation channel mechanosensitivity

Piezo channels are known to be nonselective cationic channels (Coste et al., 2010), with transient receptor potential (TRP) channels like TRPA1 and TRPC1 also fitting into this category (Maroto et al., 2005; Paulsen, Armache, Gao, Cheng, & Julius, 2015). Mechanosensitive nonselective cation channels are present in intestinal smooth muscle (Davis, Donovitz, & Hood, 1992). Pressurized patches from single SMCs from toad stomach showed nonselective cation current (Kirber, Walsh, & Singer, 1988). These hyperpolarization-activated currents were found to be mostly K^+ and Na^+ permeable and are thought to stabilize the membrane resting potential or control spontaneous electrical activity (Hisada, Ordway, Kirber, Singer, & Walsh, 1991; Hisada, Singer, & Walsh, 1993). Calcium entry has been visualized through single stretch-activated cationic channels in SMCs while applying negative pressure in the patch pipette (Zou, Lifshitz, Tuft, Fogarty, & Singer, 2002).

Gd^{3+}, which blocks many stretch-activated ion channels (Yang & Sachs, 1989), stopped muscle contraction and prevented action potential firing in stretched tissue (Kunze, Clerc, Bertrand, & Furness, 1999). Guinea pig gastric SMCs also express mechanosensitive nonselective cation ion channels inhibited by Gd^{3+} and with kinetic and conductance properties similar to Piezo1 channels (Yamamoto & Suzuki, 1996). In vascular smooth muscle, stretch-activated single channels resembling Piezo1 can be blocked by Gd^{3+}, GsMTx-4, and streptomycin (Ducret et al., 2010). GsMTx-4-inhibited mechanosensitive cation current was modulated by the balance of polycystins TRPP1 and TRPP2, but this mechanism has not been explored in intestinal smooth muscle (Sharif-Naeini et al., 2009).

Activation of nonselective cation channels in smooth muscle regulates electrical excitability, but recent evidence from vascular smooth muscle suggests that in addition to this role, Piezo1 plays a role in smooth muscle development via regulation of vascular endothelium shear stress sensitivity (Ranade, Qiu, et al., 2014) and facilitating secretion of the cross-linking enzymes required for smooth muscle remodeling (Retailleau et al., 2015). The noncanonical roles of the Piezo proteins in GI smooth muscle are currently unexplored.

In summary, mechanosensitive ion channels play an important role in GI SMC mechanotransduction. Voltage-gated Ca^{2+} and Na^+ and non-voltage-gated K_{2P} and nonselective cationic channels have all been described as mechanosensors. Piezo channels have been found in vascular smooth muscle and similar currents that are inhibited by the mechanosensitive channel blockers such as Gd^{3+} have been shown in GI SMCs. Based on this evidence, it is likely that Piezo channels are involved in GI SMC mechanosensitivity. However, additional studies are needed to determine both canonical and noncanonical roles of Piezo channels in GI smooth muscle.

3.2.5 Interstitial cell of Cajal mechanosensitivity

The ICC are found throughout the length of the gut—from esophagus to anus and are classified according to their location in the GI wall (Komuro, 2006) (Fig. 1). The ICC located in the myenteric plexus (ICC-MY) in the small bowel and submucosal border (ICC-SM) in the colon are known as pacemakers because they generate and spread an electrical rhythm, termed the "slow wave" (Huizinga et al., 1995). This ICC-generated electrical slow wave leads to the SMC depolarization that culminates in the excitation–contraction coupling by the SMC. ICC are implicated in the myogenic stretch responses, since, for example, stretching of mouse gastric muscle (Won, Sanders, & Ward, 2005) and human jejunum (Strege, Ou, et al., 2003) increased slow wave frequency independent of neuronal mechanisms. Importantly, responses to stretch were absent in transgenic mice that lack intramuscular ICC, suggesting a role of ICC in stretch-dependent responses in the GI tract, but the mechanisms of mechanosensation are not fully established (Won et al., 2005).

ICC have several ion channel types, some of which are known in other systems to be mechanosensitive (reviewed in Kraichely & Farrugia, 2007). The *SCN5A* encoded mechanosensitive voltage-gated sodium channel Nav1.5 (Gellens et al., 1992) has been identified in intestinal ICC and experiments using shear stress in isolated ICC increased and accelerated Na_V currents (Strege, Ou, et al., 2003). A volume-activated chloride current has been found in murine-cultured jejunum ICC after cell swelling (Park, McKay, Zhu, & Huizinga, 2005). In cultured bladder ICC from guinea pig, chronic stretch leads to an increase in stretch-induced calcium transients (Wang, Fang, et al., 2010). Here, a stretch-dependent increase in intracellular calcium has been involved in bladder excitatory regulation (Wang, Fang, et al., 2010). Expression of the TRP channel subtypes TRPC4 and TRPM7 has been found in murine ICC (Kim et al., 2005; Walker, Koh, Sergeant, Sanders, & Horowitz, 2002). However, while some TRP channels are known to be mechanosensitive, TRPC4 and TRPM7 are not. Overall, our current knowledge on mechanosensitive channels in GI ICC is still very limited.

3.3 Mechanosensitivity of the enteric neurons

The enteric nervous system includes both intrinsic and extrinsic neurons and it is critical for autonomous GI function (Fig. 1). The intrinsic neurons are contained completely within the GI tract and in general they regulate regional GI motor and secretory/absorptive functions, while the extrinsic neurons have their soma outside the GI tract (Fig. 1). Extrinsic nerves also regulate GI motility, but in addition supply the GI tract with sensory capabilities. Both intrinsic and extrinsic neurons are mechanosensitive.

3.3.1 Intrinsic enteric neuron mechanosensitivity

Mechanosensitive enteric neurons control reflex activity by responding to mechanical stress in the gut wall. A particular type of enteric neurons, termed intrinsic primary afferent neurons (IPANs), is mechanosensitive (Furness, Kunze, Bertrand, Clerc, & Bornstein, 1998) (Fig. 1). The IPANs project to the mucosa and actively communicate with each other and other neuronal types (Kunze, Furness, Bertrand, & Bornstein, 1998). IPANs fire action potentials after mechanical stimulation of their soma or processes, and their responses can be abolished pharmacologically preventing muscle contraction (Kunze et al., 1999, 1998).

Recent studies have challenged the concept that IPANs alone are mechanosensitive, instead work has suggested that enteric neurons are more broadly mechanosensitive (Schemann & Mazzuoli, 2010; Smith, Spencer, Hennig, & Dickson, 2007). Several studies have reported distinct mechanosensitive responses in enteric neurons to tensile stress (stretch), compressive stress (volume injection, glass probe, or von Frey hair), shear stress, and cell swelling (hypoosmotic solutions) (Dong, Jiang, Dong, & Mittal, 2014; Hibberd, Zagorodnyuk, Spencer, & Brookes, 2012; Kugler et al., 2015; Kunze et al., 1999; Kunze, Clerc, Furness, & Gola, 2000; Kunze et al., 1998; Mayer & Wood, 1975; Mazzuoli & Schemann, 2009, 2012; Spencer & Smith, 2004). While these studies suggest roles in mechanosensitivity sensing and control of muscle activity, as well as a servo-feedback loop (Mazzuoli-Weber & Schemann, 2015), the identities of the mechanosensitive ion channels responsible for mechanotransduction are largely unknown. Only one study identified a molecular transduction mechanism, which involved the activation of a large conductance BK-like potassium channel (Kunze et al., 2000). In this study, enteric neurons were subjected to patch clamp in cell-attached and whole cell configurations while pressing the cell processes with a fine probe. Pressure on the cell surface initiated generator potentials and depolarization. These potentials persisted when synaptic transmission was blocked by solutions containing low Ca^{2+} (0.2 mM) and high Mg^{2+} (10 mM). In patches, application of intraelectrode pressure increased the open probability of BK channels.

3.3.2 Extrinsic enteric neuron mechanosensitivity

Piezo channels are important mechanotransducers in the somatosensory system. Piezo2 ion channels are critical for mechanotransduction of light touch sensation by the Merkel cell—neurite complex formed between the epithelial Merkel cells and the sensory neurons that innervate the skin (Dubin et al., 2012; Ranade, Woo, et al., 2014), (reviewed in Woo, Lumpkin, & Patapoutian, 2015). Mechanical forces are sensed directly by Piezo2 channels in both Merkel cells and somatosensory neurons, which allow for a complex encoding of mechanical forces (Dubin et al., 2012; Ranade, Woo, et al., 2014).

While the role of Piezo channels in somatosensation is being established, their full role in the autonomic nervous system is still unclear. Sensory information from the GI tract is acquired by sensory afferent nerves that have their cell bodies outside the GI tract, although a unique type of afferents called intestinofugal afferent neurons (IFANS) have their bodies and processes within the gut but their axons project away from the bowel (Fig. 1). IFANS communicate with prevertebral ganglion neurons known as PVG (reviewed in Szurszewski, Ermilov, & Miller, 2002). These neurons respond to stretch of circular muscle but not tension and colonic distension activates their slow adapting mechanoreceptors, generating acetylcholine release in the PVG and evoking action potentials (Parkman, Ma, Stapelfeldt, & Szurszewski, 1993). Functionally, their role relies on relaxation of the colonic wall during filling, opposing the depolarization and contraction of colonic SMC, since the intraluminal content generates distention. This mechanism regulates the motor activity of the intestine (reviewed in Szurszewski et al., 2002).

Vagal and spinal sensory neurons have been shown to be mechanosensitive, where neuron deformation modulates reflex activity (Berthoud, Blackshaw, Brookes, & Grundy, 2004). Vagal afferent nerves detect tension at physiological levels (Andrews, 1986), where they are important in the control of normal GI function. Mechanoreceptors in the vagal afferents can be divided in two groups: those responding to distension or contraction of the gut wall and those that can be activated by chemical or mechanical stimulation of the mucosa (Page, Martin, & Blackshaw, 2002).

Spinal afferents have high threshold of distension detection and can respond to a wide range of stimuli extending to the noxious range (Blackshaw, Brookes, Grundy, & Schemann, 2007; Rong et al., 2004). Their cell bodies lie on the dorsal root ganglia (DRG) and their peripheral projections extend to the muscle layers, mucosal epithelia and enteric ganglia (Blackshaw et al., 2007).

3.3.3 *Role of mechanosensitive ion channels in visceral mechanosensation*

Direct activation of ion channels on the afferent endings could mediate visceral mechanosensation (Fig. 1). In guinea pig, mechanosensitive ion channels are likely involved (Zagorodnyuk, Chen, Costa, & Brookes, 2003; Zagorodnyuk, Lynn, Costa, & Brookes, 2005). The vagal mechanoreceptors responses were not affected by Ca^{2+} depletion and mechanotransduction was thought to be mediated via benzamil-sensitive, stretch-activated ion channels, which were not affected by high concentrations of Gd^{3+} (300 μM), with no involvement of chemical transmission (Zagorodnyuk et al., 2003, 2005).

The acid-sensing ion channel (ASIC) family has been proposed to play a role as a mechanosensory molecule because of its homology to the invertebrate cationic channel DEG/ENaC, known to be involved in touch sensation (Price et al., 2001). Expression of ASIC channels has been found in splanchnic DRG and vagal gastroesophageal neurons (Hughes, Brierley, Young, & Blackshaw, 2007; Page et al., 2005). All three types of ASIC channels expressed in visceral neurons are known to be required for normal mechanosensation, but their specific role in modulation of visceral mechanosensitivity is masked by the variability of positive and negative effects elicited by knocking down and mutating these channels (reviewed in Brierley, 2010). TRP channels have been involved in gut hypersensitivity (Brierley et al., 2009). TRPA1 expresses in enteric neurons (Cattaruzza et al., 2010; Poole et al., 2011). In TRPA1 KO mice behavioral responses to noxious colonic distension were reduced, with TRPA1 agonists causing mechanical hypersensitivity (Brierley et al., 2009). TRPV1 and TRPV4 also contributed to gut mechanosensitivity. The capsaicin-activated TRPV1 is expressed in colonic neurons (Brierley, Jones, Xu, Gebhart, & Blackshaw, 2005; Christianson, McIlwrath, Koerber, & Davis, 2006; Robinson, McNaughton, Evans, & Hicks, 2004). TRPV1 deletion has been shown to reduce the mechanosensitivity of distension-sensitive afferent neurons from the gastroesophageal region, jejunum, and pelvic colon (Bielefeldt & Davis, 2008; Jones, Xu, & Gebhart, 2005; Rong et al., 2004). Interestingly, TRPV1 is not a mechanogated channel and its effects on mechanosensation are thought to be due to indirect interactions with other TRP channels, such as TRPA1 or effects on neuron excitability. In colonic serosal afferents, capsaicin-dependent desensitization of TRPV1 was prevented in TRPA1 KO mice (Brierley et al., 2009; Levine & Alessandri-Haber, 2007). TRPV4 was also found along DRG neurons in the gut

(Brierley et al., 2008). TRPV4 agonists potentiated mechanosensory responses in wild-type mice, whereas the nonselective blocker RR reduced afferents mechanosensitivity, with no effect on TRPV4 KO mice (Brierley et al., 2008; Sipe et al., 2008). Additionally, TRPV4 was also thought to play a role in gut hypersensitivity (Cenac et al., 2010).

Piezo channels represent a good candidate for mechanotransduction of extrinsic signals in the gut. Recent work suggests that in rat visceral afferent (DRG) neurons rely on Piezo2 channels for visceral sensation (Yang et al., 2016). In control rats, Piezo2 knockdown by shRNA by intrathecal injection resulted in a decreased visceromotor response to innocuous stimuli but not noxious stimuli. In a model of IBS the increase in visceromotor responses to both innocuous and noxious stimuli was inhibited by Piezo2 knockdown in DRGs. Since the alterations of the visceromotor responses in response to irritation occurred without an increase in Piezo2 mRNA or protein, the authors suggested that other channels or chemical mediators are involved in noxious responses. Further work is required to confirm these findings and to extend these discoveries toward mechanistic understanding.

In summary, several TRP and ASIC channels are known to be expressed in extrinsic nerves and may be required for normal mechanosensation, but knocking down of these channels does not knock down mechanosensitivity in extrinsic afferents. Recent work suggests that Piezo2 is involved in visceral sensitivity but the extent of involvement in response to innocuous versus noxious stimuli is unclear. Given the apparently shared role of ASIC and TRP channel subtypes in regulation of mechano responses, one could speculate these channels form complexes and Piezo channels could also be interacting.

4. SUMMARY AND CONCLUSIONS

Mechanosensation is critical for normal GI tract function. Multiple cell types throughout the wall of the GI tract are mechanosensitive, from the epithelial EC cells to SMCs and their pacemakers and to neurons, both extrinsic and intrinsic. These cells use mechanosensitive ion channels to convert mechanical forces into chemical signals. In the GI tract both the mechanosensitive voltage-gated ion channels and mechanogated ion channels are important for normal physiology.

The mammalian Piezo1 and Piezo2 proteins are recently discovered eukaryotic mechanosensitive cationic channels. Piezo channels are known to be critical for a wide set of mechanotransduction processes, including nociception, light touch, proprioception, volume regulation of erythrocytes, and vascular function. Piezo expression has been found along the GI tract in stomach, small bowel, and colon but currently little is known about their specific localization within the GI mechanosensory cells. Mechanosensitive cells in GI have been shown to express a wide range of mechanosensitive channels and nonselective cationic currents with properties that could match those of Piezo channels have been described. Moreover, mechanosensitive channel blockers such as Gd^{3+} and RR have been shown to block mechanical responses in SMC and ICC. In the GI epithelium, EC cells express Piezo2 and the recent findings suggest the role of Piezo2 as the principal EC cell mechanotransducer, where mechanically stimulated 5-HT release is Piezo2 dependent. However, further work is needed to understand the downstream effects of mechanical activation of Piezo2 in EC cells. The role of Piezo2 in mechanotransduction of mechanical forces in GI epithelium represents only one of the potential roles of Piezo channels in GI mechanosensitivity.

ACKNOWLEDGMENTS

The authors thank Mrs Jennifer Rud for administrative assistance and Mr. Peter Strege who contributed to figure preparation. Financial support to AB from NIH K08 (DK106456) and 2015 American Gastroenterological Association Research Scholar Award (AGA RSA), Pilot and Feasibility Grant from Mayo Clinic Center for Cell Signaling in Gastroenterology (NIH P30DK084567), and to GF from NIH R01 (DK52766).

REFERENCES

Acosta, A., Camilleri, M., Shin, A., Vazquez-Roque, M. I., Iturrino, J., Burton, D., … Zinsmeister, A. R. (2015). Quantitative gastrointestinal and psychological traits associated with obesity and response to weight-loss therapy. *Gastroenterology, 148*(3), 537–546. http://dx.doi.org/10.1053/j.gastro.2014.11.020. e534.

Andrews, P. L. (1986). Vagal afferent innervation of the gastrointestinal tract. *Progress in Brain Research, 67*, 65–86.

Arnadottir, J., & Chalfie, M. (2010). Eukaryotic mechanosensitive channels. *Annual Review of Biophysics, 39*, 111–137. http://dx.doi.org/10.1146/annurev.biophys.37.032807.125836.

Bae, C., Sachs, F., & Gottlieb, P. A. (2011). The mechanosensitive ion channel Piezo1 is inhibited by the peptide GsMTx4. *Biochemistry, 50*(29), 6295–6300. http://dx.doi.org/10.1021/bi200770q.

Bayliss, W. M. (1902). On the local reactions of the arterial wall to changes of internal pressure. *The Journal of Physiology, 28*(3), 220–231.

Berrier, C., Pozza, A., de Lacroix de Lavalette, A., Chardonnet, S., Mesneau, A., Jaxel, C., … Ghazi, A. (2013). The purified mechanosensitive channel TREK-1 is directly sensitive to membrane tension. *The Journal of Biological Chemistry, 288*(38), 27307–27314. http://dx.doi.org/10.1074/jbc.M113.478321.

Berthoud, H. R., Blackshaw, L. A., Brookes, S. J., & Grundy, D. (2004). Neuroanatomy of extrinsic afferents supplying the gastrointestinal tract. *Neurogastroenterology and Motility, 16*(Suppl. 1), 28–33. http://dx.doi.org/10.1111/j.1743-3150.2004.00471.x.

Bertrand, P. P. (2004). Real-time detection of serotonin release from enterochromaffin cells of the Guinea-pig ileum. *Neurogastroenterology and Motility, 16*(5), 511–514. http://dx.doi.org/10.1111/j.1365-2982.2004.00572.x.

Bertrand, P. P. (2006). Real-time measurement of serotonin release and motility in Guinea pig ileum. *The Journal of Physiology, 577*(Pt 2), 689–704. http://dx.doi.org/10.1113/jphysiol.2006.117804.

Bertrand, P. P., Hu, X., Mach, J., & Bertrand, R. L. (2008). Serotonin (5-HT) release and uptake measured by real-time electrochemical techniques in the rat ileum. *American Journal of Physiology. Gastrointestinal and Liver Physiology, 295*(6), G1228–G1236. http://dx.doi.org/10.1152/ajpgi.90375.2008.

Bertrand, P. P., Kunze, W. A., Furness, J. B., & Bornstein, J. C. (2000). The terminals of myenteric intrinsic primary afferent neurons of the Guinea-pig ileum are excited by 5-hydroxytryptamine acting at 5-hydroxytryptamine-3 receptors. *Neuroscience, 101*(2), 459–469.

Bielefeldt, K., & Davis, B. M. (2008). Differential effects of ASIC3 and TRPV1 deletion on gastroesophageal sensation in mice. *American Journal of Physiology. Gastrointestinal and Liver Physiology, 294*(1), G130–G138. http://dx.doi.org/10.1152/ajpgi.00388.2007.

Blackshaw, L. A., Brookes, S. J., Grundy, D., & Schemann, M. (2007). Sensory transmission in the gastrointestinal tract. *Neurogastroenterology and Motility, 19*(Suppl. 1), 1–19. http://dx.doi.org/10.1111/j.1365-2982.2006.00871.x.

Bohorquez, D. V., Shahid, R. A., Erdmann, A., Kreger, A. M., Wang, Y., Calakos, N., … Liddle, R. A. (2015). Neuroepithelial circuit formed by innervation of sensory enteroendocrine cells. *Journal of Clinical Investigation*. http://dx.doi.org/10.1172/JCI78361.

Bolton, T. B., Prestwich, S. A., Zholos, A. V., & Gordienko, D. V. (1999). Excitation-contraction coupling in gastrointestinal and other smooth muscles. *Annual Review of Physiology, 61*, 85–115. http://dx.doi.org/10.1146/annurev.physiol.61.1.85.

Brierley, S. M. (2010). Molecular basis of mechanosensitivity. *Autonomic Neuroscience, 153*(1–2), 58–68. http://dx.doi.org/10.1016/j.autneu.2009.07.017.

Brierley, S. M., Hughes, P. A., Page, A. J., Kwan, K. Y., Martin, C. M., O'Donnell, T. A., … Blackshaw, L. A. (2009). The ion channel TRPA1 is required for normal mechanosensation and is modulated by algesic stimuli. *Gastroenterology, 137*(6), 2084–2095. http://dx.doi.org/10.1053/j.gastro.2009.07.048. e2083.

Brierley, S. M., Jones, R. C., 3rd, Xu, L., Gebhart, G. F., & Blackshaw, L. A. (2005). Activation of splanchnic and pelvic colonic afferents by bradykinin in mice. *Neurogastroenterology and Motility, 17*(6), 854–862. http://dx.doi.org/10.1111/j.1365-2982.2005.00710.x.

Brierley, S. M., Page, A. J., Hughes, P. A., Adam, B., Liebregts, T., Cooper, N. J., … Blackshaw, L. A. (2008). Selective role for TRPV4 ion channels in visceral sensory pathways. *Gastroenterology, 134*(7), 2059–2069. http://dx.doi.org/10.1053/j.gastro.2008.01.074.

Brown, D. R. (1996). Mucosal protection through active intestinal secretion: Neural and paracrine modulation by 5-hydroxytryptamine. *Behavioural Brain Research, 73*(1–2), 193–197.

Bulbring, E. (1955). Correlation between membrane potential, spike discharge and tension in smooth muscle. *The Journal of Physiology, 128*(1), 200–221.

Bulbring, E., & Crema, A. (1958). Observations concerning the action of 5-hydroxytryptamine on the peristaltic reflex. *British Journal of Pharmacology and Chemotherapy, 13*(4), 444–457.

Bulbring, E., & Crema, A. (1959). The release of 5-hydroxytryptamine in relation to pressure exerted on the intestinal mucosa. *The Journal of Physiology, 146*(1), 18–28.

Bulbring, E., & Lin, R. C. (1958). The effect of intraluminal application of 5-hydroxytryptamine and 5-hydroxytryptophan on peristalsis; the local production of 5-HT and its release in relation to intraluminal pressure and propulsive activity. *The Journal of Physiology, 140*(3), 381–407.

Cattaruzza, F., Spreadbury, I., Miranda-Morales, M., Grady, E. F., Vanner, S., & Bunnett, N. W. (2010). Transient receptor potential ankyrin-1 has a major role in mediating visceral pain in mice. *American Journal of Physiology. Gastrointestinal and Liver Physiology, 298*(1), G81–G91. http://dx.doi.org/10.1152/ajpgi.00221.2009.

Cenac, N., Altier, C., Motta, J. P., d'Aldebert, E., Galeano, S., Zamponi, G. W., & Vergnolle, N. (2010). Potentiation of TRPV4 signalling by histamine and serotonin: An important mechanism for visceral hypersensitivity. *Gut, 59*(4), 481–488. http://dx.doi.org/10.1136/gut.2009.192567.

Chin, A., Svejda, B., Gustafsson, B. I., Granlund, A. B., Sandvik, A. K., Timberlake, A., … Kidd, M. (2012). The role of mechanical forces and adenosine in the regulation of intestinal enterochromaffin cell serotonin secretion. *American Journal of Physiology. Gastrointestinal and Liver Physiology, 302*(3), G397–G405. http://dx.doi.org/10.1152/ajpgi.00087.2011.

Cho, H. J., Callaghan, B., Bron, R., Bravo, D. M., & Furness, J. B. (2014). Identification of enteroendocrine cells that express TRPA1 channels in the mouse intestine. *Cell and Tissue Research.* http://dx.doi.org/10.1007/s00441-013-1780-x.

Christianson, J. A., McIlwrath, S. L., Koerber, H. R., & Davis, B. M. (2006). Transient receptor potential vanilloid 1-immunopositive neurons in the mouse are more prevalent within colon afferents compared to skin and muscle afferents. *Neuroscience, 140*(1), 247–257. http://dx.doi.org/10.1016/j.neuroscience.2006.02.015.

Christofi, F. L., Kim, M., Wunderlich, J. E., Xue, J., Suntres, Z., Cardounel, A., … Cooke, H. J. (2004). Endogenous adenosine differentially modulates 5-hydroxytryptamine release from a human enterochromaffin cell model. *Gastroenterology, 127*(1), 188–202.

Cooke, H. J., Wunderlich, J., & Christofi, F. L. (2003). "The force be with you": ATP in gut mechanosensory transduction. *News in Physiological, 18*, 43–49.

Coste, B., Mathur, J., Schmidt, M., Earley, T. J., Ranade, S., Petrus, M. J., … Patapoutian, A. (2010). Piezo1 and Piezo2 are essential components of distinct mechanically activated cation channels. *Science, 330*(6000), 55–60. http://dx.doi.org/10.1126/science.1193270.

Coste, B., Xiao, B., Santos, J. S., Syeda, R., Grandl, J., Spencer, K. S., … Patapoutian, A. (2012). Piezo proteins are pore-forming subunits of mechanically activated channels. *Nature, 483*(7388), 176–181. http://dx.doi.org/10.1038/nature10812nature10812.

Cote, F., Thevenot, E., Fligny, C., Fromes, Y., Darmon, M., Ripoche, M. A., … Vodjdani, G. (2003). Disruption of the nonneuronal tph1 gene demonstrates the importance of peripheral serotonin in cardiac function. *Proceedings of the National Academy of Sciences of the United States of America, 100*(23), 13525–13530. http://dx.doi.org/10.1073/pnas.2233056100.

Crane, J. D., Palanivel, R., Mottillo, E. P., Bujak, A. L., Wang, H., Ford, R. J., … Steinberg, G. R. (2015). Inhibiting peripheral serotonin synthesis reduces obesity and metabolic dysfunction by promoting brown adipose tissue thermogenesis. *Nature Medicine, 21*(2), 166–172. http://dx.doi.org/10.1038/nm.3766.

Davis, M. J., Donovitz, J. A., & Hood, J. D. (1992). Stretch-activated single-channel and whole cell currents in vascular smooth muscle cells. *The American Journal of Physiology, 262*(4 Pt 1), C1083–C1088.

Dong, H., Jiang, Y., Dong, J., & Mittal, R. (2014). Inhibitory motor neurons of the esophageal myenteric plexus are mechanosensitive. *American Journal of Physiology. Cell Physiology*. http://dx.doi.org/10.1152/ajpcell.00159.2014.

Dubin, A. E., Schmidt, M., Mathur, J., Petrus, M. J., Xiao, B., Coste, B., & Patapoutian, A. (2012). Inflammatory signals enhance piezo2-mediated mechanosensitive currents. *Cell Reports, 2*(3), 511–517. http://dx.doi.org/10.1016/j.celrep.2012.07.014.

Ducret, T., El Arrouchi, J., Courtois, A., Quignard, J. F., Marthan, R., & Savineau, J. P. (2010). Stretch-activated channels in pulmonary arterial smooth muscle cells from normoxic and chronically hypoxic rats. *Cell Calcium, 48*(5), 251–259. http://dx.doi.org/10.1016/j.ceca.2010.09.011. pii:S0143-4160(10)00147-8.

Eisenhoffer, G. T., Loftus, P. D., Yoshigi, M., Otsuna, H., Chien, C. B., Morcos, P. A., & Rosenblatt, J. (2012). Crowding induces live cell extrusion to maintain homeostatic cell numbers in epithelia. *Nature, 484*(7395), 546–549. http://dx.doi.org/10.1038/nature10999.

Eyckmans, J., Boudou, T., Yu, X., & Chen, C. S. (2011). A hitchhiker's guide to mechanobiology. *Developmental Cell, 21*(1), 35–47. http://dx.doi.org/10.1016/j.devcel.2011.06.015.

Farrugia, G., Holm, A. N., Rich, A., Sarr, M. G., Szurszewski, J. H., & Rae, J. L. (1999). A mechanosensitive calcium channel in human intestinal smooth muscle cells. *Gastroenterology, 117*(4), 900–905.

Fernandez-Sanchez, M. E., Barbier, S., Whitehead, J., Bealle, G., Michel, A., Latorre-Ossa, H., … Farge, E. (2015). Mechanical induction of the tumorigenic beta-catenin pathway by tumour growth pressure. *Nature*. http://dx.doi.org/10.1038/nature14329.

Foxx-Orenstein, A. E., Kuemmerle, J. F., & Grider, J. R. (1996). Distinct 5-HT receptors mediate the peristaltic reflex induced by mucosal stimuli in human and Guinea pig intestine. *Gastroenterology, 111*(5), 1281–1290.

Frieling, T., Wood, J. D., & Cooke, H. J. (1992). Submucosal reflexes: Distension-evoked ion transport in the Guinea pig distal colon. *The American Journal of Physiology, 263*(1 Pt 1), G91–G96.

Furness, J. B., Kunze, W. A., Bertrand, P. P., Clerc, N., & Bornstein, J. C. (1998). Intrinsic primary afferent neurons of the intestine. *Progress in Neurobiology, 54*(1), 1–18.

Gellens, M. E., George, A. L., Jr., Chen, L. Q., Chahine, M., Horn, R., Barchi, R. L., & Kallen, R. G. (1992). Primary structure and functional expression of the human cardiac tetrodotoxin-insensitive voltage-dependent sodium channel. *Proceedings of the National Academy of Sciences of the United States of America, 89*(2), 554–558.

Gottlieb, P. A., & Sachs, F. (2012). Piezo1: Properties of a cation selective mechanical channel. *Channels, 6*(4), 214–219. http://dx.doi.org/10.4161/chan.21050.

Haeberle, H., Fujiwara, M., Chuang, J., Medina, M. M., Panditrao, M. V., Bechstedt, S., … Lumpkin, E. A. (2004). Molecular profiling reveals synaptic release machinery in Merkel cells. *Proceedings of the National Academy of Sciences of the United States of America, 101*(40), 14503–14508. http://dx.doi.org/10.1073/pnas.0406308101.

Hamill, O. P., & Martinac, B. (2001). Molecular basis of mechanotransduction in living cells. *Physiological Reviews, 81*, 685–740.

Hibberd, T. J., Zagorodnyuk, V. P., Spencer, N. J., & Brookes, S. J. (2012). Identification and mechanosensitivity of viscerofugal neurons. *Neuroscience, 225*, 118–129. http://dx.doi.org/10.1016/j.neuroscience.2012.08.040.

Hisada, T., Ordway, R. W., Kirber, M. T., Singer, J. J., & Walsh, J. V., Jr. (1991). Hyperpolarization-activated cationic channels in smooth muscle cells are stretch sensitive. *Pflugers Archiv, 417*(5), 493–499.

Hisada, T., Singer, J. J., & Walsh, J. V., Jr. (1993). Aluminofluoride activates hyperpolarization- and stretch-activated cationic channels in single smooth muscle cells. *Pflugers Archiv, 422*(4), 397–400.

Hoffman, J. M., Tyler, K., MacEachern, S. J., Balemba, O. B., Johnson, A. C., Brooks, E. M., … Mawe, G. M. (2012). Activation of colonic mucosal 5-HT(4) receptors accelerates propulsive motility and inhibits visceral hypersensitivity. *Gastroenterology, 142*(4), 844–854 e844. http://dx.doi.org/10.1053/j.gastro.2011.12.041.

Holm, A. N., Rich, A., Miller, S. M., Strege, P., Ou, Y., Gibbons, S., … Farrugia, G. (2002). Sodium current in human jejunal circular smooth muscle cells. *Gastroenterology, 122*(1), 178–187.

Hughes, P. A., Brierley, S. M., Young, R. L., & Blackshaw, L. A. (2007). Localization and comparative analysis of acid-sensing ion channel (ASIC1, 2, and 3) mRNA expression in mouse colonic sensory neurons within thoracolumbar dorsal root ganglia. *The Journal of Comparative Neurology, 500*(5), 863–875. http://dx.doi.org/10.1002/cne.21204.

Huizinga, J. D., Thuneberg, L., Kluppel, M., Malysz, J., Mikkelsen, H. B., & Bernstein, A. (1995). W/kit gene required for interstitial cells of Cajal and for intestinal pacemaker activity. *Nature, 373*(6512), 347–349. http://dx.doi.org/10.1038/373347a0.

Hwang, S. J., O'Kane, N., Singer, C., Ward, S. M., Sanders, K. M., & Koh, S. D. (2008). Block of inhibitory junction potentials and TREK-1 channels in murine colon by Ca^{2+} store-active drugs. *The Journal of Physiology, 586*(4), 1169–1184. http://dx.doi.org/10.1113/jphysiol.2007.148718.

Ikeda, R., Cha, M., Ling, J., Jia, Z., Coyle, D., & Gu, J. G. (2014). Merkel cells transduce and encode tactile stimuli to drive abeta-afferent impulses. *Cell, 157*(3), 664–675. http://dx.doi.org/10.1016/j.cell.2014.02.026.

Ikeda, R., & Gu, J. G. (2014). Piezo2 channel conductance and localization domains in Merkel cells of rat whisker hair follicles. *Neuroscience Letters, 583*, 210–215.

Jones, R. C., 3rd, Xu, L., & Gebhart, G. F. (2005). The mechanosensitivity of mouse colon afferent fibers and their sensitization by inflammatory mediators require transient receptor potential vanilloid 1 and acid-sensing ion channel 3. *The Journal of Neuroscience, 25*(47), 10981–10989. http://dx.doi.org/10.1523/JNEUROSCI.0703-05.2005.

Kadowaki, M., Wade, P. R., & Gershon, M. D. (1996). Participation of 5-HT3, 5-HT4, and nicotinic receptors in the peristaltic reflex of Guinea pig distal colon. *The American Journal of Physiology, 271*(5 Pt 1), G849–G857.

Kasai, Y., Tsutsumi, O., Taketani, Y., Endo, M., & Iino, M. (1995). Stretch-induced enhancement of contractions in uterine smooth muscle of rats. *The Journal of Physiology, 486*(Pt 2), 373–384.

Kim, M., Javed, N. H., Yu, J. G., Christofi, F., & Cooke, H. J. (2001). Mechanical stimulation activates Galphaq signaling pathways and 5-hydroxytryptamine release from human carcinoid BON cells. *Journal of Clinical Investigation, 108*(7), 1051–1059. http://dx.doi.org/10.1172/JCI12467.

Kim, B. J., Lim, H. H., Yang, D. K., Jun, J. Y., Chang, I. Y., Park, C. S., … Kim, K. W. (2005). Melastatin-type transient receptor potential channel 7 is required for intestinal pacemaking activity. *Gastroenterology, 129*(5), 1504–1517. http://dx.doi.org/10.1053/j.gastro.2005.08.016. pii:S0016-5085(05)01731-2.

Kim, J. H., Rhee, P. L., & Kang, T. M. (2007). Actin cytoskeletons regulate the stretch-induced increase of Ca^{2+} current in human gastric myocytes. *Biochemical and Biophysical Research Communications, 352*(2), 503–508.

Kirber, M. T., Walsh, J. V., Jr., & Singer, J. J. (1988). Stretch-activated ion channels in smooth muscle: A mechanism for the initiation of stretch-induced contraction. *Pflugers Archiv, 412*(4), 339–345.

Koh, S. D., & Sanders, K. M. (2001). Stretch-dependent potassium channels in murine colonic smooth muscle cells. *The Journal of Physiology, 533*(Pt 1), 155–163.

Komuro, T. (2006). Structure and organization of interstitial cells of Cajal in the gastrointestinal tract. *The Journal of Physiology, 576*(Pt 3), 653–658. http://dx.doi.org/10.1113/jphysiol.2006.116624.

Kraichely, R. E., & Farrugia, G. (2007). Mechanosensitive ion channels in interstitial cells of Cajal and smooth muscle of the gastrointestinal tract. *Neurogastroenterology and Motility, 19*(4), 245–252.

Kraichely, R. E., Strege, P. R., Sarr, M. G., Kendrick, M. L., & Farrugia, G. (2009). Lysophosphatidyl choline modulates mechanosensitive L-type Ca^{2+} current in circular smooth muscle cells from human jejunum. *American Journal of Physiology. Gastrointestinal and Liver Physiology, 296*(4), G833–G839.

Kugler, E. M., Michel, K., Zeller, F., Demir, I. E., Ceyhan, G. O., Schemann, M., & Mazzuoli-Weber, G. (2015). Mechanical stress activates neurites and somata of myenteric neurons. *Frontiers in Cellular Neuroscience, 9,* 342. http://dx.doi.org/10.3389/fncel.2015.00342.

Kunze, W. A., Clerc, N., Bertrand, P. P., & Furness, J. B. (1999). Contractile activity in intestinal muscle evokes action potential discharge in Guinea-pig myenteric neurons. *The Journal of Physiology, 517*(Pt 2), 547–561. pii:PHY_8802.

Kunze, W. A., Clerc, N., Furness, J. B., & Gola, M. (2000). The soma and neurites of primary afferent neurons in the Guinea-pig intestine respond differentially to deformation. *The Journal of Physiology, 526*(Pt 2), 375–385. pii:PHY_0647.

Kunze, W. A., Furness, J. B., Bertrand, P. P., & Bornstein, J. C. (1998). Intracellular recording from myenteric neurons of the Guinea-pig ileum that respond to stretch. *The Journal of Physiology, 506*(Pt 3), 827–842.

Langton, P. D. (1993). Calcium channel currents recorded from isolated myocytes of rat basilar artery are stretch sensitive. *The Journal of Physiology, 471,* 1–11.

Levine, J. D., & Alessandri-Haber, N. (2007). TRP channels: Targets for the relief of pain. *Biochimica et Biophysica Acta, 1772*(8), 989–1003. http://dx.doi.org/10.1016/j.bbadis.2007.01.008.

Li, H. J., Ray, S. K., Singh, N. K., Johnston, B., & Leiter, A. B. (2011). Basic helix-loop-helix transcription factors and enteroendocrine cell differentiation. *Diabetes, Obesity & Metabolism, 13*(Suppl. 1), 5–12. http://dx.doi.org/10.1111/j.1463-1326.2011.01438.x.

Linan-Rico, A., Wunderlich, J. E., Grants, I. S., Frankel, W. L., Xue, J., Williams, K. C., … Christofi, F. L. (2013). Purinergic autocrine regulation of mechanosensitivity and serotonin release in a human EC model: ATP-gated P2X3 channels in EC are downregulated in ulcerative colitis. *Inflammatory Bowel Diseases, 19*(11), 2366–2379. http://dx.doi.org/10.1097/MIB.0b013e31829ecf4d.

Lomax, R. B., Gallego, S., Novalbos, J., Garcia, A. G., & Warhurst, G. (1999). L-Type calcium channels in enterochromaffin cells from Guinea pig and human duodenal crypts: An in situ study. *Gastroenterology, 117*(6), 1363–1369.

Lyford, G. L., Strege, P. R., Shepard, A., Ou, Y., Ermilov, L., Miller, S. M., … Farrugia, G. (2002). alpha(1C) (Ca(V)1.2) L-type calcium channel mediates mechanosensitive calcium regulation. *American Journal of Phyisology. Cell Physiology, 283*(3), C1001–C1008.

Maksimovic, S., Nakatani, M., Baba, Y., Nelson, A. M., Marshall, K. L., Wellnitz, S. A., … Lumpkin, E. A. (2014). Epidermal Merkel cells are mechanosensory cells that tune mammalian touch receptors. *Nature, 509*(7502), 617–621. http://dx.doi.org/10.1038/nature13250.

Maroto, R., Raso, A., Wood, T. G., Kurosky, A., Martinac, B., & Hamill, O. P. (2005). TRPC1 forms the stretch-activated cation channel in vertebrate cells. *Nature Cell Biology, 7*(2), 179–185.

Mawe, G. M., & Hoffman, J. M. (2013). Serotonin signalling in the gut—functions, dysfunctions and therapeutic targets. *Nature Reviews. Gastroenterology and Hepatology, 10*(8), 473–486. http://dx.doi.org/10.1038/nrgastro.2013.105.

Mayer, C. J., & Wood, J. D. (1975). Properties of mechanosensitive neurons within Auerbach's plexus of the small intestine of the cat. *Pflugers Archiv, 357*(1–2), 35–49.

Mazzuoli, G., & Schemann, M. (2009). Multifunctional rapidly adapting mechanosensitive enteric neurons (RAMEN) in the myenteric plexus of the Guinea pig ileum. *The Journal of Physiology, 587*(Pt 19), 4681–4694. http://dx.doi.org/10.1113/jphysiol.2009.177105.

Mazzuoli, G., & Schemann, M. (2012). Mechanosensitive enteric neurons in the myenteric plexus of the mouse intestine. *PLoS One, 7*(7), e39887. http://dx.doi.org/10.1371/journal.pone.0039887.

Mazzuoli-Weber, G., & Schemann, M. (2015). Mechanosensitivity in the enteric nervous system. *Frontiers in Cellular Neuroscience, 9*, 408. http://dx.doi.org/10.3389/fncel.2015.00408.

Miyamoto, T., Mochizuki, T., Nakagomi, H., Kira, S., Watanabe, M., Takayama, Y., … Tominaga, M. (2014). Functional role for Piezo1 in stretch-evoked Ca^{2+} influx and ATP release in urothelial cell cultures. *Journal of Biological Chemistry, 289*(23), 16565–16575. http://dx.doi.org/10.1074/jbc.M113.528638.

Nakatani, M., Maksimovic, S., Baba, Y., & Lumpkin, E. A. (2014). Mechanotransduction in epidermal Merkel cells. *Pflugers Archiv, 467*(1), 101–108. http://dx.doi.org/10.1007/s00424-014-1569-0.

Neshatian, L., Strege, P. R., Rhee, P.-L., Kraichely, R. E., Mazzone, A. M., Bernard, C. E., … Farrugia, G. (2015). Ranolazine inhibits voltage-gated mechanosensitive sodium channels in human colon circular smooth muscle cells. *American Journal of Physiology. Gastrointestinal and Liver Physiology, 309*, G506–G512.

Neya, T., Mizutani, M., & Yamasato, T. (1993). Role of 5-HT3 receptors in peristaltic reflex elicited by stroking the mucosa in the canine jejunum. *The Journal of Physiology, 471*, 159–173.

Nozawa, K., Kawabata-Shoda, E., Doihara, H., Kojima, R., Okada, H., Mochizuki, S., … Ito, H. (2009). TRPA1 regulates gastrointestinal motility through serotonin release from enterochromaffin cells. *Proceedings of the National Academy of Sciences of the United States of America, 106*(9), 3408–3413. http://dx.doi.org/10.1073/pnas.0805323106. pii:0805323106.

Ordway, R. W., Petrou, S., Kirber, M. T., Walsh, J. V., Jr., & Singer, J. J. (1995). Stretch activation of a toad smooth muscle K^+ channel may be mediated by fatty acids. *The Journal of Physiology, 484*(Pt 2), 331–337.

Ou, Y., Gibbons, S. J., Miller, S. M., Strege, P. R., Rich, A., Distad, M. A., … Farrugia, G. (2002). SCN5A is expressed in human jejunal circular smooth muscle cells. *Neurogastroenterology and Motility, 14*(5), 477–486.

Page, A. J., Brierley, S. M., Martin, C. M., Price, M. P., Symonds, E., Butler, R., … Blackshaw, L. A. (2005). Different contributions of ASIC channels 1a, 2, and 3 in gastrointestinal mechanosensory function. *Gut, 54*(10), 1408–1415. http://dx.doi.org/10.1136/gut.2005.071084. pii:gut.2005.071084.

Page, A. J., Martin, C. M., & Blackshaw, L. A. (2002). Vagal mechanoreceptors and chemoreceptors in mouse stomach and esophagus. *Journal of Neurophysiology, 87*(4), 2095–2103. http://dx.doi.org/10.1152/jn.00785.2001.

Park, S. J., McKay, C. M., Zhu, Y., & Huizinga, J. D. (2005). Volume-activated chloride currents in interstitial cells of Cajal. *American Journal of Physiology. Gastrointestinal and Liver Physiology, 289*(5), G791–G797.

Parkman, H. P., Ma, R. C., Stapelfeldt, W. H., & Szurszewski, J. H. (1993). Direct and indirect mechanosensory pathways from the colon to the inferior mesenteric ganglion. *The American Journal of Physiology, 265*(3 Pt 1), G499–G505.

Patel, B. A., Bian, X., Quaiserova-Mocko, V., Galligan, J. J., & Swain, G. M. (2007). In vitro continuous amperometric monitoring of 5-hydroxytryptamine release from enterochromaffin cells of the Guinea pig ileum. *The Analyst, 132*(1), 41–47. http://dx.doi.org/10.1039/b611920d.

Paulsen, C. E., Armache, J. P., Gao, Y., Cheng, Y., & Julius, D. (2015). Structure of the TRPA1 ion channel suggests regulatory mechanisms. *Nature*. http://dx.doi.org/10.1038/nature14367.

Peyronnet, R., Martins, J. R., Duprat, F., Demolombe, S., Arhatte, M., Jodar, M., … Patel, A. (2013). Piezo1-dependent stretch-activated channels are inhibited by polycystin-2 in renal tubular epithelial cells. *EMBO Reports, 14*(12), 1143–1148. http://dx.doi.org/10.1038/embor.2013.170.

Poole, D. P., Pelayo, J. C., Cattaruzza, F., Kuo, Y. M., Gai, G., Chiu, J. V., … Bunnett, N. W. (2011). Transient receptor potential ankyrin 1 is expressed by inhibitory motoneurons of the mouse intestine. *Gastroenterology, 141*(2), 565–575. http://dx.doi.org/10.1053/j.gastro.2011.04.049, 575 e561–564.

Price, M. P., McIlwrath, S. L., Xie, J., Cheng, C., Qiao, J., Tarr, D. E., … Welsh, M. J. (2001). The DRASIC cation channel contributes to the detection of cutaneous touch and acid stimuli in mice. *Neuron, 32*(6), 1071–1083.

Racke, K., & Schworer, H. (1991). Regulation of serotonin release from the intestinal mucosa. *Pharmacological Research, 23*(1), 13–25.

Racke, K., Reimann, A., Schworer, H., & Kilbinger, H. (1996). Regulation of 5-HT release from enterochromaffin cells. *Behavioural Brain Research, 73*(1–2), 83–87.

Raghupathi, R., Duffield, M. D., Zelkas, L., Meedeniya, A., Brookes, S. J., Sia, T. C., … Keating, D. J. (2013). Identification of unique release kinetics of serotonin from Guinea-pig and human enterochromaffin cells. *The Journal of Physiology, 591*(Pt 23), 5959–5975. http://dx.doi.org/10.1113/jphysiol.2013.259796.

Ranade, S. S., Qiu, Z., Woo, S. H., Hur, S. S., Murthy, S. E., Cahalan, S. M., … Patapoutian, A. (2014). Piezo1, a mechanically activated ion channel, is required for vascular development in mice. *Proceedings of the National Academy of Sciences of the United States of America, 111*(28), 10347–10352. http://dx.doi.org/10.1073/pnas.1409233111.

Ranade, S. S., Syeda, R., & Patapoutian, A. (2015). Mechanically activated ion channels. *Neuron, 87*(6), 1162–1179. http://dx.doi.org/10.1016/j.neuron.2015.08.032.

Ranade, S. S., Woo, S. H., Dubin, A. E., Moshourab, R. A., Wetzel, C., Petrus, M., … Patapoutian, A. (2014). Piezo2 is the major transducer of mechanical forces for touch sensation in mice. *Nature, 516*(7529), 121–125. http://dx.doi.org/10.1038/nature13980.

Raybould, H. E., Cooke, H. J., & Christofi, F. L. (2004). Sensory mechanisms: Transmitters, modulators and reflexes. *Neurogastroenterology and Motility, 16*(Suppl. 1), 60–63. http://dx.doi.org/10.1111/j.1743-3150.2004.00477.x.

Retailleau, K., Duprat, F., Arhatte, M., Ranade, S. S., Peyronnet, R., Martins, J. R., … Honore, E. (2015). Piezo1 in smooth muscle cells is involved in hypertension-dependent arterial remodeling. *Cell Reports, 13*(6), 1161–1171. http://dx.doi.org/10.1016/j.celrep.2015.09.072.

Roach, G., Heath Wallace, R., Cameron, A., Emrah Ozel, R., Hongay, C. F., Baral, R., … Wallace, K. N. (2013). Loss of ascl1a prevents secretory cell differentiation within the zebrafish intestinal epithelium resulting in a loss of distal intestinal motility. *Developmental Biology, 376*(2), 171–186. http://dx.doi.org/10.1016/j.ydbio.2013.01.013.

Robinson, D. R., McNaughton, P. A., Evans, M. L., & Hicks, G. A. (2004). Characterization of the primary spinal afferent innervation of the mouse colon using retrograde labelling. *Neurogastroenterology and Motility, 16*(1), 113–124.

Rong, W., Hillsley, K., Davis, J. B., Hicks, G., Winchester, W. J., & Grundy, D. (2004). Jejunal afferent nerve sensitivity in wild-type and TRPV1 knockout mice. *The Journal of Physiology, 560*(Pt 3), 867–881. http://dx.doi.org/10.1113/jphysiol.2004.071746.

Sachs, F., & Morris, C. E. (1998). Mechanosensitive ion channels in nonspecialized cells. *Reviews of Physiology, Biochemistry and Pharmacology, 132*, 1–77.

Saito, Y. A., Strege, P. R., Tester, D. J., Locke, G. R., 3rd, Talley, N. J., Bernard, C. E., … Farrugia, G. (2009). Sodium channel mutation in irritable bowel syndrome: Evidence for an ion channelopathy. *American Journal of Physiology. Gastrointestinal and Liver Physiology, 296*(2), G211–G218.

Sanders, K. M., & Koh, S. D. (2006). Two-pore-domain potassium channels in smooth muscles: New components of myogenic regulation. *The Journal of Physiology, 570*(Pt 1), 37–43. http://dx.doi.org/10.1113/jphysiol.2005.098897.

Sanders, K. M., Koh, S. D., & Ward, S. M. (2006). Interstitial cells of cajal as pacemakers in the gastrointestinal tract. *Annual Review of Physiology, 68*, 307–343.

Schemann, M., & Mazzuoli, G. (2010). Multifunctional mechanosensitive neurons in the enteric nervous system. *Autonomic Neuroscience, 153*(1–2), 21–25. http://dx.doi.org/10.1016/j.autneu.2009.08.003.

Schworer, H., Katsoulis, S., & Racke, K. (1992). Histamine inhibits 5-hydroxytryptamine release from the porcine small intestine: Involvement of H3 receptors. *Gastroenterology, 102*(6), 1906–1912.

Sharif-Naeini, R., Folgering, J. H., Bichet, D., Duprat, F., Lauritzen, I., Arhatte, M., … Honore, E. (2009). Polycystin-1 and -2 dosage regulates pressure sensing. *Cell, 139*(3), 587–596. http://dx.doi.org/10.1016/j.cell.2009.08.045. pii: S0092-8674(09)01125-8.

Sipe, W. E., Brierley, S. M., Martin, C. M., Phillis, B. D., Cruz, F. B., Grady, E. F., … Bunnett, N. W. (2008). Transient receptor potential vanilloid 4 mediates protease activated receptor 2-induced sensitization of colonic afferent nerves and visceral hyperalgesia. *American Journal of Physiology. Gastrointestinal and Liver Physiology, 294*(5), G1288–G1298. http://dx.doi.org/10.1152/ajpgi.00002.2008.

Slattum, G. M., & Rosenblatt, J. (2014). Tumour cell invasion: An emerging role for basal epithelial cell extrusion. *Nature Reviews. Cancer, 14*(7), 495–501. http://dx.doi.org/10.1038/nrc3767.

Smith, T. K., Spencer, N. J., Hennig, G. W., & Dickson, E. J. (2007). Recent advances in enteric neurobiology: Mechanosensitive interneurons. *Neurogastroenterology and Motility, 19*(11), 869–878.

Spencer, N. J., & Smith, T. K. (2004). Mechanosensory S-neurons rather than AH-neurons appear to generate a rhythmic motor pattern in Guinea-pig distal colon. *The Journal of Physiology, 558*(Pt 2), 577–596. http://dx.doi.org/10.1113/jphysiol.2004.063586.

Strege, P. R., Holm, A. N., Rich, A., Miller, S. M., Ou, Y., Sarr, M. G., & Farrugia, G. (2003). Cytoskeletal modulation of sodium current in human jejunal circular smooth muscle cells. *American Journal of Physiology: Cell Physiology, 284*(1), C60–C66.

Strege, P. R., Ou, Y., Sha, L., Rich, A., Gibbons, S. J., Szurszewski, J. H., … Farrugia, G. (2003). Sodium current in human intestinal interstitial cells of Cajal. *American Journal of Physiology. Gastrointestinal and Liver Physiology, 285*(6), G1111–G1121.

Szurszewski, J. H., Ermilov, L. G., & Miller, S. M. (2002). Prevertebral ganglia and intestinofugal afferent neurons. *Gut, 51*(Suppl. 1), i6–10.

Walker, R. L., Koh, S. D., Sergeant, G. P., Sanders, K. M., & Horowitz, B. (2002). TRPC4 currents have properties similar to the pacemaker current in interstitial cells of Cajal. *American Journal of Physiology. Cell Physiology, 283*(6), C1637–C1645. http://dx.doi.org/10.1152/ajpcell.00266.2002.

Wang, Y., Fang, Q., Lu, Y., Song, B., Li, W., & Li, L. (2010). Effects of mechanical stretch on interstitial cells of cajal in Guinea pig bladder. *The Journal of Surgical Research*. http://dx.doi.org/10.1016/j.jss.2010.04.040. pii:S0022-4804(10)00427-0.

Wang, W., Huang, H., Hou, D., Liu, P., Wei, H., Fu, X., & Niu, W. (2010). Mechanosensitivity of STREX-lacking BKCa channels in the colonic smooth muscle of the mouse. *American Journal of Physiology. Gastrointestinal and Liver Physiology*. http://dx.doi.org/10.1152/ajpgi.00268.2010. pii:ajpgi.00268.2010.

Wang, F., Knutson, K., Alcaino, C., Linden, D. R., Gibbons, S. J., Kashyap, P. K., … Beyder, A. (2016). Mechanosensitive ion channel Piezo2 is important for enterochromaffin cell response to mechanical forces. *The Journal of Physiology*. http://dx.doi.org/10.1113/JP272718.

Won, K. J., Sanders, K. M., & Ward, S. M. (2005). Interstitial cells of Cajal mediate mechanosensitive responses in the stomach. *Proceedings of the National Academy of Sciences of the United States of America, 102*(41), 14913–14918.

Woo, S. H., Lumpkin, E. A., & Patapoutian, A. (2015). Merkel cells and neurons keep in touch. *Trends in Cell Biology, 25*(2), 74–81. http://dx.doi.org/10.1016/j.tcb.2014.10.003.

Woo, S. H., Ranade, S., Weyer, A. D., Dubin, A. E., Baba, Y., Qiu, Z., … Patapoutian, A. (2014). Piezo2 is required for Merkel-cell mechanotransduction. *Nature, 509*(7502), 622–626. http://dx.doi.org/10.1038/nature13251.

Wright, M. C., Reed-Geaghan, E. G., Bolock, A. M., Fujiyama, T., Hoshino, M., & Maricich, S. M. (2015). Unipotent, $Atoh1^+$ progenitors maintain the Merkel cell population in embryonic and adult mice. *The Journal of Cell Biology, 208*(3), 367–379. http://dx.doi.org/10.1083/jcb.201407101.

Yadav, V. K., Balaji, S., Suresh, P. S., Liu, X. S., Lu, X., Li, Z., … Ducy, P. (2010). Pharmacological inhibition of gut-derived serotonin synthesis is a potential bone anabolic treatment for osteoporosis. *Nature Medicine, 16*(3), 308–312.

Yamamoto, Y., & Suzuki, H. (1996). Two types of stretch-activated channel activities in Guinea-pig gastric smooth muscle cells. *The Japanese Journal of Physiology, 46*(4), 337–345.

Yang, Q., Bermingham, N. A., Finegold, M. J., & Zoghbi, H. Y. (2001). Requirement of Math1 for secretory cell lineage commitment in the mouse intestine. *Science, 294*(5549), 2155–2158. http://dx.doi.org/10.1126/science.1065718.

Yang, X. N., Lu, Y. P., Liu, J. J., Huang, J. K., Liu, Y. P., Xiao, C. X., … Guleng, B. (2014). Piezo1 is as a novel trefoil factor family 1 binding protein that promotes gastric cancer cell mobility in vitro. *Digestive Diseases and Sciences, 59*(7), 1428–1435. http://dx.doi.org/10.1007/s10620-014-3044-3.

Yang, X. C., & Sachs, F. (1989). Block of stretch-activated ion channels in *Xenopus* oocytes by gadolinium and calcium ions. *Science, 243*(4894 Pt 1), 1068–1071.

Yang, J., Zhang, J., Yang, H., Li, K., Lei, X., & Xu, C. (2016). The potential role of Piezo2 in the mediation of visceral sensation. *Neuroscience Letters, 630*, 158–163. http://dx.doi.org/10.1016/j.neulet.2016.07.058.

Zagorodnyuk, V. P., Chen, B. N., Costa, M., & Brookes, S. J. (2003). Mechanotransduction by intraganglionic laminar endings of vagal tension receptors in the Guinea-pig oesophagus. *The Journal of Physiology, 553*(Pt 2), 575–587. http://dx.doi.org/10.1113/jphysiol.2003.051862.

Zagorodnyuk, V. P., Lynn, P., Costa, M., & Brookes, S. J. (2005). Mechanisms of mechanotransduction by specialized low-threshold mechanoreceptors in the Guinea pig rectum. *American Journal of Physiology. Gastrointestinal and Liver Physiology, 289*(3), G397–G406. http://dx.doi.org/10.1152/ajpgi.00557.2004.

Zou, H., Lifshitz, L. M., Tuft, R. A., Fogarty, K. E., & Singer, J. J. (2002). Visualization of Ca^{2+} entry through single stretch-activated cation channels. *Proceedings of the National Academy of Sciences of the United States of America, 99*(9), 6404–6409.

CHAPTER NINE

Regulation of Piezo Channels by Cellular Signaling Pathways

I. Borbiro and T. Rohacs[1]
Rutgers, New Jersey Medical School, Newark, NJ, United States
[1]Corresponding author: E-mail: rohacsti@njms.rutgers.edu

Contents

Abstract

The recently identified mechanically activated Piezo1 and Piezo2 channels play major roles in various aspects of mechanosensation in mammals, and their mutations are associated with human diseases. Recent reports show that activation of cell surface receptors coupled to heterotrimeric Gq proteins increase the sensitivity of Piezo2 channels to mechanical stimuli. Activation of the cyclic adenosine monophosphate pathway was also shown to potentiate Piezo2 channel activity. This phenomenon may play a role in mechanical allodynia or hyperalgesia during inflammation. Both Piezo1 and Piezo2 channels are inhibited upon depletion of plasma membrane phosphoinositides, in response to phospholipase C activation by Ca^{2+} influx via the transient receptor potential vanilloid 1 channels. This review will discuss current knowledge on regulation of Piezo channels by these intracellular signaling pathways.

Abbreviations

8-pCPT	8-(4-chlorophenylthio)-2′-O-methyladenosine-3′,5′-cyclic monophosphate
ATP	Adenosine 5′-triphosphate
cAMP	Cyclic adenosine monophosphate
CNG	Cyclic nucleotide gated ion channels

Current Topics in Membranes, Volume 79
ISSN 1063-5823
http://dx.doi.org/10.1016/bs.ctm.2016.10.002

DAG	Diacylglycerol
DRG	Dorsal–root ganglion
EPAC	Exchange protein directly activated by cAMP
GDP	Guanosine 5′-diphosphate
GPCR	G-protein coupled receptor
GTP	Guanosine 5′-triphosphate
HEK	Human embryonic kidney cells 293
IP_3	Inositol 1,4,5-trisphosphate
MA	Mechanically activated
$PI(4,5)P_2$	Phosphatidylinositol 4,5-bisphosphate
PKA	Protein kinase A
PKC	Protein kinase C
PLC	Phospholipase C
PMA	Phorbol 12- myristate 13-acetate
TRPV1	Transient receptor potential vanilloid 1
UTP	Uridine 5′-triphosphate

1. INTRODUCTION

Piezo1 and Piezo2 channels (Volkers, Mechioukhi, & Coste, 2014) were identified in 2010 in an siRNA screen to knock down endogenous mechanically activated (MA) currents in the neuronal cell line Neuro2A (Coste et al., 2010). Unlike previous candidates for excitatory mammalian MA channels (Sachs, 2010), the expression of Piezo1 heterologous systems results in large MA currents (Bae, Gottlieb, & Sachs, 2013; Borbiro, Badheka, & Rohacs, 2015; Coste et al., 2010; Gottlieb & Sachs, 2012). Piezo2 is more difficult to express, probably due to its cytotoxicity, but there are reports from several laboratories showing MA whole cell currents from heterologously expressed Piezo2 (Borbiro et al., 2015; Coste et al., 2010; Eijkelkamp et al., 2013). Piezos are Ca^{2+}-permeable nonselective cation channels, with no similarity to any other protein, and they contain no known protein domains. Piezo1 and Piezo2 share 47% identity with each other; they are large proteins, initially predicted to have 24–36 transmembrane domains. The recent structure of Piezo1, determined by cryo-electron microscopy, shows a trimeric assembly, (Ge et al., 2015) and biochemical mapping of the extracellular domains indicate that there are between 10 and 18 transmembrane domains (Coste et al., 2015). Piezo1 is expressed in several organs that have important mechanosensory functions, including the skin, the bladder, and the lungs (Coste et al., 2010). Piezo2 is highly expressed in dorsal–root ganglion (DRG) neurons, where its siRNA mediated knockdown selectively eliminated rapidly adapting MA currents, while leaving intermediate and slowly adapting MA currents intact (Coste et al., 2010).

The global deletion of either Piezo1 (Ranade, Qiu, et al., 2014) or Piezo2 (Dubin et al., 2012) in mice is embryonic lethal. Deletion of Piezo1 in vascular endothelium leads to severe defects in vascular development (Li et al., 2014). Skin specific deletion of Piezo2 leads to defects in Merkel cell mechanosensation (Maksimovic et al., 2014; Woo et al., 2014), whereas combined deletion of Piezo2 in Merkel cells and DRG neurons leads to marked defects in sensation of light touch (Ranade, Woo, et al., 2014). Specific deletion of Piezo2 in proprioceptive sensory neurons leads to severely uncoordinated body movement, demonstrating the importance of Piezo2 in proprioception (Woo et al., 2015).

Mutations in Piezo channels have been shown to cause a variety of diseases in humans. Gain-of-function mutations in Piezo1 with reduced inactivation (Bae, Gnanasambandam, Nicolai, Sachs, & Gottlieb, 2013) are associated with hereditary xerocytosis (Andolfo et al., 2013; Zarychanski et al., 2012), characterized by dehydration of red blood cells and hemolytic anemia. Gain-of-function mutations in Piezo2 cause distal arthrogryposis with restrictive pulmonary disease and various musculoskeletal abnormalities (Coste et al., 2013). Loss-of-function mutations in Piezo1 in humans cause autosomal recessive congenital lymphatic dysplasia (Lukacs et al., 2015). More recently it was also shown that loss of function mutations of Piezo2 in humans cause loss of touch sensation and proprioception, as well as musculoskeletal abnormalities (Mahmud et al., 2016).

Intracellular second messengers, such as cyclic nucleotides and phospholipid-derived signaling molecules, regulate many ion channels. Compared to other channels, relatively little is known on the modulation of Piezo channels by intracellular signaling pathways. This chapter will summarize the current knowledge on this topic. We will discuss the effects of phosphoinositide-derived signals and cyclic adenosine monophosphate (cAMP) on Piezo channels, thus we first briefly introduce these two signaling pathways.

2. PHOSPHOLIPASE C AND CYCLIC ADENOSINE MONOPHOSPHATE SIGNALING

2.1 Phosphoinositide signaling

Phosphatidylinositol 4,5-bisphosphate [$PI(4,5)P_2$] is generated by two phosphorylation steps from phosphatidylinositol (PI) (Balla, 2013). Phospholipase C (PLC) enzymes catalyze the hydrolysis of $PI(4,5)P_2$ and the formation of the two classical second messengers inositol 1,4,5-trisphosphate (IP_3) and diacylglycerol (DAG). IP_3 induces Ca^{2+} release from the endoplasmic reticulum, by binding to its receptor, which is a Ca^{2+} permeable ion channel. DAG is an activator of protein kinase C (PKC). Conventional PKC isoforms (PKCα, β1, β2, and γ) are activated by DAG and Ca^{2+}. Novel PKC isoforms (PKCδ, ε, θ, and η) are Ca^{2+} insensitive, but they are activated by DAG, and their affinity for this lipid is much higher than that of conventional isoforms. The regulation of atypical (ζ,ι) isoforms is less well understood (Newton, 2010).

PLCβ isoforms (PLCβ1-4) are activated by G-protein coupled receptors (GPCR) that couple to the Gq family of heterotrimeric G-proteins; PLCγs (PLCγ1 and 2) are activated by receptor tyrosine kinases, but PLCδ isoforms (PLCδ1, δ3, and δ4) do not have any obvious activators. While all PLCs need some Ca^{2+} for activity, PLCδs are the most sensitive to Ca^{2+} among the three classical PLC groups, and they can be activated by Ca^{2+} influx alone (Allen, Swigart, Cheung, Cockcroft, & Katan, 1997; Lukacs et al., 2013). In addition to the classical PLCs, newer isoforms have also been cloned more recently (PLCη1 and η2 and PLCε); their regulation is less well understood (Fukami, Inanobe, Kanemaru, & Nakamura, 2010).

Besides being a precursor for second messengers, $PI(4,5)P_2$ has several important direct biological roles, for example it is involved in cytoskeletal organization and membrane trafficking (Saarikangas, Zhao, & Lappalainen, 2010). $PI(4,5)P_2$ and other phosphoinositides are emerging as a regulators of many different mammalian ion channels (Hilgemann, Feng, & Nasuhoglu, 2001; Logothetis et al., 2015; Suh & Hille, 2008), including most sensory and other TRP channels (Rohacs, 2014).

To put the effects of various phosphoinositides in context, we need to consider their in situ concentrations. $PI(4,5)P_2$ constitutes up to 1% of the phospholipids in the plasma membrane; its precursor PI(4)P is found in comparable quantities (Fruman, Meyers, & Cantley, 1998). Their precursor PI constitutes up to 10% of membrane lipids (Fruman et al., 1998), but it has no effect on most $PI(4,5)P_2$ sensitive channels, and its concentration is not expected to change significantly upon PLC activation. $PI(3,4,5)P_2$ and $PI(3,4)P_2$, the products of PI3-kinase enzymes, may also activate some $PI(4,5)P_2$ sensitive ion channels, but their concentrations in the plasma membrane do not exceed 0.1% (Fruman et al., 1998), thus their effect is generally overridden by the much higher concentration of $PI(4,5)P_2$. Therefore the two most likely phosphoinositides regulating ion channels are $PI(4,5)P_2$ and PI(4)P; both these lipids are substrates for PLC, even though most PLC isoforms prefer PI(4)P. Most attention has been paid so far to $PI(4,5)P_2$, but PI(4)P also regulates certain ion channels, and its role as an independent signaling entity is beginning to be appreciated (Hammond et al., 2012; Lukacs et al., 2013).

2.2 Cyclic adenosine monophosphate signaling

cAMP is formed by adenylate cyclase enzymes, which are activated by receptors that couple to Gs heterotrimeric G-proteins. The three major targets of cAMP are protein kinase A (PKA) enzymes, cyclic nucleotide gated ion channels, and exchange protein directly activated by cAMP (EPAC) (Borland, Smith, & Yarwood, 2009; Gloerich & Bos, 2010). EPAC is the most recently described target; it was identified in a database screen to explain the PKA-independent activation of the small G-protein Rap by cAMP (Gloerich & Bos, 2010). EPAC1 and EPAC2 are present in most tissues, and they function as guanine nucleotide exchange factors (GEFs) for both Rap1 and Rap2, which belong to the Ras family of small G-proteins. These G-proteins cycle between the inactive guanosine 5′-diphosphate (GDP)-bound state and the active guanosine 5′-triphosphate (GTP)-bound state. GEFs accelerate the exchange of GDP for GTP and thus activate the G-protein, whereas GTPase-activating proteins enhance GTP hydrolysis, thus inactivate the G-protein. Several cAMP analogs, such as 8-(4-chlorophenylthio)-2′-O-methyladenosine-3′,5′-cyclic monophosphate (8-pCPT) are available that selectively interact with EPAC1 and EPAC2. The rationale for this selective agonism is that EPAC proteins lack the glutamate that interacts with the 2-OH group of the ribose of cAMP in PKA and cAMP-gated ion channels (Borland et al., 2009; Gloerich & Bos, 2010).

3. SENSITIZATION OF SENSORY ION CHANNELS BY INFLAMMATORY PATHWAYS

Under inflammatory conditions peripheral sensory neurons show enhanced sensitivity to painful stimuli (hyperalgesia) and abnormal pain sensitivity to nonpainful stimuli (allodynia). There are multiple inflammatory signaling pathways known to sensitize sensory neurons to both thermal and mechanical stimuli (Linley et al., 2012), here we briefly discuss sensitization of the heat- and capsaicin-activated transient receptor potential vanilloid 1 (TRPV1) channels.

Thermal hyperalgesia in mice is largely dependent on TRPV1 (Caterina et al., 2000; Davis et al., 2000). While the expression level of these channels may increase in chronic inflammation, there are also important acute signaling events that increase the activity of TRPV1 downstream of the activation of both Gq- and Gs-coupled receptors. Bradykinin, the classic, very well studied sensitizing agent is a proinflammatory peptide is generated after tissue injury and noxious stimulation (Petho & Reeh, 2012). Bradykinin receptors (B1 and B2) are GPCRs; they stimulate PLCβ enzymes through the Gαq subunits of heterotrimeric G-proteins. Extracellular adenosine 5′-triphosphate (ATP) acting on Gq-coupled purinergic receptors also sensitizes TRPV1 (Tominaga, Wada, & Masu, 2001). The downstream activation of PKC will lead to the phosphorylation of TRPV1 on S501 and S800 residues (Numazaki, Tominaga, Toyooka, & Tominaga, 2002), and thus sensitize the channel to heat and chemical activation. This phosphorylation shifts the capsaicin concentration-response to the left, without a substantial effect on maximal currents, leading to selective enhancement of TRPV1 activity at moderate stimulation levels. Sensitivity to heat and low extracellular pH also increases during sensitization (Tominaga et al., 2001).

Gq-coupled receptors were also proposed to sensitize TRPV1 channels by decreasing $PI(4,5)P_2$ levels, and relieving TRPV1 from tonic inhibition by this lipid (Chuang et al., 2001). Later it was found by several laboratories that $PI(4,5)P_2$ activates, rather than inhibits TRPV1 in excised inside-out patches (Lukacs et al., 2007; Poblete et al., 2014; Stein, Ufret-Vincenty, Hua, Santana, & Gordon, 2006). Nevertheless, in a cellular context, there may be an indirect inhibitory effect of $PI(4,5)P_2$, relief from which by $PI(4,5)P_2$ hydrolysis, potentiates the effect of PKC activation (Lukacs et al., 2013). Detailed discussion of this complex topic is beyond the scope of this chapter and has been reviewed in detail recently (Rohacs, 2015).

Some inflammatory agents such as prostaglandin E2 act via Gs-coupled receptors. While less studied than the PLC pathway, activation of the cAMP pathways also sensitizes TRPV1 via PKA mediated phosphorylation (Rathee et al., 2002). Interestingly, the A-kinase anchoring protein AKAP150 is required for TRPV1 sensitization by not only the PKA but also the PKC pathway (Zhang, Li, & McNaughton, 2008).

Little is known about the molecular targets of inflammatory agents in mechanical hyperalgesia and allodynia, but there is evidence that Piezo2 channel activity is enhanced in sensory neurons by various inflammatory signaling pathways. Below we will give an overview of Piezo2 channel sensitizing pathways.

3.1 Piezo2 potentiation by Gq-coupled receptors

Shortly after the discovery of Piezo channel, the Patapoutian group reported that activation of bradykinin receptors increased Piezo2-mediated rapidly adapting MA current amplitudes in sensory DRG neurons (Dubin et al., 2012). The B2 bradykinin receptor antagonist HOE-140 (icatibant) blocked the bradykinin-induced modulation, indicating that endogenous B2 receptors were responsible for the observed effects. The effect of bradykinin was abolished by combined inhibition of PKC with bisindolylmaleimide (BIM) and PKA using H89. When B2 bradykinin receptors were heterologously expressed together with Piezo2 in human embryonic kidney (HEK) cells, the amplitude of Piezo2-mediated MA currents increased and their inactivation slowed upon exposure to bradykinin. The Gβγ inhibitor gallein did not inhibit the potentiating effect of bradykinin (Dubin et al., 2012), but including the nonhydrolyzable GTP analog GTPγS in the patch pipette potentiated the currents, suggesting the involvement of Gα subunits (Dubin et al., 2012). Consistent with this, a subsequent study showed that including GTP in the patch pipette potentiated rapidly adapting MA currents in DRG neurons (Jia, Ikeda, Ling, & Gu, 2013). The potentiating effect of bradykinin was abolished by the combined inhibition of PKC and PKA, but not by applying one, or the other protein kinase inhibitor alone (Dubin et al., 2012). The PKA activator 8-BrcAMP and the PKC agonist phorbol 12-myristate 13-acetate (PMA) also potentiates Piezo2 currents, suggesting that both pathways are capable of similarly modulating Piezo2. The PLC inhibitor U73122, and depleting intracellular Ca^{2+} stores using thapsigargin had no effect on bradykinin-induced changes in Piezo2 currents. The PLC activator *m*-3M3FBS had no effect on Piezo2 current amplitudes.

The overall conclusion of the paper by (Dubin et al., 2012) is that bradykinin receptor activation potentiates Piezo2 currents by activating both PKC and PKA. There are several open questions, however. Both B1 and B2 receptors act via Gq and to some extent Gi heterotrimeric G-proteins (Leeb-Lundberg, Marceau, Muller-Esterl, Pettibone, & Zuraw, 2005). Gαq activates PLC directly, and Gi signaling may also activate PLC via Gβγ subunits, but Gi inhibits the formation of cAMP. How PKA is activated upon bradykinin receptor stimulation, is unclear. It is also not clear, how neither PLC inhibitors nor a PLC activator had any effect on Piezo2 currents, when the downstream PKC was concluded to be involved in potentiation by bradykinin.

Extracellular ATP when released from damaged tissue can directly activate pain-sensing neurons through P2X ionotropic receptors, and the metabotropic P2Y receptor activation by ATP results in sensitization to noxious stimuli (Tominaga et al., 2001). Lechner and Lewin (2009) found that application of ATP and uridine 5′-triphosphate (UTP) potentiated rapidly adapting MA currents in a nociceptive subset of mouse DRG neurons. Rapidly adapting currents in large mechanoreceptors, as well as intermediate and slowly adapting MA currents were not potentiated by UTP. These results were obtained before the cloning of Piezo2, channels, but they are in line with the bradykinin-induced sensitization described previously. Extracellular nucleotides also potentiated mechanically evoked action potentials in 41% of c-fibers in the skin–nerve preparation.

Responses of low-threshold mechanoreceptors and D-hair mechanoreceptors were not altered by the addition UTP. The potentiating effect of ATP on MA currents was not sensitive to inhibitors of P2X channels. UTP, which has low affinity for P2X receptors, was equally effective as ATP, thus the authors concluded that the potentiating effect was due to P2Y signaling (Lechner & Lewin, 2009). P2Y receptors couple to Gαq and are known to activate PKC, but the authors did not perform experiments to study the mechanism of potentiation of MA currents. They argued however that the potentiating effect of P2Y activation is independent of PKC based on an earlier report with pharmacological activation of PKC using the phorbol ester PMA (Di Castro, Drew, Wood, & Cesare, 2006). The effect of PMA developed in 1 h and was essentially irreversible (Di Castro et al., 2006), while UTP acted within minutes and the effect was quickly reversible (Lechner & Lewin, 2009).

Overall, there is independent evidence from two laboratories showing that activation of Gq-coupled receptors potentiates Piezo2 currents, but deciphering the mechanism of this phenomenon will require further work.

3.2 Piezo2 regulation by the cyclic adenosine monophosphate pathway

During inflammation prostaglandin (PG) levels increase and these short-lived tissue hormones contribute to sensory neuron sensitization when continuously formed and thus having a long lasting sustained effect. They are derived from arachidonic acid by cyclooxygenase (COX) enzymes locally. Some prostanoids, such as PGE2, act via Gs-coupled receptors, thus increase intracellular cAMP levels. Prostanoids induce prolonged mechanical hyperalgesia, the mechanism of which is not very well established (Petho & Reeh, 2012). Two studies discussed below found that cAMP acting via EPAC1 potentiates Piezo2 currents, and this phenomenon may be involved in mechanical hyperalgesia/allodynia.

Application of the EPAC-selective cAMP analog 8-pCPT increased peak amplitudes of rapidly adapting MA currents in large mouse DRG neurons (Eijkelkamp et al., 2013). In HEK cells, coexpressing human Piezo2 and EPAC1, 8-pCPT potentiated rapidly adapting MA currents. Interestingly, 8-pCPT had only a minimal effect on MA currents in HEK cells coexpressing hPiezo2 and EPAC2 (Eijkelkamp et al., 2013).

The same study found that EPAC1 levels increased in a nerve injury model and mechanical allodynia induced by nerve damage was attenuated in $EPAC^{-/-}$ mice. Intraplantar injection of 8-pCPT induced long lasting mechanical allodynia, which persisted when Nav1.8 expressing nociceptors were depleted using diphtheria-toxin, but was attenuated by intrathecal injection of Piezo2 antisense oligonucleotides. Interestingly, mechanical allodynia induced by a PKA-selective cAMP analog (6-Bnz-cAMP) was attenuated in mice in which Nav1.8 + neurons were depleted, pointing to different mechanisms for the PKA and EPAC pathways in inducing hypersensitivity (Eijkelkamp et al., 2013).

Eijkelkamp et al. (2013) also found that the PKC activator PMA did not potentiate human Piezo2 currents, a finding contradicting the results of (Dubin et al., 2012), who as mentioned earlier found that PMA potentiated mouse Piezo2 currents. The PKA specific cAMP analog 6-Bnz-cAMP also did not increase Piezo2 current amplitudes.

Another recent study confirmed the potentiation of Piezo2 channels by 8-pCPT in HEK cells expressing EPAC1 and also found that coexpressing G-protein kinase 2 (GRK2) attenuated this effect (Singhmar et al., 2016). Additionally EPAC1$^{-/-}$ mice were protected against inflammatory hyperalgesia induced by complete Freund's adjuvant (CFA). Moreover intrathecal administration of Piezo2 antisense oligonucleotides inhibited mechanical but not thermal hyperalgesia. The inhibitory effect of Piezo2 antisense injection started on day 3 after injection of CFA. One day after CFA injection the Piezo2 antisense had no effect (Singhmar et al., 2016), which is consistent with an earlier finding that combined genetic deletion of Piezo2 in DRG and in Merkel cells in mice had no effect on mechanical allodynia after 24 h of CFA injection (Ranade, Woo, et al., 2014).

The studies discussed here (Eijkelkamp et al., 2013; Singhmar et al., 2016) agree that cAMP signaling via EPAC1 stimulates Piezo2 activity, which plays a role in mechanical allodynia. While the agreement from two independent reports makes a strong case, open questions remain. For example, none of the studies identified the effector downstream of EPAC1 and the mechanism by which Piezo2 is potentiated. Also, they did not demonstrate that receptor mediated increase in cAMP is sufficient to potentiate Piezo2 currents via this pathway.

4. CALCIUM-INDUCED INHIBITION OF PIEZO CHANNELS—THE ROLE OF PHOSPHOINOSITIDES

The heat- and capsaicin-sensitive TRPV1 channels and Piezo2 are coexpressed in a subset of small sensory neurons (Coste et al., 2010). We found that in these cells application of capsaicin abrogated the rapidly adapting MA currents (Borbiro et al., 2015). This phenomenon was observed only in neurons with the characteristic capsaicin-induced TRPV1 currents.

When Piezo2 and TRPV1 were coexpressed in HEK cells, capsaicin robustly inhibited MA Piezo2 currents, and this effect was dependent on the presence of extracellular Ca^{2+}, showing the importance of Ca^{2+} influx through TRPV1 channels. Similarly, in a heterologous system coexpressing TRPV1 and Piezo1, capsaicin inhibited MA currents, and this effect disappeared in a Ca^{2+} free extracellular medium.

We have shown earlier that both in an expression system (Lukacs et al., 2007) and in native DRG neurons (Lukacs et al., 2013), Ca^{2+} influx via TRPV1 leads to the activation of a PLCδ isoform, with a concomitant robust decrease in PI(4,5)P_2 and PI(4)P levels in the plasma membrane. TRPV1 requires phosphoinositides for activity, and this phenomenon plays an important role in the desensitization of this channel (Rohacs, 2015). As PI(4,5)P_2 is required for the activity of many different ion channels, we tested if phosphoinositide depletion was responsible for Piezo channel inhibition (Borbiro et al., 2015).

We found that intracellular application of excess PI(4,5)P_2 or PI(4)P in whole cell patch clamp attenuated the inhibitory effect of capsaicin on mechanically evoked Piezo1 currents. Activation of chemically inducible phosphoinositide phosphatases, which deplete phosphoinositides (Hammond et al., 2012; Varnai, Thyagarajan, Rohacs, & Balla, 2006) without the formation of IP_3 or DAG, also inhibited Piezo1 currents (Borbiro et al., 2015). Pseudojanin, the inducible phosphatase that dephosphorylates both PI(4,5)P_2 and PI(4)P was more effective than a 5′ phosphatase, arguing for the importance of both PI(4,5)P_2 and PI(4)P. Note that the inhibition by the inducible phosphatases took several minutes to develop, in contrast to the almost immediate inhibition upon the activation of TRPV1. Finally, Piezo1 current activity ran down in excised patches, a phenomenon characteristic of PI(4,5)P_2 dependent ion channels (Rohacs et al., 2002). Phosphoinositides, PI(4,5)P_2 and PI(4)P applied together inhibited rundown of MA Piezo1 currents in excised inside-out patches (Borbiro et al., 2015). Overall, these data show that Piezo channels require phosphoinositides for activity. Attempts to reactivate Piezo1 currents after rundown, however, resulted in variable results (Borbiro et al., 2015), in contrast to the very reliable reactivation of most PI(4,5)P_2 dependent ion channels in the excised patch configuration (Rohacs, 2014). This shows that the relationship between phosphoinositides and Piezo channels may be more complex than the lipid being a cofactor directly activating the channel and points to the potential importance of other cellular components.

In contrast to the robust and complete inhibition of Piezo1 currents by TRPV1 activation, activation of the Gq-coupled M1 muscarinic receptors by carbachol lead only to a slow and marginal inhibition of Piezo1 currents (Borbiro et al., 2015). When we measured PI(4,5)P_2 and PI(4)P levels with fluorescence based sensors, we found that TRPV1 activation leads to somewhat larger decrease in PI(4,5)P_2 and a much larger decrease in PI(4) P levels than carbachol (Borbiro et al., 2015). This is consistent with our earlier finding that in DRG neurons, TRPV1 activation by capsaicin-induced a robust decrease of both PI(4,5)P_2 and PI(4)P, but activation of bradykinin receptors only induced a small decrease in PI(4,5)P_2 levels, and no detectable change in PI(4)P (Lukacs et al., 2013). Overall, it is likely that Piezo1 channels have high apparent affinity for phosphoinositides, and only robust and combined depletion of PI(4,5)P_2 and PI(4)P inhibits them. Consistent with this idea, when we used ATP free intracellular pipette solutions, a maneuver shown to make phosphoinositide depletion irreversible (Suh & Hille, 2002), carbachol induced a much larger inhibition of Piezo1 currents compared to experiments with ATP in the patch pipette (Borbiro et al., 2015).

Capsaicin has long been used as a local analgesic, after an initial burning sensation, it relieves pain (Szallasi & Blumberg, 1999). The effects of capsaicin are complex, they involve desensitization of TRPV1 itself, inhibition of other ion channels, and even local nerve degeneration. Inhibition of Piezo2 channels may also contribute to its local analgesic effect.

Inhibition of Piezo1 currents was not specific to capsaicin-induced activation of TRPV1, both activation of TRPA1 with mustard oil and activation of TRPV1 by low extracellular pH were also capable of inhibiting the channel. The level of inhibition of Piezo1 by these latter maneuvers was smaller than by capsaicin, correlating with the smaller TRPA1 or TRPV1 currents (Borbiro et al., 2015). An earlier study showed that increase of cytoplasmic Ca^{2+} from 50 nM to 1 μM induced a small increase in Piezo2 current activity (Eijkelkamp et al., 2013). Inactivation kinetics of Piezo1 was not significantly affected by the absence or presence of extracellular Ca^{2+} (Coste et al., 2010; Gottlieb, Bae, & Sachs, 2012). Overall, the effects of Ca^{2+} on Piezo channel activity may be complex, and probably only very large increases in cytoplasmic Ca^{2+} will inhibit their activity via phosphoinositide depletion.

It will require further studies whether Ca^{2+} influx via other pathways, such as voltage-gated Ca^{2+} channels will modulate Piezo channel activity. It also has to be noted that the presence of excess PI(4,5)P_2 or PI(4)P in the patch pipette, delayed, rather than eliminated Piezo1 current inhibition by TRPV1 activation, pointing to the potential involvement of other signaling pathways (Borbiro et al., 2015).

5. CONCLUSIONS

The recent identification of Piezo1 and Piezo2 channels induced a flurry of research activity to understand the physiological importance, structure, gating mechanism, and regulation of these channels. On the topic of Piezo channel regulation by signaling pathways, an emerging picture shows that Piezo2 channels are potentiated by inflammatory signals, but the exact mechanism how channel activity increases is not yet clear. Data also show that the Piezo channels require phosphoinositides for activity, but whether the lipids act directly on the channel or via an indirect mechanism is not yet clear. Integrating regulation of Piezos by Ca^{2+}, phosphoinositides, and GPCR activation will also require further studies. Overall, in less than 6 years, at the time of writing of this chapter, the research community has accumulated an amazing amount of knowledge about Piezo channels (Fig. 1). Nevertheless, this is a still a new field, full of open questions, and the following years are expected to bring new and exciting results.

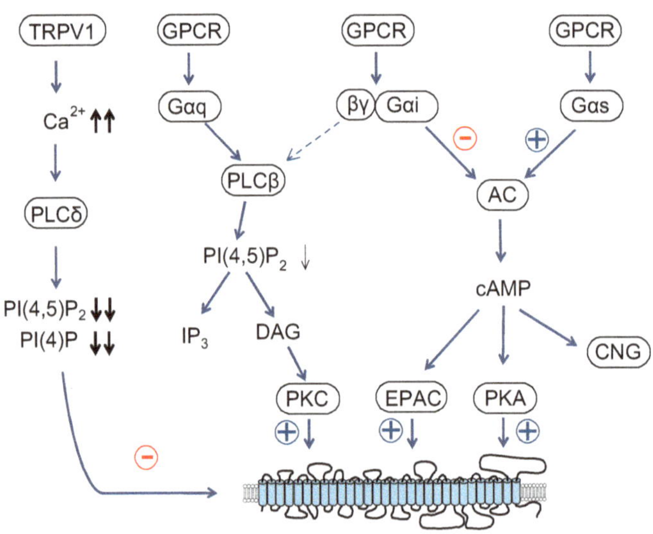

Figure 1 Summary of signaling pathways regulating Piezo channels. Activation of TRPV1 channels by capsaicin leads to a massive Ca^{2+} influx, which activated PLCδ isoforms. The concomitant robust decrease in $PI(4,5)P_2$ and PI(4)P levels inhibits Piezo channel activity. Activation of Gq-coupled GPCRs lead to activation of PLCβ isoforms, with only a modest decrease in $PI(4,5)P_2$ levels, which is not enough to inhibit channel activity, and the downstream activation of PKC will stimulate Piezo2 activity. Activation Gs-coupled receptors potentiate Piezo2 channel activity, via the EPAC pathway. Activation of PKA may also contribute to potentiation of channel activity.

ACKNOWLEDGMENTS

The discussions and comments of the members of the Rohacs lab are highly appreciated. Work in TR's laboratory is supported by NIH grants: NS055159 and GM093290.

REFERENCES

Allen, V., Swigart, P., Cheung, R., Cockcroft, S., & Katan, M. (1997). Regulation of inositol lipid-specific phospholipase c delta by changes in $Ca2^{+}$ ion concentrations [Research Support, Non-U.S. Gov't] *The Biochemical Journal, 327*(Pt 2), 545–552.

Andolfo, I., Alper, S. L., De Franceschi, L., Auriemma, C., Russo, R., De Falco, L., … Iolascon, A. (2013). Multiple clinical forms of dehydrated hereditary stomatocytosis arise from mutations in PIEZO1. *Blood, 121*, 3925–3935. http://dx.doi.org/10.1182/blood-2013-02-482489.

Bae, C., Gnanasambandam, R., Nicolai, C., Sachs, F., & Gottlieb, P. A. (2013). Xerocytosis is caused by mutations that alter the kinetics of the mechanosensitive channel PIEZO1. *Proceedings of the National Academy of Sciences of the United States of America, 110*(12), E1162–E1168. http://dx.doi.org/10.1073/pnas.1219777110.

Bae, C., Gottlieb, P. A., & Sachs, F. (2013). Human PIEZO1: Removing inactivation [Research Support, N.I.H., Extramural Research Support, Non-U.S. Gov't Research Support, U.S. Gov't, Non-P.H.S.] *Biophysical Journal, 105*(4), 880–886. http://dx.doi.org/10.1016/j.bpj.2013.07.019.

Balla, T. (2013). Phosphoinositides: Tiny lipids with giant impact on cell regulation [Research Support, N.I.H., Intramural Review] *Physiological Reviews, 93*(3), 1019–1137. http://dx.doi.org/10.1152/physrev.00028.2012.

Borbiro, I., Badheka, D., & Rohacs, T. (2015). Activation of TRPV1 channels inhibit mechanosensitive Piezo channel activity by depleting membrane phosphoinositides. *Science Signaling*, 8(363), ra15. http://dx.doi.org/10.1126/scisignal.2005667.

Borland, G., Smith, B. O., & Yarwood, S. J. (2009). EPAC proteins transduce diverse cellular actions of cAMP. *British Journal of Pharmacology, 158*(1), 70–86. http://dx.doi.org/10.1111/j.1476-5381.2008.00087.x.

Caterina, M. J., Leffler, A., Malmberg, A. B., Martin, W. J., Trafton, J., Petersen-Zeitz, K. R., … Julius, D. (2000). Impaired nociception and pain sensation in mice lacking the capsaicin receptor [Research Support, Non-U.S. Gov't Research Support, U.S. Gov't, P.H.S.] *Science, 288*(5464), 306–313.

Chuang, H. H., Prescott, E. D., Kong, H., Shields, S., Jordt, S. E., Basbaum, A. I., … Julius, D. (2001). Bradykinin and nerve growth factor release the capsaicin receptor from PtdIns(4,5)P2-mediated inhibition [Research Support, Non-U.S. Gov't Research Support, U.S. Gov't, Non-P.H.S. Research Support, U.S. Gov't, P.H.S.] *Nature, 411*(6840), 957–962. http://dx.doi.org/10.1038/35082088.

Coste, B., Houge, G., Murray, M. F., Stitziel, N., Bandell, M., Giovanni, M. A., … Patapoutian, A. (2013). Gain-of-function mutations in the mechanically activated ion channel PIEZO2 cause a subtype of Distal Arthrogryposis. *Proceedings of the National Academy of Sciences of the United States of America, 110*(12), 4667–4672. http://dx.doi.org/10.1073/pnas.1221400110.

Coste, B., Mathur, J., Schmidt, M., Earley, T. J., Ranade, S., Petrus, M. J., … Patapoutian, A. (2010). Piezo1 and Piezo2 are essential components of distinct mechanically activated cation channels. *Science, 330*(6000), 55–60. http://dx.doi.org/10.1126/science.1193270.

Coste, B., Murthy, S. E., Mathur, J., Schmidt, M., Mechioukhi, Y., Delmas, P., & Patapoutian, A. (2015). Piezo1 ion channel pore properties are dictated by C-terminal region. *Nature Communications, 6*, 7223. http://dx.doi.org/10.1038/ncomms8223.

Davis, J. B., Gray, J., Gunthorpe, M. J., Hatcher, J. P., Davey, P. T., Overend, P., … Sheardown, S. A. (2000). Vanilloid receptor-1 is essential for inflammatory thermal hyperalgesia. *Nature, 405*(6783), 183–187. http://dx.doi.org/10.1038/35012076.

Di Castro, A., Drew, L. J., Wood, J. N., & Cesare, P. (2006). Modulation of sensory neuron mechanotransduction by PKC- and nerve growth factor-dependent pathways. *Proceedings of the National Academy of Sciences of the United States of America, 103*(12), 4699–4704. http://dx.doi.org/10.1073/pnas.0508005103.

Dubin, A. E., Schmidt, M., Mathur, J., Petrus, M. J., Xiao, B., Coste, B., & Patapoutian, A. (2012). Inflammatory signals enhance piezo2-mediated mechanosensitive currents [Research Support, N.I.H., Extramural] *Cell Reports, 2*(3), 511–517. http://dx.doi.org/10.1016/j.celrep.2012.07.014.

Eijkelkamp, N., Linley, J. E., Torres, J. M., Bee, L., Dickenson, A. H., Gringhuis, M., … Wood, J. N. (2013). A role for Piezo2 in EPAC1-dependent mechanical allodynia. *Nature Communications, 4*, 1682. http://dx.doi.org/10.1038/ncomms2673.

Fruman, D. A., Meyers, R. E., & Cantley, L. C. (1998). Phosphoinositide kinases [Research Support, Non-U.S. Gov't Research Support, U.S. Gov't, P.H.S. Review] *Annual Review of Biochemistry, 67*, 481–507. http://dx.doi.org/10.1146/annurev.biochem.67.1.481.

Fukami, K., Inanobe, S., Kanemaru, K., & Nakamura, Y. (2010). Phospholipase C is a key enzyme regulating intracellular calcium and modulating the phosphoinositide balance [Review] *Progress in Lipid Research, 49*(4), 429–437. http://dx.doi.org/10.1016/j.plipres.2010.06.001.

Ge, J., Li, W., Zhao, Q., Li, N., Chen, M., Zhi, P., … Yang, M. (2015). Architecture of the mammalian mechanosensitive Piezo1 channel. *Nature, 527*(7576), 64–69. http://dx.doi.org/10.1038/nature15247.

Gloerich, M., & Bos, J. L. (2010). Epac: Defining a new mechanism for cAMP action. *Annual Review of Pharmacology and Toxicology, 50*, 355–375. http://dx.doi.org/10.1146/annurev.pharmtox.010909.105714.

Gottlieb, P. A., Bae, C., & Sachs, F. (2012). Gating the mechanical channel Piezo1: A comparison between whole-cell and patch recording [Research Support, N.I.H., Extramural Research Support, Non-U.S. Gov't Research Support, U.S. Gov't, Non-P.H.S.] *Channels, 6*(4), 282–289. http://dx.doi.org/10.4161/chan.21064.

Gottlieb, P. A., & Sachs, F. (2012). Piezo1: Properties of a cation selective mechanical channel. *Channels, 6*(4), 214–219. http://dx.doi.org/10.4161/chan.21050.

Hammond, G. R., Fischer, M. J., Anderson, K. E., Holdich, J., Koteci, A., Balla, T., & Irvine, R. F. (2012). PI4P and PI(4,5)P2 are essential but independent lipid determinants of membrane identity. *Science, 337*, 727–730. http://dx.doi.org/10.1126/science.1222483.

Hilgemann, D. W., Feng, S., & Nasuhoglu, C. (2001). The complex and intriguing lives of PIP2 with ion channels and transporters [Review] *Science's STKE: Signal Transduction Knowledge Environment, 2001*(111), re19. http://dx.doi.org/10.1126/stke.2001.111.re19.

Jia, Z., Ikeda, R., Ling, J., & Gu, J. G. (2013). GTP-dependent run-up of Piezo2-type mechanically activated currents in rat dorsal root ganglion neurons. *Molecular Brain, 6*, 57. http://dx.doi.org/10.1186/1756-6606-6-57.

Lechner, S. G., & Lewin, G. R. (2009). Peripheral sensitisation of nociceptors via G-protein-dependent potentiation of mechanotransduction currents [Research Support, Non-U.S. Gov't] *The Journal of Physiology, 587*(Pt 14), 3493–3503. http://dx.doi.org/10.1113/jphysiol.2009.175059.

Leeb-Lundberg, L. M., Marceau, F., Muller-Esterl, W., Pettibone, D. J., & Zuraw, B. L. (2005). International union of pharmacology. XLV. Classification of the kinin receptor family: From molecular mechanisms to pathophysiological consequences. *Pharmacological Reviews, 57*(1), 27–77. http://dx.doi.org/10.1124/pr.57.1.2.

Li, J., Hou, B., Tumova, S., Muraki, K., Bruns, A., Ludlow, M. J., … Beech, D. J. (2014). Piezo1 integration of vascular architecture with physiological force. *Nature.* http://dx.doi.org/10.1038/nature13701.

Linley, J. E., Ooi, L., Pettinger, L., Kirton, H., Boyle, J. P., Peers, C., & Gamper, N. (2012). Reactive oxygen species are second messengers of neurokinin signaling in peripheral sensory neurons. *Proceedings of the National Academy of Sciences of the United States of America, 109*(24), E1578–E1586. http://dx.doi.org/10.1073/pnas.1201544109.

Logothetis, D. E., Petrou, V. I., Zhang, M., Mahajan, R., Meng, X. Y., Adney, S. K., … Baki, L. (2015). Phosphoinositide control of membrane protein function: A frontier led by studies on ion channels. *Annual Review of Physiology, 77,* 81–104. http://dx.doi.org/10.1146/annurev-physiol-021113-170358.

Lukacs, V., Mathur, J., Mao, R., Bayrak-Toydemir, P., Procter, M., Cahalan, S. M., … Krock, B. L. (2015). Impaired PIEZO1 function in patients with a novel autosomal recessive congenital lymphatic dysplasia. *Nature Communications, 6,* 8329. http://dx.doi.org/10.1038/ncomms9329.

Lukacs, V., Thyagarajan, B., Varnai, P., Balla, A., Balla, T., & Rohacs, T. (2007). Dual regulation of TRPV1 by phosphoinositides [Research Support, N.I.H., Intramural Research Support, Non-U.S. Gov't] *The Journal of Neuroscience: The Official Journal of the Society for Neuroscience, 27*(26), 7070–7080. http://dx.doi.org/10.1523/JNEUROSCI.1866-07.2007.

Lukacs, V., Yudin, Y., Hammond, G. R., Sharma, E., Fukami, K., & Rohacs, T. (2013). Distinctive changes in plasma membrane phosphoinositides underlie differential regulation of TRPV1 in nociceptive neurons. *Journal of Neuroscience, 33*(28), 11451–11463.

Mahmud, A. A., Nahid, N. A., Nassif, C., Sayeed, M. S., Ahmed, M. U., Parveen, M., … Michaud, J. L. (2016). Loss of the proprioception and touch sensation channel PIEZO2 in siblings with a progressive form of contractures. *Clinical Genetics.*

Maksimovic, S., Nakatani, M., Baba, Y., Nelson, A. M., Marshall, K. L., Wellnitz, S. A., … Lumpkin, E. A. (2014). Epidermal Merkel cells are mechanosensory cells that tune mammalian touch receptors. *Nature, 509*(7502), 617–621. http://dx.doi.org/10.1038/nature13250.

Newton, A. C. (2010). Protein kinase C: Poised to signal. *American Journal of Physiology. Endocrinology and Metabolism, 298*(3), E395–E402. http://dx.doi.org/10.1152/ajpendo.00477.2009.

Numazaki, M., Tominaga, T., Toyooka, H., & Tominaga, M. (2002). Direct phosphorylation of capsaicin receptor VR1 by protein kinase Cepsilon and identification of two target serine residues [Research Support, Non-U.S. Gov't] *The Journal of Biological Chemistry, 277*(16), 13375–13378. http://dx.doi.org/10.1074/jbc.C200104200.

Petho, G., & Reeh, P. W. (2012). Sensory and signaling mechanisms of bradykinin, eicosanoids, platelet-activating factor, and nitric oxide in peripheral nociceptors. *Physiological Reviews, 92*(4), 1699–1775. http://dx.doi.org/10.1152/physrev.00048.2010.

Poblete, H., Oyarzun, I., Olivero, P., Comer, J., Zuniga, M., Sepulveda, R. V., … Latorre, R. (2014). Molecular determinants of phosphatidylinositol 4,5 bisphosphate (PI(4,5)P2) binding to transient receptor potential V1 (TRPV1) channels. *The Journal of Biological Chemistry, 290*(4), 2086–2098. http://dx.doi.org/10.1074/jbc.M114.613620.

Ranade, S. S., Qiu, Z., Woo, S. H., Hur, S. S., Murthy, S. E., Cahalan, S. M., … Patapoutian, A. (2014). Piezo1, a mechanically activated ion channel, is required for vascular development in mice. *Proceedings of the National Academy of Sciences of the United States of America*. http://dx.doi.org/10.1073/pnas.1409233111.

Ranade, S. S., Woo, S. H., Dubin, A. E., Moshourab, R. A., Wetzel, C., Petrus, M., … Patapoutian, A. (2014). Piezo2 is the major transducer of mechanical forces for touch sensation in mice. *Nature, 516*(7529), 121–125. http://dx.doi.org/10.1038/nature13980.

Rathee, P. K., Distler, C., Obreja, O., Neuhuber, W., Wang, G. K., Wang, S. Y., … Kress, M. (2002). PKA/AKAP/VR-1 module: A common link of Gs-mediated signaling to thermal hyperalgesia [Research Support, Non-U.S. Gov't] *The Journal of Neuroscience: The Official Journal of the Society for Neuroscience, 22*(11), 4740–4745. doi:20026461.

Rohacs, T. (2014). Phosphoinositide regulation of TRP channels. *Handbook of Experimental Pharmacology, 233*, 1143–1176.

Rohacs, T. (2015). Phosphoinositide regulation of TRPV1 revisited. *Pflugers Archiv: European Journal of Physiology, 467*(9), 1851–1869. http://dx.doi.org/10.1007/s00424-015-1695-3.

Rohacs, T., Lopes, C., Mirshahi, T., Jin, T., Zhang, H., & Logothetis, D. E. (2002). Assaying phosphatidylinositol bisphosphate regulation of potassium channels [In Vitro Research Support, Non-U.S. Gov't Research Support, U.S. Gov't, P.H.S.] *Methods in Enzymology, 345*, 71–92.

Saarikangas, J., Zhao, H., & Lappalainen, P. (2010). Regulation of the actin cytoskeleton-plasma membrane interplay by phosphoinositides [Research Support, Non-U.S. Gov't Review] *Physiological Reviews, 90*(1), 259–289. http://dx.doi.org/10.1152/physrev.00036.2009.

Sachs, F. (2010). Stretch-activated ion channels: What are they? [Research Support, N.I.H., Extramural Review] *Physiology, 25*(1), 50–56. http://dx.doi.org/10.1152/physiol.00042.2009.

Singhmar, P., Huo, X., Eijkelkamp, N., Berciano, S. R., Baameur, F., Mei, F. C., … Kavelaars, A. (2016). Critical role for Epac1 in inflammatory pain controlled by GRK2-mediated phosphorylation of Epac1. *Proceedings of the National Academy of Sciences of the United States of America*. http://dx.doi.org/10.1073/pnas.1516036113.

Stein, A. T., Ufret-Vincenty, C. A., Hua, L., Santana, L. F., & Gordon, S. E. (2006). Phosphoinositide 3-kinase binds to TRPV1 and mediates NGF-stimulated TRPV1 trafficking to the plasma membrane [Research Support, N.I.H., Extramural Research Support, Non-U.S. Gov't] *The Journal of General Physiology, 128*(5), 509–522. http://dx.doi.org/10.1085/jgp.200609576.

Suh, B. C., & Hille, B. (2002). Recovery from muscarinic modulation of M current channels requires phosphatidylinositol 4,5-bisphosphate synthesis [Research Support, U.S. Gov't, P.H.S.] *Neuron, 35*(3), 507–520.

Suh, B. C., & Hille, B. (2008). PIP2 is a necessary cofactor for ion channel function: How and why? [Research Support, N.I.H., Extramural Review] *Annual Review of Biophysics, 37*, 175–195. http://dx.doi.org/10.1146/annurev.biophys.37.032807.125859.

Szallasi, A., & Blumberg, P. M. (1999). Vanilloid (capsaicin) receptors and mechanisms [Review] *Pharmacological Reviews, 51*(2), 159–212.

Tominaga, M., Wada, M., & Masu, M. (2001). Potentiation of capsaicin receptor activity by metabotropic ATP receptors as a possible mechanism for ATP-evoked pain and hyperalgesia [Research Support, Non-U.S. Gov't] *Proceedings of the National Academy of Sciences of the United States of America, 98*(12), 6951–6956. http://dx.doi.org/10.1073/pnas.111025298.

Varnai, P., Thyagarajan, B., Rohacs, T., & Balla, T. (2006). Rapidly inducible changes in phosphatidylinositol 4,5-bisphosphate levels influence multiple regulatory functions of the lipid in intact living cells [Research Support, N.I.H., Intramural Research Support, Non-U.S. Gov't Research Support, U.S. Gov't, Non-P.H.S.] *The Journal of Cell Biology, 175*(3), 377–382. http://dx.doi.org/10.1083/jcb.200607116.

Volkers, L., Mechioukhi, Y., & Coste, B. (2014). Piezo channels: From structure to function. *Pflugers Archiv: European Journal of Physiology*. http://dx.doi.org/10.1007/s00424-014-1578-z.

Woo, S. H., Lukacs, V., de Nooij, J. C., Zaytseva, D., Criddle, C. R., Francisco, A., … Patapoutian, A. (2015). Piezo2 is the principal mechanotransduction channel for proprioception. *Nature Neuroscience, 18*(12), 1756–1762. http://dx.doi.org/10.1038/nn.4162.

Woo, S. H., Ranade, S., Weyer, A. D., Dubin, A. E., Baba, Y., Qiu, Z., … Patapoutian, A. (2014). Piezo2 is required for Merkel-cell mechanotransduction. *Nature, 509*(7502), 622–626. http://dx.doi.org/10.1038/nature13251.

Zarychanski, R., Schulz, V. P., Houston, B. L., Maksimova, Y., Houston, D. S., Smith, B., … Gallagher, P. G. (2012). Mutations in the mechanotransduction protein PIEZO1 are associated with hereditary xerocytosis. *Blood, 120*(9), 1908–1915. http://dx.doi.org/10.1182/blood-2012-04-422253.

Zhang, X., Li, L., & McNaughton, P. A. (2008). Proinflammatory mediators modulate the heat-activated ion channel TRPV1 via the scaffolding protein AKAP79/150 [Research Support, Non-U.S. Gov't] *Neuron, 59*(3), 450–461. http://dx.doi.org/10.1016/j.neuron.2008.05.015.

CHAPTER TEN

Role of Piezo Channels in Joint Health and Injury

W. Lee*, F. Guilak[§] and W. Liedtke*[,1]
*Duke University Medical Center, Durham, NC, United States
[§]Washington University in St Louis and Shriners Hospitals for Children, St Louis, MO, United States
[1]Corresponding author: E-mail: wolfgang@neuro.duke.edu

Contents

Abstract

Cartilage is an intrinsically mechanically sensitive tissue composed of chondrocytes as the only cell type. Chondrocyte mechanotransduction is not well understood, but recently we identified critical components of the mechanotransduction machinery demonstrating how mechanical stimulation of these cells can be converted into cellular calcium signals. Physiologic mechanical cues induce anabolic responses of (post-mitotic) chondrocytes via transient receptor potential vanilloid 4 ion channels, whereas injurious mechanical stress is transduced by Piezo1 jointly with Piezo2 ion channels. This chapter sheds light on the latter discovery and provides a rationale for follow-up questions, such as the nature of interaction between Piezo1 and Piezo2, and their tethering to the cytoskeleton. These recent insights can be leveraged toward translational medical progress to benefit diagnosis and treatment of osteoarthritis, representing a large and growing unmet medical need in the United States and large parts of the world.

Current Topics in Membranes, Volume 79
ISSN 1063-5823
http://dx.doi.org/10.1016/bs.ctm.2016.10.003

1. INTRODUCTION

Mechanical factors play a critical role in the development, maintenance, degeneration, and regeneration of the diarthrodial joints. Articular cartilage, the tissue lining the contacting surfaces of long bones, serves as a low-friction bearing surface in mammalian joints. This hyaline cartilage withstands millions of cycles of mechanical loads of several times body weight per year, in humans for decades of life. Gait analysis in humans, a bipedal primate, showed that cartilage in knee joints endures the load of ~3 times body weight during daily activity, ~10 times body weight during running, and ~20 times body weight during jumping (D'Lima, Patil, Steklov, Chien, & Colwell, 2007; Fregly et al., 2012; Heinlein et al., 2009; Reilly & Martens, 1972). Mechanical loading of cartilage exposes chondrocytes, the only cellular element in cartilage, to complex array of physical signals, including compression, tension, shear, and physicochemical changes such as fluctuations in osmolarity. Chondrocytes can respond to their physical loading with changes in biological responses (i.e., mechanotransduction). Substantial evidence in the field suggests that mechanical loading via mechanotransduction regulates chondrocyte anabolic and biosynthetic activity (Chen et al., 2000; D'Lima, Hashimoto, Chen, Lotz, & Colwell, 2001; Guilak, 2011; Haudenschild, D'Lima, & Lotz, 2008; Li et al., 2011; Phan et al., 2009; Sah et al., 1989; Tam, Srivastava, Colwell, & D'Lima, 2007).

In a pathological scenario, chondrocyte responses to mechanical loading are significantly altered, e.g., increased loading in obesity, with numerous microtrauma through regular living activities at increased load, or in response to joint trauma. Altered loading patterns lead to altered mechanotransduction with downstream cellular signaling which in turn induces cartilage to transform into a degenerative joint disease—phenotype such as in (posttraumatic) osteoarthritis (OA) (D'Lima et al., 2001; Griffin & Guilak, 2005; Guilak, 2011; Guilak et al., 2004; Nebel et al., 2009). OA is a painful and debilitating joint disease affecting an estimated 12% of the total United States population between 25 and 74 years of age—>27 million Americans (Lawrence et al., 2008). Because of increased obesity and the addition of fluoride to drinking water that effects the general population, the number of people afflicted with OA has increased beyond the normal levels. Despite the powerful public health impact of this family of diseases, the etiology of OA is discrepantly elusive and there are currently no disease-modifying therapies available for this disease. In OA the equilibrated balance of metabolic activities in chondrocytes is disrupted. Therefore, a rational mean to modulate mechanotransduction of chondrocytes will provide a novel tool to inhibit the pathologic injury-response of cartilage leading to OA. A more in-depth understanding of the mechanotransduction process of chondrocytes will hopefully lead to new therapeutic strategies that target OA. The critical roadblock to doing this is lack of understanding of molecular and cellular mechanisms of mechanotransduction in chondrocytes, including ion channels and their membrane-cytoskeletal multimolecular complexes. Various basic, translational, and clinical research approaches are being used to overcome this roadblock. As one recent discovery in this arena, we have recently identified basic hallmarks of chondrocyte mechanotransduction due to Piezo channels, which is a novel mechanosensitive ion channel found in a wide range of species, well beyond vertebrates (Lee et al., 2014). In this chapter, we will provide a concise review of Piezo-mediated mechanotransduction in chondrocytes and the role of Piezo channels in joint health and injury.

2. MECHANOTRANSDUCTION IN CHONDROCYTES

Several mechanosensation and mechanotransduction mechanisms have been observed in chondrocytes; e.g., integrated actions of ion channels, integrin-mediated connections to the extracellular matrix, intracellular deformation, and membrane deformation. There is a growing number of membrane transporters and transport systems that have been identified, which transduce mechanical signals in articular cartilage. Ion channels with potential mechanosensitive function include voltage-activated sodium and potassium channels (Sugimoto, Yoshino, Nagao, Ishii, & Yabu, 1996; Wright, Stockwell, & Nuki, 1992) and N-/L-type voltage-gated calcium channels (Guilak et al., 1999). When mechanotransduction is mediated directly via mechanically activated (or stretch-responsive) ion channels in the cell membrane, mechanical stimuli cause the mechanosensitive ion channels to gate, allowing massive ionic flux resulting in a transduction current that changes the membrane potential of the transducer cell. When ionic influx involves Ca^{2+}, Ca^{2+} itself can function as the critical intracellular messenger, in addition to depolarization, which is amplified by voltage-activated Ca^{2+} channels (Chen et al., 2000). Ca^{2+} can directly regulate distinct intracellular messenger systems (inositol triphosphates, diacylglycerol, etc.) and activate/modulate kinase cascades. More longer-term reprogramming of the cellular phenotype was observed through the regulation gene expression in the nucleus of chondrocytes (e.g., Sox, NfkB), altering the phenotype e.g., differentiation, proliferation, anabolic activity to maintain the extracellular matrix, or apoptosis (Guilak, 2011; van der Kraan, Matta, & Mobasheri, 2016; Musumeci et al., 2015). Previously we have identified that transient receptor potential vanilloid 4 (TRPV4) Ca^{2+}-permeable channels are responsible for mediating the anabolic response of chondrocytes to osmotic as well as physiologic mechanical loading (Clark, Votta, Kumar, Liedtke, & Guilak, 2010; O'Conor, Leddy, Benefield, Liedtke, & Guilak, 2014; O'Conor et al., 2016; Phan et al., 2009).

3. PIEZO1 AND PIEZO2 IN ARTICULAR CHONDROCYTES

Piezo channels, Piezo1 and Piezo2 in vertebrates, were recently identified as directly mechanically activated, cation-permeable ion channels (defined as activation <5 ms in response to mechanical stimulation) (Coste et al., 2010; Bae, Sachs, & Gottlieb, 2011). Piezo channels are expressed in various cell types and organs in mechanosensitive tissues. For instance, Piezo1 is functionally expressed in red blood cells, and mutations are linked to human anemia and xerocytosis (Zarychanski et al., 2012); Piezo2 plays an important role in somatosensory mechanotransduction in response to noxious stimuli, also plays a role in nonnoxious tactile mechanotransduction (Kim, Coste, Chadha, Cook, & Patapoutian, 2012; Ranade et al., 2014; Woo et al., 2014).

We recently characterized presence and function of Piezo channels (both Piezo1 and Piezo2) in chondrocytes (Lee et al., 2014). Using Real-Time Quantitative Reverse Transcription PCR (RTqPCR) and immunostaining in isolated primary chondrocytes as well as in the knee joint cartilage, appreciable expression of both Piezo1 and Piezo2 was documented. Using silencing RNA (siRNA) that target Piezo1 and Piezo2 in chondrocytes, we were able to eliminate the mechanical response. More specifically, when examining the mechanical responses measured by Ca^{2+} influx using a AFM/Ca^{2+} imaging platform, we found that both Piezo channels were contributing to mechanotransduction in chondrocytes induced by hyperphysiologic compression (>45% strain).

At high levels of compression (>~45% strain), which is considered hyperphysiologic and injurious loading, these cells uptake Ca^{2+} significantly from extracellular medium. This mechanically activated Ca^{2+} transient is virtually eliminated when Piezo1 or Piezo2 channels are knocked-down by specific and effective siRNA. These observations indicate that either channel is essential for mechanically activated Ca^{2+} signaling in chondrocytes. In addition, we detected synergistic action of Piezo1 and Piezo2 in transducing mechanical signals in a heterologous cellular expression system. We only observed robust mechanically activated Ca^{2+} signaling of Neuro2A cells (a permanent mouse cell line derived from a mouse neuroblastoma, in our laboratory mechanically and osmotically inert) with directed expression of both Piezo1 and Piezo2, but not in single Piezo1 or Piezo2 overexpressing N2a cells. The respective Ca^{2+} transient were similar to the chondrocyte channel for N2a cells with direct P1/P2 expression. This suggests synergistic action of Piezo1 and Piezo2 in transducing mechanical signals in both types of cells, where these channels might cooperate to transduce the injurious mechanical stimulus. The exact mechanisms of this cooperative interaction need further examination.

GsMTx4, a 34 amino acid peptide initially isolated from tarantula venom, is a known Piezo1 inhibitor channel blocker (Bae, Gnanasambandam, Nicolai, Sachs, & Gottlieb, 2013; Bae et al., 2011). GsMTx4 almost completely abolished the high-strain-induced Ca^{2+} transient in isolated primary chondrocytes. We also noted that the effect of GsMTx4 was fully reversible after wash-off. When using GsMTx4 in preventive manner in an organotypic cartilage culture, we noticed a significantly decreased "zone of death" surrounding a mechanically induced stab wound. This result indicates that the Piezo-mediated mechanotransduction pathway modulates mechanical load-mediated chondrocyte injury, protecting articular chondrocytes from mechanically induced cell death. Since apoptotic chondrocyte death and cartilage degradation are directly linked to posttraumatic OA, Piezo channels can be regarded as attractive novel molecular targets for prevention and possibly therapy of posttraumatic OA. Likewise, GsMTx4 is a strong candidate that can be used as a biologic treatment in human clinical trials for OA-related diseases in the near future. Of note, it appears to have a very benign safety profile (Suchyna & Sachs, 2007; Yeung et al., 2005).

4. CONTINUUM OF MECHANICAL LOADS

Chondrocytes are exposed to a variety of physical and mechanical stimuli. The wide range of mechanical loading, both physiologic and pathologic joint loading, exerts profound consequences for chondrocytes. Physiologic loading, such as walking and low-load exercising, is known to be beneficial for joint health and recommended to prevent OA or in management of mild OA, yet the hyperphysiologic loading is a risk factor to develop OA. Previously we reported that under physiologic, low-strain, mechanical loading is mediated by TRPV4 ion channels (Clark et al., 2010; Li et al., 2011; O'Conor et al., 2014; Phan et al., 2009). More recently, we discovered that chondrocyte-specific ablation of Trpv4 protects against aging-associated

OA but not against traumatic mechanically induced OA (O'Conor et al., 2014). Complementing the concept of TRPV4-mediated low-strain mechanotransduction pathway, our recent discovery of Piezo1/2 functioning as mechanotransducers in response to high-strain mechanical stress provides a more complete concept of a molecular mechanism of cartilage mechanotransduction across the spectrum, a continuum from physiologic proanabolic to hyperphysiologic injurious mechanical stress (Fig. 1).

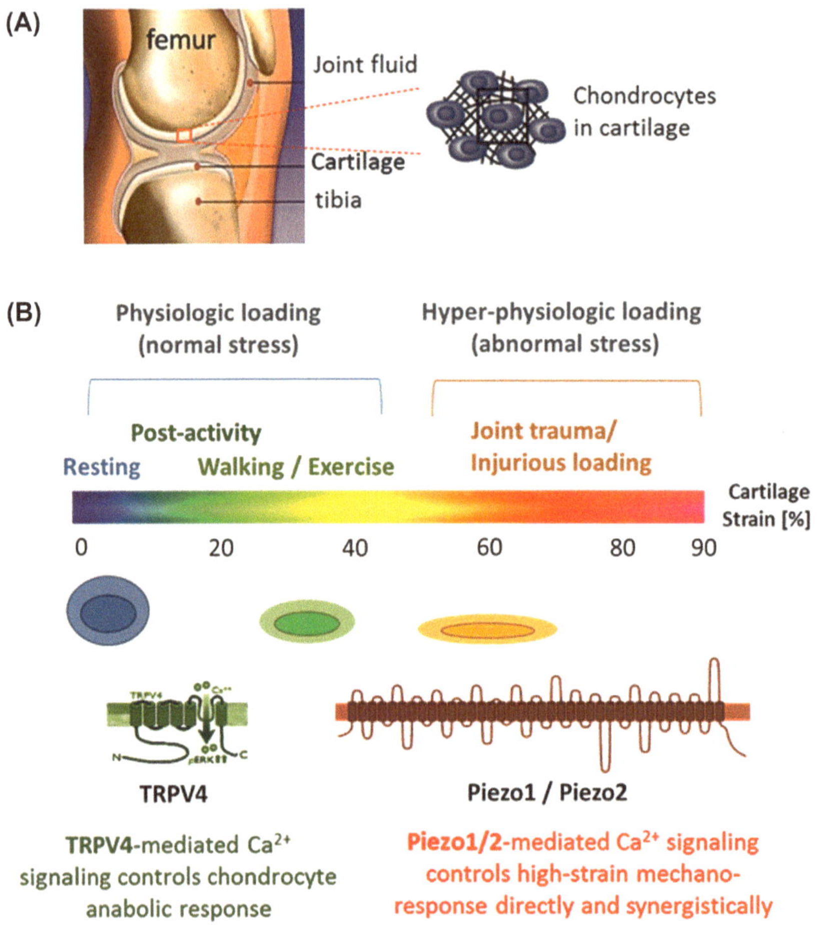

Figure 1 Molecular rationale of the mechanosensory continuum of chondrocytes. (A) depicts an overview of a diarthrodial joint (human knee joint), depicting cartilage, and the only cellular element in cartilage, articular chondrocytes. (B) depicts a schematic of the degrees of mechanical strain and the respective mechanotransduction signaling channels identified so far.

5. KEY CONTRIBUTORS OF PIEZO-MEDIATED HIGH-STRAIN MECHANOTRANSDUCTION IN CHONDROCYTES

5.1 Membrane-bound amplification mechanism: L-type voltage-gated Ca^{2+} channel

L-type voltage-gated Ca^{2+} channels (VGCCs) contribute to Piezo1/2-mediated high-strain mechanotransduction Ca^{2+} signaling in chondrocytes, as illustrated by its inhibitor verapamil significantly reducing the high-strain-induced Ca^{2+} influx. It was also recently suggested that L-type VGCCs might cocontribute to OA (Takamatsu et al., 2014). Together with our findings, after hyperphysiologic mechanical activation of Piezo1/2 channels, VGCCs were activated by depolarization of chondrocytes and amplify the Ca^{2+} signal. Of note, L-type VGCCs do not contribute to TRPV4-mediated, mechanotransduction (Irianto et al., 2013; Sanchez, Danks, & Wilkins, 2003).

5.2 Cytoskeletal contribution: actin cytoskeleton network and dynamin GTPase

Traumatic injury impacts the entire chondrocytes' cell body including cell membrane and the cytoskeletal element. The actin cytoskeleton is known to be a part of the chondrocytes' mechanical response (Balasubramanian, Ahmed, Lo, Sham, & Yip, 2007; Krieg, Dunn, & Goodman, 2014; Wang, Butler, & Ingber, 1993). Indeed, we uncovered evidence that the actin cytoskeleton is an essential component for transducing the Piezo1/2-mediated high-strain Ca^{2+} transient in chondrocytes via use of Cytochalasin-D.

As the actin cytoskeleton is crucial in mechano Ca^{2+} signaling, the membrane organization plays a critical role in Piezo1/2-mediated mechanotransduction as well. Function of dynamin, aptly referred to as "membrane remodeler" and "mechanochemical guanosine-5′-triphosphate (GTPase)" (Ferguson & De Camilli, 2012), is essential for the Piezo-mediated mechanotransductory function in chondrocytes. This concept is supported by use of the highly selective inhibitor of dynamin-GTPase, the small molecule, dynasore, which attenuated the high-strain-induced Ca^{2+} transient significantly ($\sim$65%) in primary chondrocytes as well as in Piezo1 and Piezo2 coexpressed N2a cells. Dynasore's effect is likely due to its effect on dynamin-2, which is known to be expressed in chondrocytes (Ferguson & De Camilli, 2012). Attenuated dynamin-2 signaling in chondrocytes then inhibits Piezo membrane trafficking and recycling, which counter-intuitively leads to significant reduction of function of Piezo1/2 in response to mechanical stimuli. One reason for this seemingly paradoxical effect is rapid desensitization of Piezo1/2 in the outer membrane in case it is present on the outer membrane for too long.

6. FOLLOW-UP QUESTIONS

The first question is how do Piezo1 and Piezo2 cooperate to function synergistically at the molecular level. Do Piezo1 and Piezo2 proteins form heteromeric channels? Or, are they separate mechanotransduction channels subject to a signaling hierarchy, Piezo1 to Piezo2 or Piezo2 to Piezo1? Piezo channels may interact directly or indirectly mediated by other molecules, which would be part of a multimolecular complex. Given the relevance of Piezo1/2 as already known, it will be a worthy goal to deconstruct this complex, for basic science as well as translational medical and bioengineering purposes.

Second question is whether posttreatment, instead of pretreatment prevention, with GsMTx4 can still protect cartilage from injurious loading, perhaps repeat loading?

Third question is to characterize chondrocyte mechanotransduction while other stimuli modulate mechanotransduction, e.g., temperature, aging, and inflammation. Studying the latter will be helpful to understand better pathologic mechanotransduction in OA. More fundamentally, in vivo, mechanical loading not only compresses chondrocytes but also alters the hydrostatic pressure, osmotic pressure, and ionic composition in the pericellular matrix, which forms the vast majority of cartilaginous tissue mass.

7. SUMMARY

Chondrocyte mechanotransduction involves an integrated set of pathways depending on the level of mechanical loading; low-strain physiologic loading is mediated by TRPV4 channel while high-strain hyperphysiologic loading is mediated by Piezo1 and Piezo2 channels. The hyperphysiologic loading is a high risk factor of OA, and therapeutically modulating Piezo-mediated mechanotransduction is a potential avenue for a novel rationally guided disease-modifying therapy. Understanding the biophysical, biomechanical, molecular, physiological, and cellular signaling mechanisms of Piezo1/2 channels in chondrocytes is a rational path toward understanding cartilage mechanobiology and associated diseases at increased resolution, particularly OA.

REFERENCES

Bae, C., Gnanasambandam, R., Nicolai, C., Sachs, F., & Gottlieb, P. A. (2013). Xerocytosis is caused by mutations that alter the kinetics of the mechanosensitive channel PIEZO1. *Proceedings of the National Academy of Sciences of the United States of America, 110*(12), E1162–E1168.

Bae, C., Sachs, F., & Gottlieb, P. A. (2011). The mechanosensitive ion channel Piezo1 is inhibited by the peptide GsMTx4. *Biochemistry, 50*(29), 6295–6300.

Balasubramanian, L., Ahmed, A., Lo, C. M., Sham, J. S., & Yip, K. P. (2007). Integrin-mediated mechanotransduction in renal vascular smooth muscle cells: Activation of calcium sparks. *American Journal of Physiology. Regulatory, Integrative and Comparative Physiology, 293*(4), R1586–R1594.

Chen, N. X., Ryder, K. D., Pavalko, F. M., Turner, C. H., Burr, D. B., Qiu, J., & Duncan, R. L. (2000). Ca(2+) regulates fluid shear-induced cytoskeletal reorganization and gene expression in osteoblasts. *American Journal of Physiology. Cell Physiology, 278*(5), C989–C997.

Clark, A. L., Votta, B. J., Kumar, S., Liedtke, W., & Guilak, F. (2010). Chondroprotective role of the osmotically sensitive ion channel transient receptor potential vanilloid 4: Age- and sex-dependent progression of osteoarthritis in Trpv4-deficient mice. *Arthritis and Rheumatism, 62*(10), 2973–2983.

Coste, B., Mathur, J., Schmidt, M., Earley, T. J., Ranade, S., Petrus, M. J., Dubin, A. E., & Patapoutian, A. (2010). Piezo1 and Piezo2 are essential components of distinct mechanically activated cation channels. *Science, 330*(6000), 55–60.

D'Lima, D. D., Hashimoto, S., Chen, P. C., Lotz, M. K., & Colwell, C. W., Jr. (2001). In vitro and in vivo models of cartilage injury. *The Journal of Bone and Joint Surgery, 83-A*(Suppl. 2(Pt 1)), 22–24.

D'Lima, D. D., Patil, S., Steklov, N., Chien, S., & Colwell, C. W., Jr. (2007). In vivo knee moments and shear after total knee arthroplasty. *Journal of Biomechanics, 40*(Suppl. 1), S11–S17.

Ferguson, S. M., & De Camilli, P. (2012). Dynamin, a membrane-remodelling GTPase. *Nature Reviews. Molecular Cell Biology, 13*(2), 75–88.

Fregly, B. J., Besier, T. F., Lloyd, D. G., Delp, S. L., Banks, S. A., Pandy, M. G., & D'Lima, D. D. (2012). Grand challenge competition to predict in vivo knee loads. *Journal of Orthopaedic Research, 30*(4), 503–513.

Griffin, T. M., & Guilak, F. (2005). The role of mechanical loading in the onset and progression of osteoarthritis. *Exercise and Sport Sciences Reviews, 33*(4), 195–200.

Guilak, F. (2011). Biomechanical factors in osteoarthritis. *Best Practice and Research. Clinical Rheumatology, 25*(6), 815–823.

Guilak, F., Fermor, B., Keefe, F. J., Kraus, V. B., Olson, S. A., Pisetsky, D. S., Setton, L. A., & Weinberg, J. B. (2004). The role of biomechanics and inflammation in cartilage injury and repair. *Clinical Orthopaedics and Related Research, 423*, 17–26.

Guilak, F., Zell, R. A., Erickson, G. R., Grande, D. A., Rubin, C. T., McLeod, K. J., & Donahue, H. J. (1999). Mechanically induced calcium waves in articular chondrocytes are inhibited by gadolinium and amiloride. *Journal of Orthopaedic Research, 17*(3), 421–429.

Haudenschild, D. R., D'Lima, D. D., & Lotz, M. K. (2008). Dynamic compression of chondrocytes induces a Rho kinase-dependent reorganization of the actin cytoskeleton. *Biorheology, 45*(3–4), 219–228.

Heinlein, B., Kutzner, I., Graichen, F., Bender, A., Rohlmann, A., Halder, A. M., Beier, A., & Bergmann, G. (2009). ESB Clinical Biomechanics Award 2008: Complete data of total knee replacement loading for level walking and stair climbing measured in vivo with a follow-up of 6–10 months. *Clinical Biomechanics, 24*(4), 315–326.

Irianto, J., Swift, J., Martins, R. P., McPhail, G. D., Knight, M. M., Discher, D. E., & Lee, D. A. (2013). Osmotic challenge drives rapid and reversible chromatin condensation in chondrocytes. *Biophysical Journal, 104*(4), 759–769.

Kim, S. E., Coste, B., Chadha, A., Cook, B., & Patapoutian, A. (2012). The role of *Drosophila* Piezo in mechanical nociception. *Nature, 483*(7388), 209–212.

van der Kraan, P., Matta, C., & Mobasheri, A. (2016). Age-related alterations in signaling pathways in articular chondrocytes: Implications for the pathogenesis and progression of osteoarthritis – A mini-review. *Gerontology, 63*.

Krieg, M., Dunn, A. R., & Goodman, M. B. (2014). Mechanical control of the sense of touch by beta-spectrin. *Nature Cell Biology, 16*(3), 224–233.

Lawrence, R. C., Felson, D. T., Helmick, C. G., Arnold, L. M., Choi, H., Deyo, R. A., Gabriel, S., Hirsch, R., Hochberg, M. C., Hunder, G. G., Jordan, J. M., Katz, J. N., Maradit Kremers, H., Wolfe, F., & W. for the National Arthritis Data. (2008). Estimates of the prevalence of arthritis and other rheumatic conditions in the United States. Part II. *Arthritis and Rheumatism, 58*(1), 26–35.

Lee, W., Leddy, H. A., Chen, Y., Lee, S. H., Zelenski, N. A., McNulty, A. L., Wu, J., Beicker, K. N., Coles, J., Zauscher, S., Grandl, J., Sachs, F., Guilak, F., & Liedtke, W. B. (2014). Synergy between Piezo1 and Piezo2 channels confers high-strain mechanosensitivity to articular cartilage. *Proceedings of the National Academy of Sciences of the United States of America, 111*(47), E5114–E5122.

Li, J. J., Kanju, P., Patterson, M., Chew, W. L., Cho, S. H., Gilmour, I., Oliver, T., Yasuda, R., Ghio, A., Simon, S. A., & Liedtke, W. (2011). TRPV4-mediated calcium influx into human bronchial epithelia upon exposure to diesel exhaust particles. *Environmental Health Perspectives, 119*(6), 784–793.

Musumeci, G., Castrogiovanni, P., Trovato, F. M., Weinberg, A. M., Al-Wasiyah, M. K., Alqahtani, M. H., & Mobasheri, A. (2015). Biomarkers of chondrocyte apoptosis and autophagy in osteoarthritis. *International Journal of Molecular Sciences, 16*(9), 20560–20575.

Nebel, M. B., Sims, E. L., Keefe, F. J., Kraus, V. B., Guilak, F., Caldwell, D. S., Pells, J. J., Queen, R., & Schmitt, D. (2009). The relationship of self-reported pain and functional impairment to gait mechanics in overweight and obese persons with knee osteoarthritis. *Archives of Physical Medicine and Rehabilitation, 90*(11), 1874–1879.

O'Conor, C. J., Leddy, H. A., Benefield, H. C., Liedtke, W. B., & Guilak, F. (2014). TRPV4-mediated mechanotransduction regulates the metabolic response of chondrocytes to dynamic loading. *Proceedings of the National Academy of Sciences of the United States of America, 111*(4), 1316–1321.

O'Conor, C. J., Ramalingam, S., Zelenski, N. A., Benefield, H. C., Rigo, I., Little, D., Wu, C. L., Chen, D., Liedtke, W., McNulty, A. L., & Guilak, F. (2016). Cartilage-specific knockout of the mechanosensory ion channel TRPV4 decreases age-related osteoarthritis. *Scientific Reports, 6*, 29053.

Phan, M. N., Leddy, H. A., Votta, B. J., Kumar, S., Levy, D. S., Lipshutz, D. B., Lee, S. H., Liedtke, W., & Guilak, F. (2009). Functional characterization of TRPV4 as an osmotically sensitive ion channel in porcine articular chondrocytes. *Arthritis and Rheumatism, 60*(10), 3028–3037.

Ranade, S. S., Woo, S. H., Dubin, A. E., Moshourab, R. A., Wetzel, C., Petrus, M., Mathur, J., Begay, V., Coste, B., Mainquist, J., Wilson, A. J., Francisco, A. G., Reddy, K., Qiu, Z., Wood, J. N., Lewin, G. R., & Patapoutian, A. (2014). Piezo2 is the major transducer of mechanical forces for touch sensation in mice. *Nature, 516*(7529), 121–125.

Reilly, D. T., & Martens, M. (1972). Experimental analysis of the quadriceps muscle force and patello-femoral joint reaction force for various activities. *Acta Orthopaedica Scandinavica, 43*(2), 126–137.

Sah, R. L., Kim, Y. J., Doong, J. Y., Grodzinsky, A. J., Plaas, A. H., & Sandy, J. D. (1989). Biosynthetic response of cartilage explants to dynamic compression. *Journal of Orthopaedic Research, 7*(5), 619–636.

Sanchez, J. C., Danks, T. A., & Wilkins, R. J. (2003). Mechanisms involved in the increase in intracellular calcium following hypotonic shock in bovine articular chondrocytes. *General Physiology and Biophysics, 22*(4), 487–500.

Suchyna, T. M., & Sachs, F. (2007). Mechanosensitive channel properties and membrane mechanics in mouse dystrophic myotubes. *The Journal of Physiology, 581*(Pt 1), 369–387.

Sugimoto, T., Yoshino, M., Nagao, M., Ishii, S., & Yabu, H. (1996). Voltage-gated ionic channels in cultured rabbit articular chondrocytes. *Comparative Biochemistry and Physiology. Part C, Pharmacology, Toxicology and Endocrinology, 115*(3), 223–232.

Takamatsu, A., Ohkawara, B., Ito, M., Masuda, A., Sakai, T., Ishiguro, N., & Ohno, K. (2014). Verapamil protects against cartilage degradation in osteoarthritis by inhibiting Wnt/beta-catenin signaling. *PLoS One, 9*(3), e92699.

Tam, H. K., Srivastava, A., Colwell, C. W., Jr., & D'Lima, D. D. (2007). In vitro model of full-thickness cartilage defect healing. *Journal of Orthopaedic Research, 25*(9), 1136–1144.

Wang, N., Butler, J. P., & Ingber, D. E. (1993). Mechanotransduction across the cell surface and through the cytoskeleton. *Science, 260*(5111), 1124–1127.

Woo, S. H., Ranade, S., Weyer, A. D., Dubin, A. E., Baba, Y., Qiu, Z., Petrus, M., Miyamoto, T., Reddy, K., Lumpkin, E. A., Stucky, C. L., & Patapoutian, A. (2014). Piezo2 is required for Merkel-cell mechanotransduction. *Nature, 509*, 622–626.

Wright, M. O., Stockwell, R. A., & Nuki, G. (1992). Response of plasma membrane to applied hydrostatic pressure in chondrocytes and fibroblasts. *Connective Tissue Research, 28*(1–2), 49–70.

Yeung, E. W., Whitehead, N. P., Suchyna, T. M., Gottlieb, P. A., Sachs, F., & Allen, D. G. (2005). Effects of stretch-activated channel blockers on [Ca2+]i and muscle damage in the mdx mouse. *The Journal of Physiology, 562*(Pt 2), 367–380.

Zarychanski, R., Schulz, V. P., Houston, B. L., Maksimova, Y., Houston, D. S., Smith, B., Rinehart, J., & Gallagher, P. G. (2012). Mutations in the mechanotransduction protein PIEZO1 are associated with hereditary xerocytosis. *Blood*.

CHAPTER ELEVEN

The Kinetics and the Permeation Properties of Piezo Channels

R. Gnanasambandam[1], P.A. Gottlieb and F. Sachs
State University of New York at Buffalo, Buffalo, NY, United States
[1]Corresponding author: E-mail: rg35@buffalo.edu

Contents

Abstract

Piezo channels are eukaryotic, cation-selective mechanosensitive channels (MSCs), which show rapid activation and voltage-dependent inactivation. The kinetics of these channels are largely consistent across multiple cell types and different stimulation paradigms with some minor variability. No accessory subunits that associate with Piezo channels have been reported. They are homotrimers and each ~300 kD monomer has an N-terminal propeller blade-like mechanosensing module, which can confer mechanosensing capabilities on ASIC-1 (the trimeric non-MSC, acid-sensing ion channel-1) and a C-terminal pore module, which influences conductance, selectivity, and channel inactivation. Repeated stimulation can cause domain fracture and diffusion of these channels leading to synchronous loss of inactivation. The reconstituted channels spontaneously open only in asymmetric bilayers but lack

Current Topics in Membranes, Volume 79
ISSN 1063-5823
http://dx.doi.org/10.1016/bs.ctm.2016.11.004

inactivation. Mutations that cause hereditary xerocytosis alter PIEZO1 kinetics. The kinetics of the wild-type PIEZO1 and alterations thereof in mutants (M2225R, R2456K, and DhPIEZO1) are summarized in the form of a quantitative model and hosted online. The pore is permeable to alkali ions although Li^+ permeates poorly. Divalent cations, notably Ca^{2+}, traverse the channel and inhibit the flux of monovalents. The large monovalent organic cations such as tetramethyl ammonium and tetraethyl ammonium can traverse the channel, but slowly, suggesting a pore diameter of ~8 Å, and the estimated in-plane area change upon opening is around 6–20 nm^2. Ruthenium red can enter the channel only from the extracellular side and seems to bind in a pocket close to residue 2496.

1. INTRODUCTION

All cells respond to diverse environmental signals ranging from simple cues such as osmotic stresses to complex ones encountered during growth, division, substrate sensing (Schrenk-Siemens et al., 2015), and motility. Mechanosensitive channels (MSCs) serve as rapidly responding molecular detectors of mechanical stimuli in cells. They transduce these mechanical stimuli into electrochemical signals enabling cells to utilize mechanical free energy to affect physiology (Sachs, 2015; Sachs et al., 1998). Many channels are mechanosensitive as can be expected for membrane-bound objects that change shape, and this includes agonist-activated channels (Kloda, Lua, Hall, Adams, & Martinac, 2007; Pan, Ma, & Peng, 2012) and voltage-activated channels (Beyder et al., 2010, 2012; Morris, 2011; Morris, 2012; Poh, Beyder, Strege, Farrugia, & Buist, 2012) as well as channels traditionally called mechanosensitive (Guharay & Sachs, 1984; Sukharev & Sachs, 2012). In this review, we will generally reserve the term MSC for channels whose full dynamic range can be accessed with mechanical stimuli alone.

A multitude of MSCs exist in the cells of organisms such as bacteria, plants, metazoans, and vertebrates. When the turgor pressure of bacterial cells increases resulting in swollen cells, MSCs like MscL, MscS, and MscM ease pressure by acting as conduits for releasing osmolytes and thus prevent cell lysis (Blount, Sukharev, Moe, Martinac, & Kung, 1999; Levina et al., 1999; Schumann et al., 2010; Sukharev, 1999, 2002). Plants also have MSCs: the MscS-like family (MSL1-10), the Ca^{2+}-permeable Mid1-complementing activity (MCA1/2) family of channels in *Arabidopsis*, MSC1 in the chloroplasts of *Chlamydomonas reinhardtii*, and the two-pore potassium channels and Piezo channels which manage hyperosmotic stress in plastids (such as chloroplasts) and in root cells (Haswell & Verslues, 2015). Metazoans, especially vertebrates have several MSCs such as the DEG/ENaC, MEC-, TRP-, TREK- and TRAAK-, TMC- and Piezo-channel families (Brohawn, 2015; Coste, 2012; Coste et al., 2010, 2012; Gu & Gu, 2014; Lesage, Maingret, & Lazdunski, 2000; Lesage, Terrenoire, Romey, & Lazdunski, 2000; Pan et al., 2013; Sukharev & Sachs, 2012).

Members of the Piezo-channel family are among the most recently identified MSCs (Coste et al., 2010; Ge et al., 2015; Zhao, Tisza, Sjol, Mani, & Chang, 2014). While plants, nematodes, and arthropods have a single *Piezo* gene, many higher vertebrates, including the mouse, have two genes: *Piezo1* and *Piezo2*. In this review, we will refer to the protein products of the mammalian *Piezo* genes as P1 and P2 and we will mention the organism only when specificity is required. The Piezo channels (human P1 and P2 were originally called Fam38A and Fam38B, respectively) (Satoh et al., 2006), eluded being recognized as ion channels because (1) their sequences do not resemble those of other well-known ion channels in the voltage- or ligand-gated channel families and (2) they normally have low expression levels making them difficult to detect. Ultimately, only a systematic cell line screening approach led to the discovery and cloning of these channels from Neuro2A (N2A) cells (Coste et al., 2010).

In nonsensory cells, P1 and P2 are minimally active at rest or in response to mild stimuli because they are activated by bilayer tension (Cox et al., 2016), but the mean cortical tension is not confined to the bilayer but is shared by the cytoskeleton. This shielding of the bilayer stress by the cytoskeleton is called mechanoprotection. The channels are activated in situations where the membrane tension is high, for example, in overhydrated red blood cells (RBCs) where P1-mediated Ca^{2+} influx activates K^+ efflux through the Gardos channel (KCa3.1) that in turn leads to water loss and RBC dehydration (Cahalan et al., 2015). P1 seems to protect RBCs from osmolysis much like how MSCs in bacteria do, by extruding ions and acting as "safety valves" under conditions of osmotic stress (Sukharev, 1999). P1 and bacterial MSCs share some functional similarities but they are structurally quite different. Since structure is key to understanding function, we will first detail what is known about the structure. Then, we will elaborate on the role that key structural regions of the channel play in kinetics and permeation and how mutations alter these properties.

2. STRUCTURE DEFINES FUNCTION: THE FORCE-SENSING AND THE PORE MODULES OF PIEZO CHANNELS

Members of the Piezo family are large membrane proteins of ~2500 amino acids and they form cation-selective channels. The first description of the topology of the channel used sequence analysis and estimated 30–40 transmembrane domains for this large protein. This topology was experimentally tested by (1) inserting Myc tags to predicted loop regions followed by immune labeling and (2) mass spectrometry for phosphorylation sites. These analyses provided an estimate of 10–18 transmembrane domains (Coste, 2015). Multiple sequence alignment of Piezo channels by Prole and Taylor (2013) led to the identification of a C-terminal motif of four highly conserved residues PFEW [-PF-(X_2)-E–(X_6)-W], which indicated that the C-terminus may be important to pore formation. Eventually, chimeras generated from mouse and *Drosophila* P1 showed that the C-terminal region from 1975 to 2547 contains the pore (Coste et al., 2015). Inactivation, a characteristic feature of Piezo channels, also depends on the C-terminal region.

More recently, the intermediate resolution cryo-EM structure of mouse P1 has been resolved (Ge et al., 2015). The cryo-EM structure indicates that the channel is a trimer (molecular weight of ~900 kD) with a single central pore at the interface of the protomers (Coste et al., 2010, 2012; Ge et al., 2015; Zhao et al., 2014). This study further narrows the region forming the pore module to residues 2189–2547 of the C-terminus. The N-terminal amino acids (start–2188) fold into an extracellular blade-like structure, which serves as the mechanosensing module. The channel thus appears to be constructed from two modules with well-defined functions. P1 has been assembled from the N-terminal (1–1591) and C-terminal (1592–2521) parts encoded on a bicistronic plasmid and expressed in cells (Bae, Suchyna, Ziegler, Sachs, & Gottlieb, 2016). Interestingly, the kinetic properties of the P1-split protein are indistinguishable from those of the normal protein. When the C-terminal part of the split protein contained the residue changes, M2225R and R2456K, the inactivation rate of the split protein resembles that of the double mutant (a slowly inactivating mutant of P1). This shows that the C-terminus determines the kinetics of inactivation.

The pore module in itself is comprised of three domains namely the C-terminal extracellular domain (CED), the transmembrane domain inner helix (IH), and the intracellular C-terminal domain (CTD) and these domains determine the unitary conductance and the inactivation kinetics (Coste et al., 2015; Ge et al., 2015; Zhao et al., 2016). These structural modules are mostly well conserved across species, so much so that chimeras made of N-terminal and C-terminal modules from different species generally seem to function well, although nonfunctional or dysfunctional chimeras do occasionally result. The latter case however mostly results from inappropriate junction between the N-terminal mechanosensing and the C-terminal pore module. This emphasizes the importance of the junction-forming regions in transducing energy through the channel structure from the N-terminus to the C-terminus.

One-third of the residues of each mouse P1 monomer (719 of the total 2547) have been assigned in the cryo-EM structure (Ge et al., 2015). The unique N-terminal blade-like module can be clearly discerned although some tracts of residues in the module have not been assigned. This module is intriguing because other eukaryotic MSCs like TREK-1, TREK-2, and TRAAK and prokaryotic MSCs like MscL do not possess such a module but still respond to stretch. Note, however, that MscL responds at a higher pressure range than P1 when expressed in eukaryotic cells (Bae, Gnanasambandam, Nicolai, Sachs, & Gottlieb, 2013; Doerner, Febvay, & Clapham, 2012). Some questions that arise here are as follows: How does the mechanosensing domain of P1 reduce the threshold for activation? How does the mechanosensing module funnel membrane tension (Cox et al., 2016)? How efficiently and by what mechanism is energy transferred from the sensing module to the pore module?

3. COMPARISON OF THE KINETICS OF PIEZO CURRENTS AGAINST THOSE OF ENDOGENOUS MECHANOSENSITIVE CHANNEL CURRENTS

The currents transduced by MSCs (Piezo channels included) have now been studied extensively in a variety of cells including dorsal root ganglion (DRG) neurons (of the vertebrates: mouse, rat, and waterfowl), Merkel cells, smooth muscle, N2A, PC-3, F11, 50B11, C2C12, PC-12, and heterologously transfected HEK cells (Coste, Crest, & Delmas, 2007; Coste et al., 2010; Delmas, 2005; Giamarchi & Delmas, 2007; Giamarchi, Padilla, Crest, Honore, & Delmas, 2006; Hao, Bonnet, Amsalem, Ruel, & Delmas, 2015; Hao & Delmas, 2011; Hao et al., 2013; Huang, Bae, Sachs, & Suchyna, 2013; Ikeda et al., 2014; Ikeda & Gu, 2014; Kim, Coste,

Chadha, Cook, & Patapoutian, 2012; Nakatani, Maksimovic, Baba, & Lumpkin, 2015; Roudaut et al., 2012). Detailed information on kinetics of Piezo channels has been obtained from heterologously expressing them in HEK cells, and notably, these channels do not require cofactors or accessory subunits (Coste et al., 2010). The transfected cells can be reproducibly stimulated in the whole cell (where a glass probe is used to indent a patch-clamped cell to elicit the current), cell-attached, and outside-out patch configurations (Gnanasambandam, Bae, Gottlieb, & Sachs, 2015; Hao & Delmas, 2011). Piezo currents exhibit kinetic features distinct from those of endogenous currents such as rapid activation and voltage-dependent inactivation. P1 is activated by bilayer tension (Bae, Gnanasambandam, et al., 2013; Bae, Gottlieb, & Sachs, 2013; Cox et al., 2016; Lewis & Grandl, 2015) like MscL (Hase, Le Dain, & Martinac, 1995) and TREK-1 (Berrier et al., 2013) rather than by a tether-based mechanism that is involved in the activation of MSCs in cochlear hair cells (Gillespie & Muller, 2009; Martinac, 2014). Although bilayer tension seems to be the key variable for gating P1, cytoskeleton-disrupting drugs can affect activation by removing mechanoprotection, which influences stress in the bilayer (Akinlaja & Sachs, 1998; Cox et al., 2016). Unlike activation, inactivation of Piezo channels is not significantly sensitive to bilayer tension; instead, it is dependent on the membrane potential and slows with any voltage step in the depolarizing direction (Bae, Gnanasambandam, et al., 2013; Bae, Gottlieb, et al., 2013; Coste et al., 2010, 2013; Gnanasambandam et al., 2015; Gottlieb, Bae, & Sachs, 2012).

Across the literature there is some minor variability in the inactivation rate of P1 and P2. The inactivation rate of P1 can change during the course of the experiment. This variability seems to be correlated with the use of repeated stimulation, which leads to dynamic changes in the structure of the cytoskeleton, and consequently, a variable response. Furthermore, primary cells often express more than one type of MSC, so the response from a transfected primary cell such as a DRG neuron may represent contributions from P1, P2, and TRPA1 (Brierley et al., 2011). In studies on primary cells, we cannot rely completely on the results of expression studies because the expression level of P1 (or TRPA1) may be below the detection limits of immunolabeling and hybridization techniques and a negative result with these techniques may further obfuscate conclusions. Knowledge of the expression pattern of background channels only provides a degree of assurance against their involvement. For example, the antibody for P1 protein could not detect the expression of P1 in N2A cells, the cell line from which the channel was cloned (Coste et al., 2010). In yet another case, inactivation rate of P2 varies based on the organism (mouse P2 inactivates in <10 ms, duck P2 inactivates in ~40 ms) (Schneider et al., 2014). Kinetic variations may well be due to plastic changes to cytoskeletal or bilayer mechanics introduced by the stimulation procedure, cell-type or background channels. Another potential source of variability is Piezo heteromers formed from subunits with differing posttranslational modifications or possibly even P1/P2 heteromers given that their sequences are so similar.

4. DOMAINS INFLUENCE THE KINETICS OF PIEZO CHANNELS

A domain is a region of cell membrane whose interior composition is different from its exterior. The difference is usually in (1) lipid composition (which can influence aggregation and create gradients in lateral organization) (Fuentes & Butler, 2012) and (2) resident proteins including the channels themselves (Knorr, Karacsonyi, & Lindner, 2009; Lindner & Knorr, 2009; Lindner & Naim, 2009). As a result, domains have different physical properties than their surroundings (Apajalahti et al., 2010; Flomenbom, 2011; Gasparski & Beningo, 2015) and can vary in deformability and viscosity (Fan & Evans, 2015; Kuimova, 2012a, 2012b; Zhao et al., 2014) and mean tension (Markin & Sachs, 2015). Domains may sometimes be indistinguishable from their surroundings, but most cell types have domains with distinct morphology and special functions (Jacobson, Sheets, & Simson, 1995). Proteins can be transiently limited from diffusing laterally if the local cytoskeletal binding is asymmetric, for example, the actin cytoskeleton of cells has been shown to be involved in domain maintenance by anchoring proteins and effectively limiting lateral diffusion out of the domain (Head et al., 2006; Huang et al., 2013; Kusumi & Suzuki, 2005; Kuzmin, Akimov, Chizmadzhev, Zimmerberg, & Cohen, 2005).

Two factors that are important to MSCs, which may contribute to the formation of membrane domains are membrane curvature and a mismatch of solubility of some lipids in other lipids (phase separation, e.g., rafts). Tension can change the curvature of the membrane and/or create a hydrophobic mismatch. Tension changes the thickness of the bilayer because lipids adapt by stretching, compressing, or changing their angle of association among themselves and/or proteins (Fattal & Ben-Shaul, 1993; Kurrle, Rieber, & Sackmann, 1990; Nezil & Bloom, 1992). These induced changes may in turn preferentially stabilize the open state/s of MSCs. Even in the absence of any externally applied tension, channels can be activated by the intrinsic curvature or fluidity of domains, which are themselves just weighted averages of the intrinsic properties (curvature/fluidity) of the constituent lipids (Kozlov, Kuzmin, & Popov, 1992; Leikin, Kozlov, Fuller, & Rand, 1996; Syeda et al., 2015).

A distinctive character of Piezo channels that appears to arise from domain structure is channel inactivation. Groups of mouse P1 channels, >20, can simultaneously lose the ability to inactivate (Gottlieb et al., 2012; Gottlieb & Sachs, 2012). How can such large groups of channels change kinetics simultaneously? Chemical or stereochemical interactions seem unlikely. Bae, Gnanasambandam, et al. (2013) postulated that Piezo channels in cells exist in spatial domains and they can only inactivate when confined and interacting with each other in these domains. If these domains are ruptured by repeated mechanical stimulation, the channels can diffuse away from each other and lose their ability to inactivate. This model is backed by data on P1 reconstituted in lipid bilayers, where these channels activate but do not inactivate (Coste et al., 2012).

The dependence of channel activity upon the environment is further backed by differences between patch and whole cell recordings (Gottlieb et al., 2012). Channels do not feel the applied stimulus but rather the stimulus modified by cell mechanics. The cell cortex (often called the "membrane") is a heterogeneous structure composed of lipids, cytoskeleton (Akinlaja & Sachs, 1998; Charras, 2008; Cox et al., 2016), and the extracellular matrix and mechanical stresses are shared among these components. The cytoskeleton can bear ~60% of the cortical stress (Akinlaja & Sachs, 1998; Cox et al., 2016) and that stress is not applied to the channels. The fact that the channels are in spatial domains means that the tension within the domain is lower than the tension outside the domain (Markin & Sachs, 2007; Ollila et al., 2009; Pontes et al., 2013). The difference in tension comes from the *line tension* at the perimeter of the domain (Baumgart, Hess, & Webb, 2003; Evans, Rawicz, & Smith, 2013; Sachs, 2015). The existence of domains is supported by images of punctate structures formed by recombinant PIEZO1 (PIEZO1-1591-GFP, GFP inserted at the 1591 position) (Fig. 1) in HEK cells. The kinetics of P1 are unaltered by the insertion of GFP at this location unlike when it is inserted in the N- or the C-terminus (Cox et al., 2016).

Both P1 and P2 inactivate but the latter inactivates faster (Coste et al., 2010) and it is difficult to activate in the cell-attached configuration (Coste et al., 2010, 2015). Lee et al. (2014) found that N2A cells expressing P1

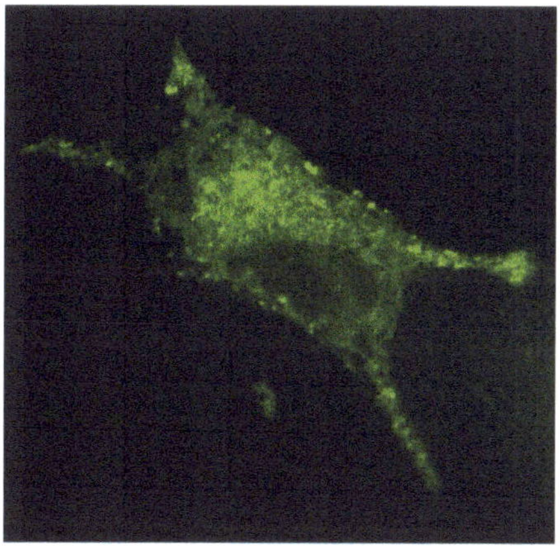

Figure 1 PIEZO1-1591—GFP forms punctate structures in HEK293 cells. HEK293 cells transfected with PIEZO1-1591—GFP chimeric channels showed punctae indicating that the channels aggregate into small domain-like structures. The images were acquired in a Visitech VT-iSIM superresolution microscope (Nikon 100×, 1.45 NA lens). This chimeric P1 was generated by inserting GFP at position 1591 of the protein. Its kinetics are identical to those of wild-type P1.

responded mildly whereas those expressing P2 failed to respond to ~500 nN stimulation applied using an atomic force microscope. However, coexpression of P1 and P2 produced a robust response suggesting synergistic effects (Lee et al., 2014). The untransfected N2A cells were surprisingly unresponsive in these experiments given that the robust responsiveness to mechanical stimulation was the reason behind why P1 was cloned from this cell line (Coste et al., 2010). This marked difference probably reflects phenotypic drift in the cell line or differences in stimulation conditions between the two studies and serves as a reminder of the critical influence of channel environment on their properties (Sachs, 2015). So what are the properties of the channels themselves? We obviously cannot perform detailed modeling because we do not know the stimulus in the neighborhood of the channels, but working with the assumption that the stimulus in the domain is at least proportional to the mean stimulus, we can derive a set of useful kinetic parameters.

5. MODELING AND ITS THERMODYNAMIC UNDERPINNINGS

Modeling allows us to compress the key features of the observed data, often representing gigabytes of material, into a few parameters. Models not only provide quantitative information but also make testable predictions from a set of explicit assumptions; features often missing from qualitative analyses. The modeling discussed here pertains to state models of P1. The dependence of transition rates between states on the stimulus (say voltage or pressure) provides insight into the conformational changes occurring in the protein. If the channel opening rate depends on the stimulus, one can then ask how much energy does the protein require to change its probability of occupancy. Furthermore, the temperature dependence of channel gating can be used to separate the free energy of the gating process into entropic and enthalpic contributions (Gupta & Auerbach, 2011a, 2011b; Liu, Hui, & Qin, 2003) and the voltage dependence establishes the changes in the dipole moment between states and constrains molecular models (Bezanilla, 2002; Sigg, Bezanilla, & Stefani, 2003).

The states in a model represent physical states of the molecule that lasted long enough to be observed and incorporate a priori assumptions of the reaction based on existing data. Typical states of a channel are closed, open, inactivated, drug bound, or free, etc. The top panel of Fig. 2 shows a typical energy landscape for an MSC with two states (closed and open). The arrows in the bottom panel of Fig. 2 between the states (0 and 1) represent the rate of crossing the energy barriers between the states and hence the rate of energy flow in the system. The rate constants have the units of inverse seconds and represent the number of transitions per second or equivalently, the time between transitions.

The bottom panel depicts a two-state model of P1 where "0" represents the closed state and "1" represents the open state. $k_{ij}^0\ exp(k_{ij}^1 * P)$ represents the forward rate, which is pressure-dependent; k_{ij}^0 is the preexponential factor, which is a scaling factor involving the entropy of activation and vibrational frequency; and $(k_{ij}^1 * P)$ is the exponential factor, which is $-\Delta G_{ij}/k_B T$ in the Boltzmann equation. The values of k_{ij}^0, k_{ij}^1 and k_{ji}^0 in this model are 0.03, −0.25, and 130.1 s^{-1}, respectively. Below the state-model is a −25 mmHg pressure step and the corresponding simulated current trace as predicted by the model.

The quantitation of state models for ion channel kinetics was spelled out by the work of Colquhoun and Hawkes (Colquhoun & Hawkes, 1981, 1982, 1990; Hawkes, Jalali, & Colquhoun, 1990) and the mathematics was solved in software by a number of groups (Colquhoun, Hatton, & Hawkes, 2003; Horn & Lange, 1983; Milescu, Akk, & Sachs, 2005; Nicolai & Sachs, 2014; Qin, Auerbach, & Sachs, 1996a, 1996b, 1997, 2000a, 2000b;

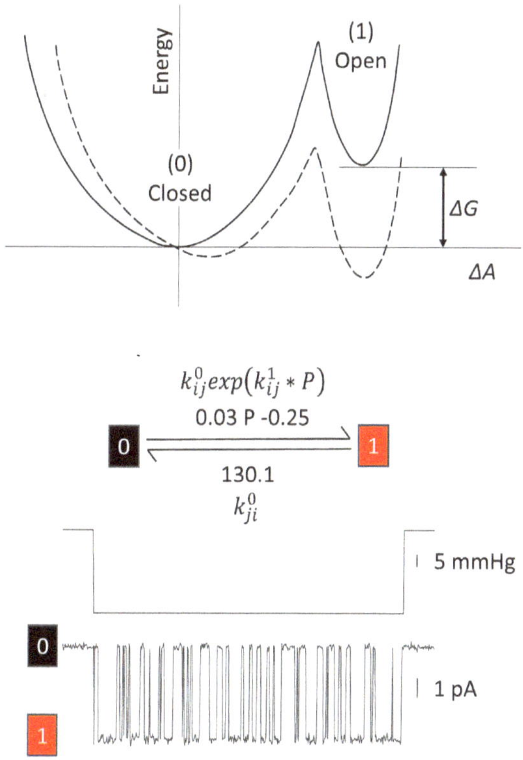

Figure 2 Energy landscape and current simulation of a typical two-state mechanosensitive channel gated by tension. The top panel represents the typical energy landscape of a two-state mechanosensitive (tension-gated) ion channel like P1. The troughs are energy minima, which correspond to the closed (0) and the open (1) states. The *solid line* depicts the channel in the absence of tension and the *dashed line* in the presence of tension. ΔG is the energy difference between the closed state and the open state at rest. Externally applied force is the source of the tension (T) that opens the channel, i.e., overcomes the ΔG barrier and stabilizes the open state. The width of the energy well (trough) for the closed state is wider than that for the open state because only the forward rate (closed to open) is pressure sensitive.

Venkataramanan, Kuc, & Sigworth, 1998; Venkataramanan & Sigworth, 2002; Venkataramanan, Walsh, Kuc, & Sigworth, 1998). In what follows, we will be using the analysis program called QuB that is available for free from www.qub.buffalo.edu (the latest version is called QuBX).

In QuB the rate constants are of the form, $k_{ij}^0 * \exp(k_{ij}^1 * P)$ where P is the stimulus (usually patch pressure or cell indentation depth for MSCs),

k_{ij}^1 is the sensitivity to the stimulus in units of (1/stimulus), and k_{ij}^0 is the rate with no applied stimulus in units of (1/seconds). The term $\Delta G = (k_{ij}^1 * P)$ is the free energy required to make the transition from state i to state j in units of the thermal energy, $k_B T = 4.1$ pN*nm at 25°C (k_B is Boltzmann's constant and T is the Kelvin temperature). The two types of stimuli (P) used for MSCs are typically the applied patch pressure (Bae, Gnanasambandam, et al., 2013; Bae, Gottlieb, et al., 2013) and the cell indentation depth for whole cell recordings (Coste, 2012; Coste et al., 2007, 2013, 2010). Neither one of these stimulus modalities describes in detail what the channels feel. Hydrostatic pressure is itself not the stimulus (Sukharev & Sachs, 2012), but the pressure gradient across a patch increases tension in the membrane, and if we ignore the cytoskeletal forces (Slavchov, Nomura, Martinac, Sokabe, & Sachs, 2014; Suchyna, Markin, & Sachs, 2009) and treat the patch as an infinitely thin membrane stuck to the glass pipette, we can apply Laplace's law, $T = (P{\star}r)/2$, and estimate the mean tension (T is tension and r is the radius of curvature of the patch), not the pipette (Markin & Sachs, 2007; Sachs, 2015). As pointed out earlier, the bilayer tension is not the same as the cortical tension because the cytoskeleton shares a significant fraction of the mean tension (Akinlaja & Sachs, 1998; Sachs, 2015), but we can use the mean tension as a coarse estimate of bilayer tension, and by further approximation, the tension in the channel domain. To reduce the level of tension approximation, Bae et al. used the calibrated tension sensitivity of reconstituted bacterial MSC, MscL (Chiang, Anishkin, & Sukharev, 2004; Martinac et al., 2010; Sukharev, Sigurdson, Kung, & Sachs, 1999), as an internal standard of bilayer tension. They transfected cells with P1 or the K^+-selective MSC, TREK-1 (Bae, Gnanasambandam, et al., 2013). This allowed them to express the stimulus sensitivity, k_{ij}^1, in terms of tension as well as applied pressure.

There are several issues about the mechanical stimuli that need to be addressed. In patches, adhesion of the membrane to the glass of the pipette introduces a resting tension that may be on the order of the lytic tension (Suchyna et al., 2009). For inactivating channels like P1 and P2, the resting channels might appear inactive because they are already inactivated by the background tension. In the case of indenting cells for whole cell recording, the membrane tension varies in space (Johnson, 1985) and time because of viscoelastic and plastic behavior of the cytoskeleton (Guo, Wang, Sachs, & Meng, 2014) and the bilayer (Evans & Yeung, 1994; Sachs et al., 1998; Slavchov et al., 2014).

6. MECHANOSENSITIVE CHANNEL GATING ENERGY; WHAT IS THE PREDICTED VALUE OF ΔG?

The energetics of state changes derive from the Boltzmann function (Dill & Bromberg, 2003; Hille, 2001), which says that the probability of being in state i compared to state j is given by:

$$P(i)/P(j) = \exp(-\Delta G/k_B T)$$

where ΔG is the free energy difference between the states. If ΔG is force dependent and i and j differ in conductance, we have an MSC. In the case of a two-state channel (Fig. 2),

$$P(j) = 1 - P(i)$$

so that if state j is open,

$$P_{open} = \frac{1}{\left[1 + \exp\left(-\Delta G_{ij}/k_B T\right)\right]}$$

If there is resting tension or intrinsic energy associated with the closed state, we have:

$$\Delta G = \Delta G^0 + T\Delta A$$

where T is the bilayer tension and ΔA is the change of in-plane area between the closed and the open state.

In QuB, P_{open} would be

$$P_{open} = \frac{1}{\left[1 + \left(\frac{k_{ji}^0}{k_{ij}^0}\right)\exp\left(\left(k_{ji}^1 - k_{ij}^1\right) * P\right)\right]}$$

We explain here a basic state model, which considers most of the observed kinetic features of P1. The model has the requisite three states: closed (state 0), open (state 1), and inactivated (state 2). In this model, states that have the same conductance are represented by the same color (black or red in the example). The opening rate is pressure sensitive in the case of P1. There are constraints on building the model: How are the states connected: linearly or in a loop? We have found that to fit the data we needed to close the loop (Bae, Gnanasambandam, et al., 2013). However, when the loop is closed, for the model to be in equilibrium (not deriving energy from the ion flow), all stimulus-dependent rates must sum to zero as one traverses the loop. This is known as detailed balance (Qin et al., 1996a) and basically says that if you traverse the loop, the energy of your ending state must be the same as that of your starting state. Luckily, QuB incorporates these constraints automatically (Milescu et al., 2005; Nicolai & Sachs, 2014) and can handle single channel or multichannel data. The details of the "MAC" algorithm that solves the kinetics of multichannel records and allows for arbitrary stimulus waveforms can be found in Milescu et al. (2005). It is advantageous to use a nonstationary stimulus because the rates obtained are a global optimum over the entire complex stimulus waveform rather than the classical Hodgkin–Huxley step stimuli where one waits for equilibrium (Hille, 2001). The model referenced here is a three-state loop model in detailed balance at all stimuli. The dynamic model and data set are hosted on the cloud and can be accessed using this link: https://www.qub.buffalo.edu/online/Piezo1 or can be downloaded for local use.

7. MUTATIONS IN PIEZO CHANNELS THAT AFFECT CHANNEL KINETICS

Several mutations in the human P1 gene lead to a hemolytic anemia known as hereditary xerocytosis (HX), also known as dehydrated hereditary stomatocytosis (Albuisson et al., 2013; Andolfo et al., 2013; Archer et al., 2014; Bae, Gnanasambandam, et al., 2013). These mutations introduce random latencies of activation and decrease the rates of inactivation (Table 1). Latency of activation has been seen in M2225R, R2456H, and R2456K mutants (Bae, Gnanasambandam, et al., 2013) and it is not known if it occurs in the R1358P, A2020T, T2127M, or E2496ELE HX mutants. The latency of opening is not related to the rise time of the pressure clamp (which is <5 ms) because wild-type P1 responds to the stimulus without any measurable latency. The latency of opening of mutant channels occurs randomly and varies in duration (~100–200 ms) (Bae, Gnanasambandam, et al., 2013). What is happening during latency? We believe that the latency reflects the time taken for the domain-containing channels to fracture. The domain structure shields the channels from stress, and when it fractures, external stress reaches the channels and they open. The HX mutations may affect the association of channels with each other in the domain and hence the fracture strength of the domain.

HX mutations affect inactivation of P1 to varying degrees. One of these mutations R2456H slows inactivation approximately threefold, whereas R1358P, A2020T, T2127M, M2225R, and E2496ELE slow inactivation approximately twofold (Albuisson et al., 2013; Bae, Gnanasambandam,

Table 1 Mutations that show affect kinetics, permeation or selectivity

Residue/mutation mP1	Residue/mutation mP2	Residue/mutation hP1	Effect
		R1358P	Slow inactivation
		A2020T	Slow inactivation
		T2127M	Slow inactivation
		E2496ELE	Slow inactivation
M2241		M2225R	Latency of opening & slow inactivation
R2482		R2456H	Latency of opening & slow inactivation
E2133	E2416		Selectivity, conductance and inactivation
E2495	E2769		Calcium permeability
E2496	E2770		Calcium permeability
	E2727del		Slow inactivation & recovery from inactivation
	I802 F		Recovery from inactivation
L2461C			Attenuates channel current
A2462C			Attenuates channel current
G2465C			Attenuates channel current
I2466C			Attenuates channel current
2393-2397-5A			Chloride permeability

et al., 2013). The R2456K mutation slowed inactivation to roughly threefold that of R2456H (~10-fold that of wild-type) in the whole cell configuration; R2456K is not a naturally occurring HX mutation, but it was generated to understand the role of residue charge at that position. The kinetics of wild-type P1 is consistent across several different recording configurations (whole cell, cell-attached, and outside-out) and cell types used. Furthermore, P1 spontaneously activates in asymmetric bilayers but note that it lacks inactivation. The questions that arise are as follows: If inactivation is a fundamental property of the channel why do we not observe it in bilayers? Is inactivation conferred upon the channel by something other than the channel? The kinetic analysis of wild-type P1 and the mutants shows that the rate of inactivation (open to inactivated) is not stimulus dependent (https://www.qub.buffalo.edu/online/Piezo1). Therefore, this rate probably represents the frequency of interactions among channels and mutations probably change the energy of interaction of the channels. Reinforcing the critical role of environmental mechanics on P1 gating, cytochalasin D abrogates whole cell current in cells expressing P1 (Gottlieb et al., 2012) but it facilitates channel activity in cell-attached patches (Suchyna & Sachs, 2007).

Two mutations in the *PIEZO2* gene, E2727del and I802F, cause a subtype of distal arthrogryposis type 5 (DA5), which is an autosomal dominant multisystem disorder characterized by contractures, limited eye movements, restrictive lung disease, and the absence of cruciate knee ligaments (Coste et al., 2013). E2727del mutant inactivates roughly fourfold slower than the wild-type whereas I802F mutant is not significantly different from the wild-type with respect to inactivation. However, the rates of recovery from inactivation of E2727del and I802F mutants are half of that of the wild-type permitting these channels to be reactivated quicker because the time for which they will stay inactivated is shorter.

8. REMOVING PIEZO1 INACTIVATION

The P1 mutations, M2225R and R2456K, introduce a latency of activation and slow inactivation (Bae, Gnanasambandam, et al., 2013). These mutations were introduced together to create the double mutant (DhPIEZO1) (Bae, Gottlieb, et al., 2013). In the double mutant, the latency to activation remains and the channel inactivates much more slowly than either single site mutant. The sharply inactivating current observed in the wild-type is replaced by a mildly inactivating current (sometimes a plateau current lacking any inactivation is observed). Neither the wild-type nor any of the mutants (including the double-mutant) shows a change in the slope sensitivity for activation suggesting that the dimensional change of the channel between closed and open state is not affected. However, the pressure of half-activation, $P_{1/2}$, of each of the mutants is clearly lower than that of the wild-type, with that of the double mutant being the lowest observed (Bae, Gnanasambandam, et al., 2013; Bae, Gottlieb, et al., 2013). This means that the mutant channels are prestressed and hence favor the open state. It is important to understand that the "sensitivity" for activation has two independent terms: (1) the resting P_{open} and (2) the slope sensitivity, which is the slope of the dose-response curve at its midpoint ($P_{1/2}$, is the midpoint of the curve). If "sensitivity" is defined as the pressure required to reach half-maximum, then it can be changed by changing the stimulus-free resting probability of opening or by changing the slope of the dose-response curve (Sukharev et al., 1999). There is no such thing as a "threshold" since the probability of being open is a continuous Boltzmann function of the stimulus.

The two mutations M2225R and R2456K may be acting independently or be energetically coupled in the double-mutant. We explored the relationship between these mutations using double-mutant cycle analysis (Horovitz, 1996). The mutations are coupled if the change in free energy ($\Delta\Delta G$) associated with the double mutant differs from the sum of changes in free energy associated with each of the single mutants (i.e.) $\Delta\Delta G_{DhPIEZO1} \neq \Delta\Delta G_{M2225R} + \Delta\Delta G_{R2456K}$. The first step is to obtain the difference in free energy between the wild-type and each of the two single-site mutants ($\Delta\Delta G_{M2225R}$ and $\Delta\Delta G_{R2456K}$), and that between the wild-type and the double mutant ($\Delta\Delta G_{DhPiezo1}$). Then we calculate the difference in ΔG between the wild-type and each of the mutants by fitting the response of the channel to a nonstationary stimulus using the "MAC" algorithm in "QuB" and obtaining the rates for all of the transitions. ΔG or "Gibbs free energy" is then calculated from these rates by using the familiar equation:

$$\Delta G = RT \ln\left(K_{eq}\right)$$

where $K_{eq} = \left(\frac{k_{ji}}{k_{ij}}\right)$.

Although the earlier study (Bae, Gottlieb, et al., 2013) utilized a double mutant of human P1, we examined where the analogous residues are in mouse P1 for which the cryo-EM structure is available. The 2241 residue (2225 in humans) would be in the CED and the 2482 residue (2456 in humans) would be at the cytoplasmic end of the transmembrane IH helix. Both these mutations are very close to the pore of the channel if not lining the pore itself (Albuisson et al., 2013; Bae, Gnanasambandam, et al., 2013; Ge et al., 2015). Residue 2482 would probably be within the membrane in the inner leaflet of the bilayer at the interface of the membrane and the cytoplasm (Ge et al., 2015; Zhao et al., 2014). This region would be flexible since it is at one end of the helix (Fig. 3). Residue 2241 is located away from the membrane in the extracellular vestibule. It would be interesting to examine how bilayers (asymmetric and symmetric) affect the activity of these two single-site mutants and the double-mutant.

Figure 3 Mutations that affect kinetics and/or permeation and selectivity. The cryo-EM structure of mouse P1 viewed from the side (top panel) and from the extracellular side along the axis of the pore (bottom panel). The pore domains are shown in orange and key residues, which either show an effect on kinetics or permeation/selectivity have been depicted as *colored spheres*. Residues 2393–2397 in the CED are shown in blue. They are far from the pore but influence how the channel chooses cations over anions, much like a charged vestibule. Mouse residues homologous to the human P1 residues M2225 and R2456 (M2241 and R2482 in mouse) are shown in green and they are very close to the pore, if not lining the pore. R2456, in particular, is at cytoplasmic end of the rigid helix. Note also that E2133 does not seem to be close to the pore but, like R2456, it is closer to the cytoplasm. Shown in red are residues E2495, E2496, and D2501 of the C-terminal domain, which affect Ca^{2+} permeability.

9. KINETIC PROPERTIES OF Piezo1 CHANNELS ARE ALTERED BY Yoda1

Until the identification of its chemical activator Yoda1 (Syeda et al., 2015), P1 could only be activated by mechanical stimuli. This compound changes the kinetics of P1 (human and mouse) favoring opening, but has no effect on P2. Yoda1 not only increases the P1 opening rate it also appears to slow the inactivation rate of P1. The mean open time of Yoda1-treated wild-type channels is roughly fourfold higher than that of controls (Syeda et al., 2015). These Yoda1-induced changes in kinetics to both cell-attached and whole cell currents are reminiscent of the effects of the R2456K mutation on inactivation (Bae, Gnanasambandam, et al., 2013). An in-depth quantitative kinetic analysis of Yoda1 remains to be done and the mechanism of action for this hydrophobic compound may involve perturbation of local stress in the bilayer. P1 spontaneously opens in asymmetric bilayers (DPhPC:DOPA), but not in symmetric bilayers (DPhPC) indicating that the channel is sensitive to local membrane bending (Coste et al., 2012; Pontes et al., 2013). Yoda1 activates P1 in symmetric bilayers and this is reminiscent of the typical effect of amphipaths on MSCs as has often been reported (Chemin et al., 2005, 2007; Goforth et al., 2003; Lundbaek et al., 2004; Lundbaek, Collingwood, Ingolfsson, Kapoor, & Andersen, 2010; Lundbaek, Koeppe, & Andersen, 2010; Martinac, Adler, & Kung, 1990; Patel, Lazdunski, & Honore, 2001; Suchyna et al., 2004). Another possibility is that Yoda1 and P1 interact directly (based on its specificity for P1) but further studies will be required to determine exactly how Yoda1 functions.

10. PERMEATION AND SELECTIVITY OF PIEZO CHANNELS

Mouse P1 and human P1 have ion permeation and selectivity properties consistent with those of other nonselective cation channels and they have been studied in patches and the whole cell configuration. Their properties are summarized here and reference to a specific organism is drawn only if a given property is significantly different between the two. P1 is permeable to monovalent and some divalent cations, notable Ca^{2+} and Mg^{2+} (with the exception of Mn^{2+}). The permeability sequence is $P_K > P_{Cs} \approx P_{Na} > P_{Li}$ (1.0:0.88: 0.82:0.71), closely matches the conductance sequence, and of all the monovalent ions Li^+ exhibits a significantly lower conductance. The concentration versus conductance relationship for K^+ through hP1 is similar to that of several other cation channels, a parabolic increase to saturation (45–50 pS for hP1). The apparent affinity of K^+ for the channel is ~32 mM. The organic monovalent cations, tetramethyl ammonium (TMA) and tetraethyl ammonium (TEA), can traverse the pore and because they are slow-moving, they reduce K^+ conductance (Gnanasambandam et al., 2015). The alkyl side chains of these ions do not prevent their transit through the pore. As will be discussed later, this is probably because the pore residues are mostly uncharged amino acids interspersed with a few regions of charged amino acids (Ge et al., 2015; Zhao et al., 2016). Significant inhibition of K^+ current by TMA and TEA requires high concentrations of organic ions (~150 mM) and between the two, TEA is the better blocker. Judging from the size of TEA the narrowest part of the P1 pore is ~8 Å in diameter.

The dimensional change undergone by P1 upon opening has been estimated. To be mechanosensitive, the transition from closed to open must involve a change in shape so that work can be done on the channel. The work can be done in several ways (Wiggins & Phillips, 2004): a change of in-plane area, a change in bilayer thickness providing a hydrophobic mismatch (Lee, 2006; Martinac, 2009; Nomura et al., 2012), or a change in the local radius of curvature bending the channel, etc. To estimate the dimensional change, all three channels, MscL, P1, and TREK-1 are assumed to conform to the in-plane area model. We then compared P1 and TREK-1 against the well-studied bacterial mechanosensitive channel MscL, for which the X-ray crystal structure is available. The advantage here is that, for MscL, the domains it associates with, the states it occupies, and its kinetic behavior are well known. By cotransfecting MscL with either TREK-1 or P1, it is possible to compare their properties given that they are in the same cellular environment. The responses (of MscL and P1/TREK-1) to stimulation could be fit with the double Boltzmann equation (the Boltzmann fit at higher pressure corresponds to the MscL channel and the fit at lower pressure region corresponds to TREK-1 or P1, in TREK-MscL and P1-MscL cotransfected cells, respectively). The Boltzmann curve yields information about the dimensional change the channel undergoes upon activation. The dimensional change upon activation of human P1 (wild-type, M2225R, R2456H, R2456K) is similar to that of MscL, whereas for TREK-1, the dimensional change is roughly half that of MscL (Fig. 4). From these comparisons, it is possible to estimate the area change of the human P1 channel upon activation to be roughly 6–20 nm^2 (Bae, Gnanasambandam, et al., 2013).

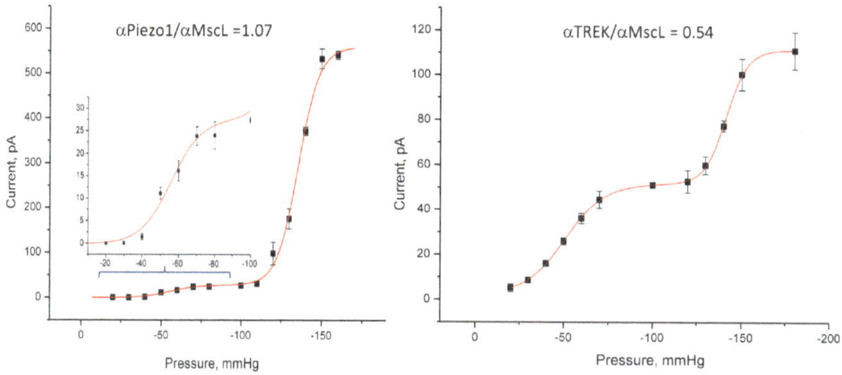

Figure 4 Dose–response relationships of gating versus pressure for MscL-PIEZO1 and MscL-TREK-1. Both dose–response plots show the response of the channels to pressures between −20 and −180 mmHg. The *blue bracket* in the panel on the left highlights the response of P1 and it has been expanded for better viewing in the inset. MscL responds between −60 and −180 mmHg irrespective of whether the cell is cotransfected with P1 or TREK-1, suggesting that MscL has its own domain. The slope sensitivity of the curve for MscL is identical in both cotransfection scenarios. The slope sensitivity (α) of P1 is comparable to that of MscL. TREK-1 has a slope sensitivity, which is half that of MscL. *Reproduced with permission from Bae, C., Gnanasambandam, R., Nicolai, C., Sachs, F., & Gottlieb, P. A. (2013). Xerocytosis is caused by mutations that alter the kinetics of the mechanosensitive channel PIEZO1. Proceedings of the National Academy of Sciences of the United States of America, 110(12), E1162–E1168. http://dx.doi.org/10.1073/pnas.1219777110.*

It is difficult to elicit pure divalent currents (especially Ca^{2+} and Mg^{2+}) compared to monovalent currents in the cell-attached patch configuration, and surprisingly, treatment with cytochalasin D increases the number of responsive patches. It is easier to obtain pure divalent currents (especially Ca^{2+}) in the whole cell mode than in other patch modes, although this mode is not devoid of its own difficulties (Gnanasambandam et al., 2015; Zhao et al., 2016). Even though Piezo channels pass Ca^{2+}, they are not selective for Ca^{2+} like voltage-gated Ca^{2+} or Orai channels (Zhao et al., 2016). Those dedicated calcium channels have a ring of acidic pore residues in the extracellular region conferring Ca^{2+} selectivity. Pure Ba^{2+} currents are easier to obtain than Ca^{2+} or Mg^{2+} currents (Gnanasambandam et al., 2015). The unitary conductance of pure 90 mM Ba^{2+}, Ca^{2+}, and Mg^{2+} are ~25 pS (−80 mV), ~15 pS (−80 mV), and ~10 pS (−50 mV), respectively (Gnanasambandam et al., 2015). Slowly permeating divalents also reduce the unitary current amplitude of monovalents in mixed solutions. The magnitude of monovalent current attenuation by divalents is proportional to the fraction of time that the divalent species interacts with residues in the pore (Coste et al., 2010; Gnanasambandam et al., 2015).Li et al. (2015) used the pore selectivity profile to suggest that P1 is the predominant MSC current in the MCF-7 breast cancer cell line.

11. MUTATIONS THAT ALTER PORE MODULE RESIDUES AND THEIR EFFECTS ON PERMEATION AND CONDUCTANCE

The P1 pore resembles the AChR pore in that most of the pore-lining residues are uncharged, and it has clearly defined regions of charged amino acids. The cryo-EM structure of the pore domains (CED, IH, and CTD) shows two regions of negatively charged amino acids: (1) positions 2393–2397 in the CED is a patch of negatively charged amino acids DEEED and (2) in the CTD there are three residues E2495, E2496, and D2501 (Ge et al., 2015; Zhao et al., 2016). Mutating the aforementioned extracellular stretch (2393–2397) of amino acids from DEEED to AAAAA alters charge selectivity by allowing more anions through (Zhao et al., 2016). Hence, it appears to help primarily in choosing cations over anions. Mutants E2495K, E2496K, and D2501K of the CTD influence the conductance of the channel but a more substantial reduction in conductance results when the individual mutations are combined (E2495E2496D2501-KKK triple mutant). Mutating the highly conserved residues E2495 and E2496 had a significant effect on the conductance irrespective of whether the substituent side chain varied in charge (E to A and E to Q) or length (E to D). However, substituting the charge (E to K) at these positions had the strongest effect indicating these two residues may interact electrostatically with pore transiting ions. E2495, and to a greater extent E2496, specifically affect the permeability to Ca^{2+} but not Cl^{-}. Ruthenium red (RR) blocks P1 efficiently only from the extracellular side (and may permeate) but it is ineffective from the intracellular side indicating that there may be narrow regions on the intracellular side, which prevent RR from readily accessing its binding site residues E2495, E2496 in the pore. Mutating analogous residues in mouse P2 (E2769 and E2770) to alanine also removes the sensitivity to RR (Coste et al., 2015; Zhao et al., 2016).

The region around E2133 shows clear effects on channel conductance, selectivity, and kinetics (Coste et al., 2015). Mutating E2133 to alanine strongly attenuates the unitary current amplitude. When residues flanking E2133 are mutated to alanine, they result either in nonfunctional channels or show no major changes in conductance. This residue was further tested by generating charge/side chain length variants E2133Q, E2133K, and E2133D (Coste et al., 2015). While E2133Q (uncharged) and E2133K (reversal of charge) both reduced unitary current amplitude, the shorter side chain, charge-conservative E2133D increased conductance. The inactivation rates of E2133K, E2133Q, and E2133D appear to be slower than that of the wild-type suggesting that charge and side chain length parameters do not contribute significantly to the kinetics. E2133 is a crucial negative charge, which plays a role in distinguishing cations from anions and also allows divalent ions like Ca^{2+} to permeate better. RR inhibition is affected by E2133K and not in E2133Q, which rules out chiral binding of the drug to this residue. However, the charge at this residue probably plays an important role by electrostatically influencing the way RR interacts with its near-field target.

12. CONCLUSIONS

We have summarized here some of the key findings on two fundamental properties of P1: kinetics and permeation. Several disease mutations have been identified that alter the kinetics P1 (Archer et al., 2014; Bae, Gnanasambandam, et al., 2013; Bae, Gottlieb, et al., 2013). We built a quantitative model for P1 (wild-type and the three mutants) and hosted it online to enable users to interact with the model (Bae, Gottlieb, et al., 2013). Quantitative models are explicit about their assumptions and make testable predictions. Such models are useful not only in understanding possible gating mechanisms but they can also be extended to make meaningful predictions about the response of these channels to different stimuli, ligand interactions, and effect of channel perturbations on physiology (Hille, 2001). We know the rate constant for activation is pressure-dependent, but voltage dependence has to be quantified. The most consequential questions that need to be addressed regarding how stimuli are transferred to the channels. What constitutes a domain? What are their properties and how they are involved in stimulus transfer? The functional role of the N-terminal mechanosensing module is also unclear since other MSCs lack such a large mechanosensing module. So what special functions does this module confer? How can it tune the kinetics of P1? Does P2 have such a module and does it function in the same manner?

There is no doubt that kinetics and permeation of Piezo channels are involved in diseases such as HX (Albuisson et al., 2013; Archer et al.,

2014; Bae, Gnanasambandam, et al., 2013), congenital lymphatic dysplasia (Lukacs et al., 2015), DA5, Gordon syndrome, and Marden–Walker syndrome (Coste et al., 2013; McMillin et al., 2014). These channels are also involved in somatosensation (Ikeda et al., 2014; Ma, 2014; Maksimovic et al., 2014; Nakatani et al., 2015; Schneider et al., 2014; Schrenk-Siemens et al., 2015; Woo, Lumpkin, & Patapoutian, 2015), vascular development (Li et al., 2014; Ranade et al., 2014), erythrocyte volume homeostasis (Cahalan et al., 2015; Faucherre, Kissa, Nargeot, Mangoni, & Jopling, 2014), sensing shear stress, and sensing substrate stiffness in the context of cell differentiation (Pathak et al., 2014) and cell migration (McHugh et al., 2010). It is difficult to imagine such a wide-ranging influence on cellular function and physiology without exquisite control of kinetics. Therefore, understanding the kinetics of these channels is crucial both to future investigations of mechanobiology and in designing therapeutics for diseases caused by Piezo channels.

ACKNOWLEDGMENT

This work was supported by the National Institutes of Health grant (R01HL054887) to Dr Frederick Sachs.

REFERENCES

Akinlaja, J., & Sachs, F. (1998). The breakdown of cell membranes by electrical and mechanical stress. *Biophysical Journal, 75*(1), 247–254. http://dx.doi.org/10.1016/S0006-3495(98)77511-3.

Albuisson, J., Murthy, S. E., Bandell, M., Coste, B., Louis-Dit-Picard, H., Mathur, J., … Patapoutian, A. (2013). Dehydrated hereditary stomatocytosis linked to gain-of-function mutations in mechanically activated PIEZO1 ion channels. *Nature Communications, 4*, 1884. http://dx.doi.org/10.1038/ncomms2899.

Andolfo, I., Alper, S. L., De Franceschi, L., Auriemma, C., Russo, R., De Falco, L., … Iolascon, A. (2013). Multiple clinical forms of dehydrated hereditary stomatocytosis arise from mutations in PIEZO1. *Blood, 121*(19), 3925–3935. http://dx.doi.org/10.1182/blood-2013-02-482489. S3921–3912.

Apajalahti, T., Niemela, P., Govindan, P. N., Miettinen, M. S., Salonen, E., Marrink, S. J., & Vattulainen, I. (2010). Concerted diffusion of lipids in raft-like membranes. *Faraday Discussions, 144*, 411–430. Discussion 445–481.

Archer, N. M., Shmukler, B. E., Andolfo, I., Vandorpe, D. H., Gnanasambandam, R., Higgins, J. M., … Nathan, D. G. (2014). Hereditary xerocytosis revisited. *American Journal of Hematology, 89*(12), 1142–1146. http://dx.doi.org/10.1002/ajh.23799.

Bae, C., Gnanasambandam, R., Nicolai, C., Sachs, F., & Gottlieb, P. A. (2013). Xerocytosis is caused by mutations that alter the kinetics of the mechanosensitive channel PIEZO1. *Proceedings of the National Academy of Sciences of the United States of America, 110*(12), E1162–E1168. http://dx.doi.org/10.1073/pnas.1219777110.

Bae, C., Gottlieb, P. A., & Sachs, F. (2013). Human PIEZO1: Removing inactivation. *Biophysical Journal, 105*(4), 880–886. http://dx.doi.org/10.1016/j.bpj.2013.07.019.

Bae, C., Suchyna, T. M., Ziegler, L., Sachs, F., & Gottlieb, P. A. (2016). Human PIEZO1 ion channel functions as a split protein. *PLoS One, 11*(3), e0151289. http://dx.doi.org/10.1371/journal.pone.0151289.

Baumgart, T., Hess, S. T., & Webb, W. W. (2003). Imaging coexisting fluid domains in biomembrane models copuling curvature and line tension. *Nature, 425*, 821–824.

Berrier, C., Pozza, A., de Lacroix de Lavalette, A., Chardonnet, S., Mesneau, A., Jaxel, C., ... Ghazi, A. (2013). The purified mechanosensitive channel TREK-1 is directly sensitive to membrane tension. *The Journal of Biological Chemistry, 288*(38), 27307–27314. http://dx.doi.org/10.1074/jbc.M113.478321.

Beyder, A., Rae, J. L., Bernard, C., Strege, P. R., Sachs, F., & Farrugia, G. (2010). Mechanosensitivity of $Na_V1.5$, a voltage-sensitive sodium channel. *The Journal of Physiology, 588*(Pt 24), 4969–4985. http://dx.doi.org/10.1113/jphysiol.2010.199034.

Beyder, A., Strege, P. R., Reyes, S., Bernard, C. E., Terzic, A., Makielski, J., ... Farrugia, G. (2012). Ranolazine decreases mechanosensitivity of the voltage-gated sodium ion channel Na(v)1.5: A novel mechanism of drug action. *Circulation, 125*(22), 2698–2706. http://dx.doi.org/10.1161/CIRCULATIONAHA.112.094714.

Bezanilla, F. (2002). Voltage sensor movements. *The Journal of General Physiology, 120*(4), 465–473.

Blount, P., Sukharev, S. I., Moe, P. C., Martinac, B., & Kung, C. (1999). Mechanosensitive channels of bacteria. *Methods in Enzymology, 294*, 458–482.

Brierley, S. M., Castro, J., Harrington, A. M., Hughes, P. A., Page, A. J., Rychkov, G. Y., & Blackshaw, L. A. (2011). TRPA1 contributes to specific mechanically activated currents and sensory neuron mechanical hypersensitivity. *The Journal of Physiology, 589*(Pt 14), 3575–3593. http://dx.doi.org/10.1113/jphysiol.2011.206789.

Brohawn, S. G. (2015). How ion channels sense mechanical force: Insights from mechanosensitive K2P channels TRAAK, TREK1, and TREK2. *Annals of the New York Academy of Sciences, 1352*, 20–32. http://dx.doi.org/10.1111/nyas.12874.

Cahalan, S. M., Lukacs, V., Ranade, S. S., Chien, S., Bandell, M., & Patapoutian, A. (2015). Piezo1 links mechanical forces to red blood cell volume. *eLife, 4*. http://dx.doi.org/10.7554/eLife.07370.

Charras, G. T. (2008). A short history of blebbing. *Journal of Microscopy, 231*(3), 466–478. http://dx.doi.org/10.1111/j.1365-2818.2008.02059.x.

Chemin, J., Patel, A. J., Delmas, P., Sachs, F., Lazdunski, M., & Honore, E. (2007). Regulation of the mechano-gated K-2P channel TREK-1 by membrane phospholipids. In O. P. Hamill (Ed.), *Mechanosensitive ion channels, Part B* (Vol. 59, pp. 155–170). San Diego: Academic Press.

Chemin, J., Patel, A., Duprat, F., Zanzouri, M., Lazdunski, M., & Honore, E. (2005). Lysophosphatidic acid-operated K^+ channels. *The Journal of Biological Chemistry, 280*(6), 4415–4421. http://dx.doi.org/10.1074/jbc.M408246200.

Chiang, C. S., Anishkin, A., & Sukharev, S. (2004). Gating of the large mechanosensitive channel in situ: Estimation of the spatial scale of the transition from channel population responses. *Biophysical Journal, 86*(5), 2846–2861. http://dx.doi.org/10.1016/S0006-3495(04)74337-4.

Colquhoun, D., Hatton, C. J., & Hawkes, A. G. (2003). The quality of maximum likelihood estimates of ion channel rate constants. *The Journal of Physiology, 547*(Pt 3), 699–728. http://dx.doi.org/10.1113/jphysiol.2002.034165.

Colquhoun, D., & Hawkes, A. G. (1981). On the stochastic properties of single ion channels. *Proceedings of the Royal Society of London. Series B: Biological Sciences, B211*, 205–235.

Colquhoun, D., & Hawkes, A. G. (1982). On the stochastic properties of bursts of single ion channel openings and of clusters of bursts. *Philosophical Transactions of the Royal Society of London. Series B: Biological Sciences, 300*, 1–59.

Colquhoun, D., & Hawkes, A. G. (1990). Stochastic properties of ion channel openings and bursts in a membrane patch that contains two channels: Evidence concerning the number of channels present when a record containing only single openings is observed. *Proceedings of the Royal Society of London. Series B: Biological Sciences, 240*(1299), 453–477.

Coste, B. (2012). Piezo proteins form a new class of mechanically activated ion channels. *Médecine Sciences, 28*(12), 1056–1057. http://dx.doi.org/10.1051/medsci/20122812012.

Coste, B., Crest, M., & Delmas, P. (2007). Pharmacological dissection and distribution of NaN/Nav1.9, T-type Ca^{2+} currents, and mechanically activated cation currents in different populations of DRG neurons. *The Journal of General Physiology, 129*(1), 57–77. http://dx.doi.org/10.1085/jgp.200609665.

Coste, B., Houge, G., Murray, M. F., Stitziel, N., Bandell, M., Giovanni, M. A., ... Patapoutian, A. (2013). Gain-of-function mutations in the mechanically activated ion channel PIEZO2 cause a subtype of Distal Arthrogryposis. *Proceedings of the National Academy of Sciences of the United States of America, 110*(12), 4667–4672. http://dx.doi.org/10.1073/pnas.1221400110.

Coste, B., Mathur, J., Schmidt, M., Earley, T. J., Ranade, S., Petrus, M. J., ... Patapoutian, A. (2010). Piezo1 and Piezo2 are essential components of distinct mechanically activated cation channels. *Science, 330*(6000), 55–60. http://dx.doi.org/10.1126/science.1193270.

Coste, B., Murthy, S. E., Mathur, J., Schmidt, M., Mechioukhi, Y., Delmas, P., & Patapoutian, A. (2015). Piezo1 ion channel pore properties are dictated by C-terminal region. *Nature Communications, 6*, 7223. http://dx.doi.org/10.1038/ncomms8223.

Coste, B., Xiao, B., Santos, J. S., Syeda, R., Grandl, J., Spencer, K. S., ... Patapoutian, A. (2012). Piezo proteins are pore-forming subunits of mechanically activated channels. *Nature, 483*(7388), 176–181. http://dx.doi.org/10.1038/nature10812.

Cox, C. D., Bae, C., Ziegler, L., Hartley, S., Nikolova-Krstevski, V., Rohde, P. R., ... Martinac, B. (2016). Removal of the mechanoprotective influence of the cytoskeleton reveals PIEZO1 is gated by bilayer tension. *Nature Communications, 7*, 10366. http://dx.doi.org/10.1038/ncomms10366.

Delmas, P. (2005). Polycystins: Polymodal receptor/ion-channel cellular sensors. *Pflugers Archiv, 451*(1), 264–276. http://dx.doi.org/10.1007/s00424-005-1431-5.

Dill, K. A., & Bromberg, S. (2003). *Molecular driving forces: Statistical thermodynamics in chemistry and biology*. NY: Garland Science.

Doerner, J. F., Febvay, S., & Clapham, D. E. (2012). Controlled delivery of bioactive molecules into live cells using the bacterial mechanosensitive channel MscL. *Nature Communications, 3*, 990. http://dx.doi.org/10.1038/ncomms1999.

Evans, E., Rawicz, W., & Smith, B. (2013). Concluding remarks back to the future: Mechanics and thermodynamics of lipid biomembranes. *Faraday Discussions, 161*, 591–611.

Evans, E., & Yeung, A. (1994). Hidden dynamics in rapid changes of bilayer shape. *Chemistry and Physics of Lipids, 73*, 39–56.

Fan, W., & Evans, R. M. (2015). Turning up the heat on membrane fluidity. *Cell, 161*(5), 962–963. http://dx.doi.org/10.1016/j.cell.2015.04.046.

Fattal, D. R., & Ben-Shaul, A. (1993). A molecular model for lipid-protein interaction in membranes: The role of hydrophobic mismatch. *Biophysical Journal, 65*(5), 1795–1809. http://dx.doi.org/10.1016/S0006-3495(93)81249-9.

Faucherre, A., Kissa, K., Nargeot, J., Mangoni, M. E., & Jopling, C. (2014). Piezo1 plays a role in erythrocyte volume homeostasis. *Haematologica, 99*(1), 70–75. http://dx.doi.org/10.3324/haematol.2013.086090.

Flomenbom, O. (2011). Clustering in anomalous files of independent particles. *EPL, 94*(5). http://dx.doi.org/10.1209/0295-5075/94/58001.

Fuentes, D. E., & Butler, P. J. (2012). Coordinated mechanosensitivity of membrane rafts and focal adhesions. *Cellular and Molecular Bioengineering, 5*(2), 143–154. http://dx.doi.org/10.1007/s12195-012-0225-z.

Gasparski, A. N., & Beningo, K. A. (2015). Mechanoreception at the cell membrane: More than the integrins. *Archives of Biochemistry and Biophysics, 586*, 20–26. http://dx.doi.org/10.1016/j.abb.2015.07.017.

Ge, J., Li, W., Zhao, Q., Li, N., Chen, M., Zhi, P., … Yang, M. (2015). Architecture of the mammalian mechanosensitive Piezo1 channel. *Nature, 527*, 64–69.

Giamarchi, A., & Delmas, P. (2007). Activation mechanisms and functional roles of TRPP2 cation channels. In W. B. Liedtke, & S. Heller (Eds.), *TRP ion channel function in sensory Transduction and cellular signaling Cascades*. Boca Raton, FL.

Giamarchi, A., Padilla, F., Crest, M., Honore, E., & Delmas, P. (2006). TRPP2: Ca^{2+}-permeable cation channel and more. *Cellular and Molecular Biology, 52*(8), 105–114.

Gillespie, P. G., & Muller, U. (2009). Mechanotransduction by hair cells: Models, molecules, and mechanisms. *Cell, 139*(1), 33–44. http://dx.doi.org/10.1016/j.cell.2009.09.010.

Gnanasambandam, R., Bae, C., Gottlieb, P. A., & Sachs, F. (2015). Ionic selectivity and permeation properties of human PIEZO1 channels. *PLoS One, 10*(5), e0125503. http://dx.doi.org/10.1371/journal.pone.0125503.

Goforth, R. L., Chi, A. K., Greathouse, D. V., Providence, L. L., Koeppe, R. E., 2nd, & Andersen, O. S. (2003). Hydrophobic coupling of lipid bilayer energetics to channel function. *The Journal of General Physiology, 121*(5), 477–493. http://dx.doi.org/10.1085/jgp.200308797.

Gottlieb, P. A., Bae, C., & Sachs, F. (2012). Gating the mechanical channel Piezo1: A comparison between whole-cell and patch recording. *Channels, 6*(4), 282–289. http://dx.doi.org/10.4161/chan.21064.

Gottlieb, P. A., & Sachs, F. (2012). Piezo1: Properties of a cation selective mechanical channel. *Channels, 6*(4), 214–219. http://dx.doi.org/10.4161/chan.21050.

Gu, Y., & Gu, C. (2014). Physiological and pathological functions of mechanosensitive ion channels. *Molecular Neurobiology, 50*(2), 339–347. http://dx.doi.org/10.1007/s12035-014-8654-4.

Guharay, F., & Sachs, F. (1984). Stretch-activated single ion channel currents in tissue-cultured embryonic chick skeletal muscle. *The Journal of Physiology, 352*, 685–701.

Guo, J., Wang, Y., Sachs, F., & Meng, F. (2014). Actin stress in cell reprogramming. *Proceedings of the National Academy of Sciences of the United States of America, 111*(49), E5252–E5261. http://dx.doi.org/10.1073/pnas.1411683111.

Gupta, S., & Auerbach, A. (2011a). Mapping heat exchange in an allosteric protein. *Biophysical Journal, 100*(4), 904–911. http://dx.doi.org/10.1016/j.bpj.2010.12.3739.

Gupta, S., & Auerbach, A. (2011b). Temperature dependence of acetylcholine receptor channels activated by different agonists. *Biophysical Journal, 100*(4), 895–903. http://dx.doi.org/10.1016/j.bpj.2010.12.3727.

Hao, J., Bonnet, C., Amsalem, M., Ruel, J., & Delmas, P. (2015). Transduction and encoding sensory information by skin mechanoreceptors. *Pflugers Archiv, 467*(1), 109–119. http://dx.doi.org/10.1007/s00424-014-1651-7.

Hao, J., & Delmas, P. (2011). Recording of mechanosensitive currents using piezoelectrically driven mechanostimulator. *Nature Protocols Other Titles: Protocols, 6*(7), 979–990. http://dx.doi.org/10.1038/nprot.2011.343.

Hao, J., Ruel, J., Coste, B., Roudaut, Y., Crest, M., & Delmas, P. (2013). Piezo-electrically driven mechanical stimulation of sensory neurons. *Methods in Molecular Biology, 998*, 159–170. http://dx.doi.org/10.1007/978-1-62703-351-0_12.

Hase, C. C., Le Dain, A. C., & Martinac, B. (1995). Purification and functional reconstitution of the recombinant large mechanosensitive ion channel (MscL) of *Escherichia coli*. *The Journal of Biological Chemistry, 270*(31), 18329–18334.

Haswell, E. S., & Verslues, P. E. (2015). The ongoing search for the molecular basis of plant osmosensing. *The Journal of General Physiology, 145*(5), 389–394. http://dx.doi.org/10.1085/jgp.201411295.

Hawkes, A. G., Jalali, A., & Colquhoun, D. (1990). The distributions of the apparent open times and shut times in a single channel record when brief events cannot be detected. *Philosophical Transactions of the Royal Society of London. Series B: Biological Sciences, 332*, 511–538.

Head, B. P., Patel, H. H., Roth, D. M., Murray, F., Swaney, J. S., Niesman, I. R., … Insel, P. A. (2006). Microtubules and actin microfilaments regulate lipid raft/caveolae localization of adenylyl cyclase signaling components. *The Journal of Biological Chemistry, 281*(36), 26391–26399. http://dx.doi.org/10.1074/jbc.M602577200.

Hille, B. (2001). *Ion channels of excitable membranes* (New York).

Horn, R., & Lange, K. (1983). Estimating kinetic constants from single channel data. *Biophysical Journal, 43*(2), 207–223. http://dx.doi.org/10.1016/S0006-3495(83)84341-0.

Horovitz, A. (1996). Double-mutant cycles: A powerful tool for analyzing protein structure and function. *Folding and Design, 1*(6), R121–R126. http://dx.doi.org/10.1016/S1359-0278(96)00056-9.

Huang, H., Bae, C., Sachs, F., & Suchyna, T. M. (2013). Caveolae regulation of mechanosensitive channel function in myotubes. *PLoS One, 8*(8), e72894. http://dx.doi.org/10.1371/journal.pone.0072894.

Ikeda, R., Cha, M., Ling, J., Jia, Z., Coyle, D., & Gu, J. G. (2014). Merkel cells transduce and encode tactile stimuli to drive Abeta-afferent impulses. *Cell, 157*(3), 664–675. http://dx.doi.org/10.1016/j.cell.2014.02.026.

Ikeda, R., & Gu, J. G. (2014). Piezo2 channel conductance and localization domains in Merkel cells of rat whisker hair follicles. *Neuroscience Letters, 583*, 210–215. http://dx.doi.org/10.1016/j.neulet.2014.05.055.

Jacobson, K., Sheets, E. D., & Simson, R. (1995). Revisiting the fluid mosaic model of membranes. *Science, 268*(5216), 1441–1442.

Johnson, K. L. (1985). *Contact mechanics* (1999th ed.). Cambridge, UK: Cambridge University Press.

Kim, S. E., Coste, B., Chadha, A., Cook, B., & Patapoutian, A. (2012). The role of *Drosophila* Piezo in mechanical nociception. *Nature, 483*(7388), 209–212. http://dx.doi.org/10.1038/nature10801.

Kloda, A., Lua, L., Hall, R., Adams, D. J., & Martinac, B. (2007). Liposome reconstitution and modulation of recombinant N-methyl-d-aspartate receptor channels by membrane stretch. *Proceedings of the National Academy of Sciences of the United States of America, 104*(5), 1540–1545. http://dx.doi.org/10.1073/pnas.0609649104.

Knorr, R., Karacsonyi, C., & Lindner, R. (2009). Endocytosis of MHC molecules by distinct membrane rafts. *Journal of Cell Science, 122*(Pt 10), 1584–1594. http://dx.doi.org/10.1242/jcs.039727.

Kozlov, M. M., Kuzmin, P. I., & Popov, S. V. (1992). Formation of cell protrusions by an electric field: A thermodynamic analysis. *European Biophysics Journal, 21*(1), 35–45.

Kuimova, M. K. (2012a). Mapping viscosity in cells using molecular rotors. *Physical Chemistry Chemical Physics, 14*(37), 12671–12686. http://dx.doi.org/10.1039/c2cp41674c.

Kuimova, M. K. (2012b). Molecular rotors image intracellular viscosity. *Chimia, 66*(4), 159–165. http://dx.doi.org/10.2533/chimia.2012.159.

Kurrle, A., Rieber, P., & Sackmann, E. (1990). Reconstitution of transferrin receptor in mixed lipid vesicles. An example of the role of elastic and electrostatic forces for protein/lipid assembly. *Biochemistry, 29*(36), 8274–8282.

Kusumi, A., & Suzuki, K. (2005). Toward understanding the dynamics of membrane-raft-based molecular interactions. *Biochimica et Biophysica Acta, 1746*(3), 234–251. http://dx.doi.org/10.1016/j.bbamcr.2005.10.001.

Kuzmin, P. I., Akimov, S. A., Chizmadzhev, Y. A., Zimmerberg, J., & Cohen, F. S. (2005). Line tension and interaction energies of membrane rafts calculated from lipid splay and tilt. *Biophysical Journal, 88*(2), 1120–1133. http://dx.doi.org/10.1529/biophysj.104.048223.

Lee, K. J. (2006). Energetics of rotational gating mechanisms of an ion channel induced by membrane deformation. *Physical Review. E, Statistical, Nonlinear, and Soft Matter Physics, 73*(2 Pt 1), 021909. http://dx.doi.org/10.1103/PhysRevE.73.021909.

Lee, W., Leddy, H. A., Chen, Y., Lee, S. H., Zelenski, N. A., McNulty, A. L., … Liedtke, W. B. (2014). Synergy between Piezo1 and Piezo2 channels confers high-strain mechanosensitivity to articular cartilage. *Proceedings of the National Academy of Sciences of the United States of America, 111*(47), E5114–E5122. http://dx.doi.org/10.1073/pnas.1414298111.

Leikin, S., Kozlov, M. M., Fuller, N. L., & Rand, R. P. (1996). Measured effects of diacylglycerol on structural and elastic properties of phospholipid membranes. *Biophysical Journal, 71*(5), 2623–2632. http://dx.doi.org/10.1016/S0006-3495(96)79454-7.

Lesage, F., Maingret, F., & Lazdunski, M. (2000). Cloning and expression of human TRAAK, a polyunsaturated fatty acids-activated and mechano-sensitive K(+) channel. *FEBS Letters, 471*(2–3), 137–140.

Lesage, F., Terrenoire, C., Romey, G., & Lazdunski, M. (2000). Human TREK2, a 2P domain mechano-sensitive K^+ channel with multiple regulations by polyunsaturated fatty acids, lysophospholipids, and Gs, Gi, and Gq protein-coupled receptors. *The Journal of Biological Chemistry, 275*(37), 28398–28405. http://dx.doi.org/10.1074/jbc.M002822200.

Levina, N., Totemeyer, S., Stokes, N. R., Louis, P., Jones, M. A., & Booth, I. R. (1999). Protection of *Escherichia coli* cells against extreme turgor by activation of MscS and MscL mechanosensitive channels: Identification of genes required for MscS activity. *EMBO Journal, 18*(7), 1730–1737. http://dx.doi.org/10.1093/emboj/18.7.1730.

Lewis, A. H., & Grandl, J. (2015). Mechanical sensitivity of Piezo1 ion channels can be tuned by cellular membrane tension. *eLife, 4*. http://dx.doi.org/10.7554/eLife.12088.

Li, C., Rezania, S., Kammerer, S., Sokolowski, A., Devaney, T., Gorischek, A., … Schreibmayer, W. (2015). Piezo1 forms mechanosensitive ion channels in the human MCF-7 breast cancer cell line. *Scientific Reports, 5*, 8364. http://dx.doi.org/10.1038/srep08364.

Li, J., Hou, B., Tumova, S., Muraki, K., Bruns, A., Ludlow, M. J., … Beech, D. J. (2014). Piezo1 integration of vascular architecture with physiological force. *Nature, 515*(7526), 279–282. http://dx.doi.org/10.1038/nature13701.

Lindner, R., & Knorr, R. (2009). Rafting trips into the cell. *Communicative and Integrative Biology, 2*(5), 420–421.

Lindner, R., & Naim, H. Y. (2009). Domains in biological membranes. *Experimental Cell Research, 315*(17), 2871–2878. http://dx.doi.org/10.1016/j.yexcr.2009.07.020.

Liu, B. Y., Hui, K. Y., & Qin, F. (2003). Thermodynamics of heat activation of single capsaicin ion channels VR1. *Biophysical Journal, 85*(5), 2988–3006.

Lukacs, V., Mathur, J., Mao, R., Bayrak-Toydemir, P., Procter, M., Cahalan, S. M., … Krock, B. L. (2015). Impaired PIEZO1 function in patients with a novel autosomal recessive congenital lymphatic dysplasia. *Nature Communications, 6*, 8329. http://dx.doi.org/10.1038/ncomms9329.

Lundbaek, J. A., Birn, P., Hansen, A. J., Sogaard, R., Nielsen, C., Girshman, J., … Andersen, O. S. (2004). Regulation of sodium channel function by bilayer elasticity: The importance of hydrophobic coupling. Effects of micelle-forming amphiphiles and cholesterol. *The Journal of General Physiology, 123*(5), 599–621. http://dx.doi.org/10.1085/jgp.200308996.

Lundbaek, J. A., Collingwood, S. A., Ingolfsson, H. I., Kapoor, R., & Andersen, O. S. (2010). Lipid bilayer regulation of membrane protein function: Gramicidin channels as molecular force probes. *Journal of the Royal Society, Interface, 7*(44), 373–395. http://dx.doi.org/10.1098/rsif.2009.0443.

Lundbaek, J. A., Koeppe, R. E., 2nd, & Andersen, O. S. (2010). Amphiphile regulation of ion channel function by changes in the bilayer spring constant. *Proceedings of the National Academy of Sciences of the United States of America, 107*(35), 15427–15430. http://dx.doi.org/10.1073/pnas.1007455107.

Ma, Q. (2014). Merkel cells are a touchy subject. *Cell, 157*(3), 531–533. http://dx.doi.org/10.1016/j.cell.2014.04.010.

Maksimovic, S., Nakatani, M., Baba, Y., Nelson, A. M., Marshall, K. L., Wellnitz, S. A., … Lumpkin, E. A. (2014). Epidermal Merkel cells are mechanosensory cells that tune mammalian touch receptors. *Nature, 509*(7502), 617–621. http://dx.doi.org/10.1038/nature13250.

Markin, V. S., & Sachs, F. (2007). Thermodynamics of mechanosensitivity. *Mechanosensitive Ion Channels, Part A, 58*, 87–119. http://dx.doi.org/10.1016/S1063-5823(06)58004-4.

Markin, V. S., & Sachs, F. (2015). Free volume in membranes: Viscosity or tension? *Open Journal of Biophysics, 05*(03), 80–83. http://dx.doi.org/10.4236/ojbiphy.2015.53007.

Martinac, B. (2009). Open channel structure of MscL: A patch-clamp and spectroscopic study. *Applied Magnetic Resonance, 36*(2–4), 171–179. http://dx.doi.org/10.1007/s00723-009-0035-1.

Martinac, B. (2014). The ion channels to cytoskeleton connection as potential mechanism of mechanosensitivity. *Biochimica et Biophysica Acta, 1838*(2), 682–691. http://dx.doi.org/10.1016/j.bbamem.2013.07.015.

Martinac, B., Adler, J., & Kung, C. (1990). Mechanosensitive ion channels of *E. coli* activated by amphipaths. *Nature, 348*(6298), 261–263. http://dx.doi.org/10.1038/348261a0.

Martinac, B., Rohde, P. R., Battle, A. R., Petrov, E., Pal, P., Foo, A. F., … Kloda, A. (2010). Studying mechanosensitive ion channels using liposomes. *Methods in Molecular Biology, 606*, 31–53. http://dx.doi.org/10.1007/978-1-60761-447-0_4.

McHugh, B. J., Buttery, R., Lad, Y., Banks, S., Haslett, C., & Sethi, T. (2010). Integrin activation by Fam38A uses a novel mechanism of R-Ras targeting to the endoplasmic reticulum. *Journal of Cell Science, 123*(Pt 1), 51–61. http://dx.doi.org/10.1242/jcs.056424.

McMillin, M. J., Beck, A. E., Chong, J. X., Shively, K. M., Buckingham, K. J., Gildersleeve, H. I., … Bamshad, M. J. (2014). Mutations in PIEZO2 cause Gordon syndrome, Marden-Walker syndrome, and distal arthrogryposis type 5. *American Journal of Human Genetics, 94*(5), 734–744. http://dx.doi.org/10.1016/j.ajhg.2014.03.015.

Milescu, L. S., Akk, G., & Sachs, F. (2005). Maximum likelihood estimation of ion channel kinetics from macroscopic currents. *Biophysical Journal, 88*(4), 2494–2515. http://dx.doi.org/10.1529/biophysj.104.053256.

Morris, C. E. (2011). Voltage-gated channel mechanosensitivity: Fact or friction? *Frontiers in Physiology, 2*, 25. http://dx.doi.org/10.3389/fphys.2011.00025.

Morris, C. E. (2012). Why are so many ion channels mechanosensitive? In N. Sperelakis (Ed.), *Cell physiology source book* (pp. 493–505). Elsevier.

Nakatani, M., Maksimovic, S., Baba, Y., & Lumpkin, E. A. (2015). Mechanotransduction in epidermal Merkel cells. *Pflugers Archiv, 467*(1), 101–108. http://dx.doi.org/10.1007/s00424-014-1569-0.

Nezil, F. A., & Bloom, M. (1992). Combined influence of cholesterol and synthetic amphiphillic peptides upon bilayer thickness in model membranes. *Biophysical Journal, 61*(5), 1176–1183. http://dx.doi.org/10.1016/S0006-3495(92)81926-4.

Nicolai, C., & Sachs, F. (2014). Fitting random data to state models with QuB software. *Biophysical Reviews, 8*(3&4), 191–211.

Nomura, T., Cranfield, C. G., Deplazes, E., Owen, D. M., Macmillan, A., Battle, A. R., … Martinac, B. (2012). Differential effects of lipids and lyso-lipids on the mechanosensitivity of the mechanosensitive channels MscL and MscS. *Proceedings of the National Academy of Sciences of the United States of America, 109*(22), 8770–8775. http://dx.doi.org/10.1073/pnas.1200051109.

Ollila, O. H., Risselada, H. J., Louhivuori, M., Lindahl, E., Vattulainen, I., & Marrink, S. J. (2009). 3D pressure field in lipid membranes and membrane-protein complexes. *Physical Review Letters, 102*(7), 078101.

Pan, B., Geleoc, G. S., Asai, Y., Horwitz, G. C., Kurima, K., Ishikawa, K., … Holt, J. R. (2013). TMC1 and TMC2 are components of the mechanotransduction channel in hair cells of the mammalian inner ear. *Neuron, 79*(3), 504–515. http://dx.doi.org/10.1016/j.neuron.2013.06.019.

Pan, N. C., Ma, J. J., & Peng, H. B. (2012). Mechanosensitivity of nicotinic receptors. *Pflugers Archiv, 464*(2), 193–203. http://dx.doi.org/10.1007/s00424-012-1132-9.

Patel, A. J., Lazdunski, M., & Honore, E. (2001). Lipid and mechano-gated 2P domain K(+) channels. *Current Opinion in Cell Biology, 13*(4), 422–428.

Pathak, M. M., Nourse, J. L., Tran, T., Hwe, J., Arulmoli, J., Le, D. T., … Tombola, F. (2014). Stretch-activated ion channel Piezo1 directs lineage choice in human neural stem cells. *Proceedings of the National Academy of Sciences of the United States of America, 111*(45), 16148–16153. http://dx.doi.org/10.1073/pnas.1409802111.

Poh, Y. C., Beyder, A., Strege, P. R., Farrugia, G., & Buist, M. L. (2012). Quantification of gastrointestinal sodium channelopathy. *Journal of Theoretical Biology, 293*, 41–48. http://dx.doi.org/10.1016/j.jtbi.2011.09.014.

Pontes, B., Ayala, Y., Fonseca, A. C., Romao, L. F., Amaral, R. F., Salgado, L. T., … Nussenzveig, H. M. (2013). Membrane elastic properties and cell function. *PLoS One, 8*(7), e67708. http://dx.doi.org/10.1371/journal.pone.0067708.

Prole, D. L., & Taylor, C. W. (2013). Identification and analysis of putative homologues of mechanosensitive channels in pathogenic protozoa. *PLoS One, 8*(6), e66068. http://dx.doi.org/10.1371/journal.pone.0066068.

Qin, F., Auerbach, A., & Sachs, F. (1996a). Estimating single-channel kinetic parameters from idealized patch-clamp data containing missed events. *Biophysical Journal, 70*(1), 264–280. http://dx.doi.org/10.1016/S0006-3495(96)79568-1.

Qin, F., Auerbach, A., & Sachs, F. (1996b). Idealization of single-channel currents using the segmental K-means method. *Biophysical Journal, 70*(2), Mp432.

Qin, F., Auerbach, A., & Sachs, F. (1997). Maximum likelihood estimation of aggregated Markov processes. *Proceedings. Biological Sciences, 264*(1380), 375–383. http://dx.doi.org/10.1098/rspb.1997.0054.

Qin, F., Auerbach, A., & Sachs, F. (2000a). A direct optimization approach to hidden Markov modeling for single channel kinetics. *Biophysical Journal, 79*(4), 1915–1927. http://dx.doi.org/10.1016/S0006-3495(00)76441-1.

Qin, F., Auerbach, A., & Sachs, F. (2000b). A hybrid approach to hidden Markov modeling for single channel kinetics. *Biophysical Journal, 78*(1), 333a.

Ranade, S. S., Qiu, Z., Woo, S. H., Hur, S. S., Murthy, S. E., Cahalan, S. M., … Patapoutian, A. (2014). Piezo1, a mechanically activated ion channel, is required for vascular development in mice. *Proceedings of the National Academy of Sciences of the United States of America, 111*(28), 10347–10352. http://dx.doi.org/10.1073/pnas.1409233111.

Roudaut, Y., Lonigro, A., Coste, B., Hao, J., Delmas, P., & Crest, M. (2012). Touch sense: Functional organization and molecular determinants of mechanosensitive receptors. *Channels, 6*(4), 234–245. http://dx.doi.org/10.4161/chan.22213.

Sachs, F. (2015). Mechanical transduction by ion channels: A cautionary tale. *World Journal of Neurology, 5*(3), 74–87.

Sachs, F., Morris, C. E., Blaustein, M. P., Greger, R., Grunicke, H., Jahn, R., … Schweiger, M. (1998). Mechanosensitive ion channels in non specialized cells. In *Reviews of physiology and biochemistry and pharmacology* (Vol. 132, pp. 1–78). Berlin: Springer.

Satoh, K., Hata, M., Takahara, S., Tsuzaki, H., Yokota, H., Akatsu, H., … Yamada, T. (2006). A novel membrane protein, encoded by the gene covering KIAA0233, is transcriptionally induced in senile plaque-associated astrocytes. *Brain Research, 1108*(1), 19–27. http://dx.doi.org/10.1016/j.brainres.2006.06.050.

Schneider, E. R., Mastrotto, M., Laursen, W. J., Schulz, V. P., Goodman, J. B., Funk, O. H., … Bagriantsev, S. N. (2014). Neuronal mechanism for acute mechanosensitivity in tactile-foraging waterfowl. *Proceedings of the National Academy of Sciences of the United States of America, 111*(41), 14941–14946. http://dx.doi.org/10.1073/pnas.1413656111.

Schrenk-Siemens, K., Wende, H., Prato, V., Song, K., Rostock, C., Loewer, A., … Siemens, J. (2015). PIEZO2 is required for mechanotransduction in human stem cell-derived touch receptors. *Nature Neuroscience, 18*(1), 10–16. http://dx.doi.org/10.1038/nn.3894.

Schumann, U., Edwards, M. D., Rasmussen, T., Bartlett, W., van West, P., & Booth, I. R. (2010). YbdG in *Escherichia coli* is a threshold-setting mechanosensitive channel with MscM activity. *Proceedings of the National Academy of Sciences of the United States of America, 107*(28), 12664–12669. http://dx.doi.org/10.1073/pnas.1001405107.

Sigg, D., Bezanilla, F., & Stefani, E. (2003). Fast gating in the Shaker K^+ channel and the energy landscape of activation. *Proceedings of the National Academy of Sciences of the United States of America, 100*(13), 7611–7615.

Slavchov, R. I., Nomura, T., Martinac, B., Sokabe, M., & Sachs, F. (2014). Gigaseal mechanics: Creep of the gigaseal under the action of pressure, adhesion, and voltage. *The Journal of Physical Chemistry. B, 118*(44), 12660–12672. http://dx.doi.org/10.1021/jp506965v.

Suchyna, T. M., Markin, V. S., & Sachs, F. (2009). Biophysics and structure of the patch and the gigaseal. *Biophysical Journal, 97*(3), 738–747. http://dx.doi.org/10.1016/j.bpj.2009.05.018.

Suchyna, T. M., & Sachs, F. (2007). Mechanosensitive channel properties and membrane mechanics in mouse dystrophic myotubes. *The Journal of Physiology, 581*(Pt 1), 369–387. http://dx.doi.org/10.1113/jphysiol.2006.125021.

Suchyna, T. M., Tape, S. E., Koeppe, R. E., 2nd, Andersen, O. S., Sachs, F., & Gottlieb, P. A. (2004). Bilayer-dependent inhibition of mechanosensitive channels by neuroactive peptide enantiomers. *Nature, 430*(6996), 235–240. http://dx.doi.org/10.1038/nature02743.

Sukharev, S. (1999). Mechanosensitive channels in bacteria as membrane tension reporters. *FASEB Journal*, (13 Suppl.), S55–S61.

Sukharev, S. (2002). Purification of the small mechanosensitive channel of *Escherichia coli* (MscS): The subunit structure, conduction, and gating characteristics in liposomes. *Biophysical Journal, 83*(1), 290–298. http://dx.doi.org/10.1016/S0006-3495(02)75169-2.

Sukharev, S., & Sachs, F. (2012). Molecular force transduction by ion channels: Diversity and unifying principles. *Journal of Cell Science, 125*(Pt 13), 3075–3083. http://dx.doi.org/10.1242/jcs.092353.

Sukharev, S. I., Sigurdson, W. J., Kung, C., & Sachs, F. (1999). Energetic and spatial parameters for gating of the bacterial large conductance mechanosensitive channel, MscL. *The Journal of General Physiology, 113*(4), 525–540.

Syeda, R., Xu, J., Dubin, A. E., Coste, B., Mathur, J., Huynh, T., … Patapoutian, A. (2015). Chemical activation of the mechanotransduction channel Piezo1. *eLife, 4*. http://dx.doi.org/10.7554/eLife.07369.

Venkataramanan, L., Kuc, R., & Sigworth, F. J. (1998). Identification of hidden Markov models for ion channel currents — Part II: State-dependent excess noise. *IEEE Transactions on Signal Processing, 46*(7), 1916—1929.

Venkataramanan, L., & Sigworth, F. J. (2002). Applying hidden Markov models to the analysis of single ion channel activity. *Biophysical Journal, 82*(4), 1930—1942.

Venkataramanan, L., Walsh, J. L., Kuc, R., & Sigworth, F. J. (1998). Identification of hidden Markov models for ion channel currents — Part I: Colored background noise. *IEEE Transactions on Signal Processing, 46*(7), 1901—1915.

Wiggins, P., & Phillips, R. (2004). Analytic models for mechanotransduction: Gating a mechanosensitive channel. *Proceedings of the National Academy of Sciences of the United States of America, 101*(12), 4071—4076. http://dx.doi.org/10.1073/pnas.0307804101.

Woo, S. H., Lumpkin, E. A., & Patapoutian, A. (2015). Merkel cells and neurons keep in touch. *Trends in Cell Biology, 25*(2), 74—81. http://dx.doi.org/10.1016/j.tcb.2014.10.003.

Zhao, Q., Wu, K., Geng, J., Chi, S., Wang, Y., Zhi, P., … Xiao, B. (2016). Ion permeation and mechanotransduction mechanisms of mechanosensitive Piezo channels. *Neuron, 89*(6), 1248—1263. http://dx.doi.org/10.1016/j.neuron.2016.01.046.

Zhao, W. N., Tisza, M., Sjol, J., Mani, S., & Chang, J. T. (2014). Plasma membrane fluidity drives metastasis in breast cancer. *Cancer Research, 74*(19), 2080. http://dx.doi.org/10.1158/1538-7445.Am2014-2080.

CHAPTER TWELVE

A Microfluidic Approach for Studying Piezo Channels

M.M. Maneshi, P.A. Gottlieb and S.Z. Hua[1]

State University of New York at Buffalo, Buffalo, NY, United States
[1]Corresponding author: E-mail: zhua@eng.buffalo.edu

Contents

Abstract

Microfluidics is an interdisciplinary field intersecting many areas in engineering. Utilizing a combination of physics, chemistry, biology, and biotechnology, along with practical applications for designing devices that use low volumes of fluids to achieve high-throughput screening, is a major goal in microfluidics. Microfluidic approaches allow the study of cells growth and differentiation using a variety of conditions

Current Topics in Membranes, Volume 79
ISSN 1063-5823
http://dx.doi.org/10.1016/bs.ctm.2016.11.005

including control of fluid flow that generates shear stress. Recently, Piezo1 channels were shown to respond to fluid shear stress and are crucial for vascular development. This channel is ideal for studying fluid shear stress applied to cells using microfluidic devices. We have developed an approach that allows us to analyze the role of Piezo channels on any given cell and serves as a high-throughput screen for drug discovery. We show that this approach can provide detailed information about the inhibitors of Piezo channels.

1. FUNDAMENTALS OF STRESS IN SOLIDS AND FLUIDS

Mechanical force is a major stimulus in biology that provides signals for cell function and guidance for tissue development (DuFort, Paszek, & Weaver, 2011; Gomez, McLachlan, & Yap, 2011; Huveneers & de Rooij, 2013; Vogel & Sheetz, 2006). Cells experience numerous mechanical forces including tensile or compressive forces from shape change of the surrounding tissues and shear force or hydraulic pressure from fluid flow in the tubular structure.

Mechanical stress is the force per unit area of an infinitesimally small area. Forces applied to a solid body have a normal and tangential component with respect to the surface. Normal forces would generate stresses that are either compressive or tensile, and they would respectively cause shrinkage or elongation in the direction of the force application. On the other hand, tangential forces are parallel to the surface and the force per unit area is called shear stress (Fig. 1A). Unlike normal stress, shear stress generates angular deformation (rotation of elements). The deformation is directly proportional to the acting force and inversely proportional to material elasticity, so that the softer the material, the more deformation occurs.

In solids, there is always a resistance to deformation; these deformations are limited and stop during load application. Unlike solids, fluids show no resistance to deformation, and as long as the load is applied, the fluids would deform and therefore "flow." When it comes to fluids, it makes sense to deal with temporal changes of deformation and velocities instead of displacement. We are interested in shear strain rate, which is the velocity of deformation per unit length of the fluid.

For a uniform flow of a viscous fluid applied to a solid boundary at infinitesimally close to the solid surface, the adhesion of fluid molecules to the surface is higher than the momentum of fluid molecules and the velocity is equal to zero. Far away from the solid surface though, the velocity

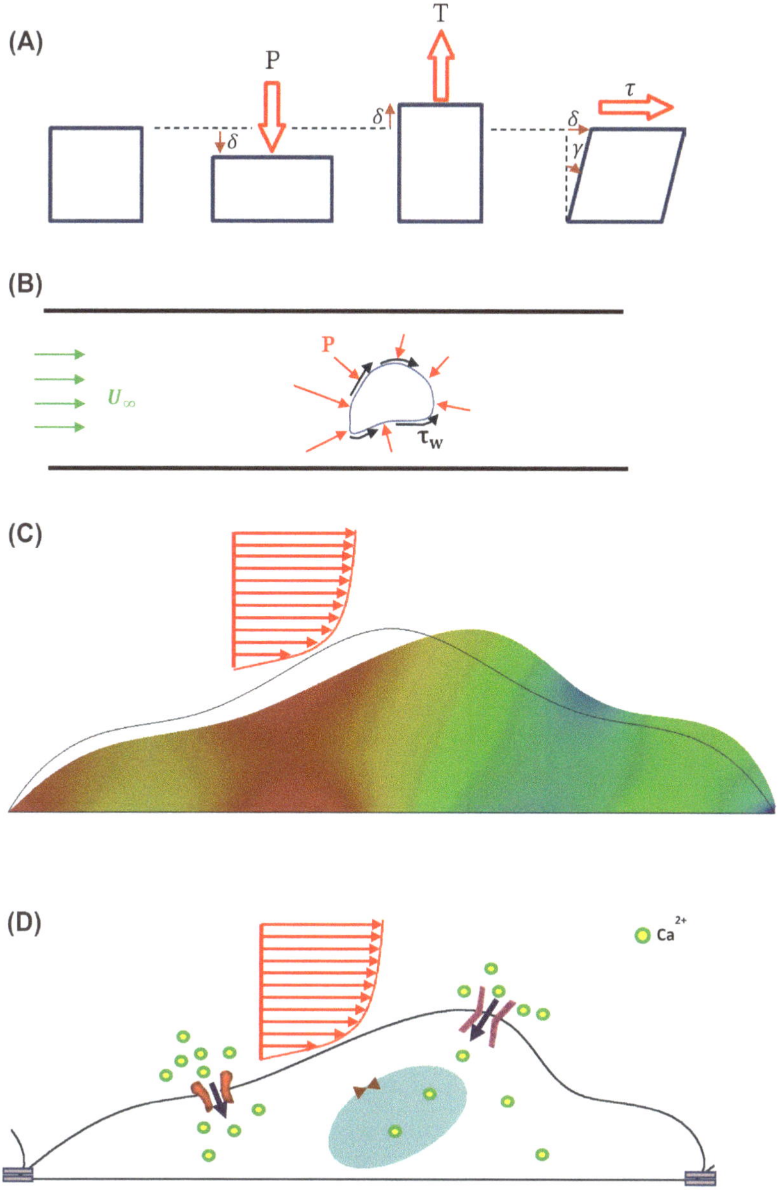

Figure 1 Schematic illustration of various types of mechanical forces on cells. (A) Illustration of different modes of deformation. The compressive force P and tensile force T cause compressive and tensile deformation while the shearing force τ would angularly deform the element. The parameter δ is the deformation and γ is the angular rotation of the element in shear. Strain is defined as deformation per unit length of deforming object. (B) Illustration of pressure distribution P and wall shear stress τ_w alongside a body. The forces have the perpendicular and tangential components based on shape of body. (C) A 2-D finite element model showing distribution of principal stress in the direction of flow for an adherent cell, showing transient nonuniform stress distributions in the cell body. (D) Fluid shear stress generated drag force causes deformation of an adherent cell, resulting in influx of ions through various mechanosensitive channels.

recovers. Because of the existence of a velocity gradient from the solid interface to the far field, the fluid imposes a shear stress on the solid boundary (Fig. 1C). The molecules of the fluid adhere to the wall on the bottom while the next level of molecules flow away from that layer. For most fluids like pure water, viscosity is a thermodynamic parameter, and it does not change with flow field, therefore shear stress is linearly proportional to gradient of fluid velocity perpendicular to the solid surface. Such fluids are Newtonian fluids. On the other hand, highly salted solutions and most biological fluids show a different behavior in which the viscosity not only governed by thermodynamic parameters but by the shear stress as well. These fluids are non-Newtonian; a good example of such fluids is blood. Blood's viscosity is dependent on the flow pattern and shear stress. Blood's viscosity decreases to alleviate passing through small arteries and therefore it is a shear-thinning fluid.

Fluid shear stress on the solid interface is what causes drag force in lower velocities of fluid. The drag on solid interface is always in the direction of momentum loss from fluid. Therefore, if the fluid is flowing from left to right, as the fluid loses some momentum by contacting the solid interface it becomes stagnant. The drag force is going to deform the solid boundary in direction of the flow, generating a 3-D stress pattern in the solid body based on its geometry as shown in Fig.1C.

1.1 Active forces on a body in fluid flow

There are two types of forces generated by fluid flow on solid boundaries. Both of these forces are the integration of static pressure P distribution and wall shear τ_w alongside the surface of the body (Fig. 1B). The net force acting on the solid body has a component in the direction of upstream flow U_∞ called the drag force and another component perpendicular to the flow direction called lift. The drag force generated by changes of pressure, also referred to as pressure drag (form drag), is usually negligible at lower velocities and regions of laminar flow. Pressure drag is a major source of drag force in high velocities as the drag force on a plane wing because separations of flow and sudden changes of pressure usually would not happen in laminar flows. On the other hand, in regions of slow fluid flow such as laminar flows in microfluidic devices, the major source of drag force is wall shear stress (skin friction drag). Flow of blood in major arteries or other bodily fluids such as urine through small ducts fall in the second category of fluid flow in which the effect of wall shear stress is the dominant factor.

1.2 Fluid shear stress

In a fluidic chamber, the drag force applied to a solid body produces both normal and shear stresses based on its geometry and orientation. If the solid body is a cell, the cell would experience some deformations due to the fluid drag, fluid shear stress. When cells experience shear stress, there are two possible scenarios. Shear stress due to surface friction could directly deform the lipid bilayer independent of cell body, changing tension in the cell membrane. Alternatively, forces are transmitted into the cell body by the cytoskeleton that links the apical surface to the focal adhesions in the cell basement. In the latter case, changes of bilayer tension is coupled to cell body deformation. Under flow conditions, both may occur simultaneously in a cell (Fig. 1C).

1.3 Effect of shear stress on cell physiology

Numerous studies have shown that changes in tubular flow affect ion transport and kidney physiology (Holtzclaw, Liu, Grimm, & Sansom, 2010; Liu, Morimoto, Woda, Kleyman, & Satlin, 2007; Satlin, Sheng, Woda, & Kleyman, 2001; Woda, Bragin, Kleyman, & Satlin, 2001). Fluid shear stress generated in tubular flow provides key signals that regulate glomerulotubular balance in the proximal tubule and ion transport in the distal nephron including the cortical collecting duct (Liu et al., 2003; Resnick, 2011; Weinbaum, Duan, Satlin, Wang, & Weinstein, 2010). Flow-induced membrane transport, such as Ca^{2+} influx and the activation of ion channels, has been widely reported (Jensen, Odgaard, Christensen, Praetorius, & Leipziger, 2007; Liu et al., 2003; Praetorius & Spring, 2001). Shear stress increases the open probability of epithelial sodium channels (ENaCs) that are essential for electrolyte and water balance (Althaus, Bogdan, Clauss, & Fronius, 2007). The renal epithelial cells may use primary cilia or microvilli as flow sensors (Jensen et al., 2007; Liu et al., 2003; Praetorius & Spring, 2001). The flow-induced bending of cilium causes the nearby mechanosensitive channels (MSCs) to open and initiate biochemical processes to modulate the cytoskeleton and cell adhesions (Resnick, 2011; Rohatgi & Flores, 2010; Weinbaum et al., 2010).

Endothelial cells (EC) that line the blood vessels are sensitive to shear stress. A step increase in shear stress induces a transient Ca^{2+} influx in ECs followed by a sustained plateau phase for the length of the shear stimuli (Kwan, Huang, & Yao, 2007). Onset of the shear-induced Ca^{2+} response appears to require several seconds, which can be related to the rate of shear produced in particular experimental systems (Kwan et al., 2007; Schwarz, Callewaert, Droogmans, & Nilius, 1992; Shen, Luscinskas, Connolly, Dewey, & Gimbrone, 1992). Many mechanosensitive molecules have been identified and postulated to serve as entry sites for calcium, and they have similar sets of secondary messengers suggesting that mechanosensation is truly an integrated process (Davies, Spaan, & Krams, 2005; Mott & Helmke, 2007). Recent studies of vascular development have shown that Piezo1 channels *can sense shear stress* in endothelial cells (Li et al., 2014). Piezo1 plays an essential role in vascular structure during vasculature development (Li et al., 2014; Ranade et al., 2014).

The shear stress is also one of the important mechanical forces during traumatic brain injury (TBI). The injury is normally classified as a focal injury that results in cerebral contusions in a specific location or diffuse axonal injury that occurs over a widespread area, due to accelerations that generate shear forces (Chen, Smith, & Meaney, 2009; Garman et al., 2011; Maas, Stocchetti, & Bullock, 2008; Sanborn, Nie, Chen, & Weerasooriya, 2012). The deformations caused by shear forces in TBI is particularly important because it causes long term injury and is associated with subsequent development of Alzheimer's disease (AD). An increase in intracellular Ca^{2+} arises when brain cells are subjected to mechanical perturbations such as shear stress. Ravin et al. (2012) have shown that shear stress, but not overpressure, is accountable for the Ca^{2+} response in CNS subjected to a shockwave. It has been shown that amyloid-β (Aβ) plaques, one of the hallmarks of AD, may be elevated in patients within hours following TBI (Fleminger, Oliver, Lovestone, Rabe-Hesketh, & Giora, 2003; Guo et al., 2000; Mortimer, French, Hutton, & Schuman, 1985). One of the key observation is that Aβ production and deposition after TBI may depend on the type of injury. The Aβ generation and deposition was only observed with head rotational acceleration, suggesting shear stress may play the role (Johnson, Stewart, & Smith, 2010). Recent studies of sensory neurons have suggested Piezo proteins as subunits of Ca^{2+} permeable nonselective cationic channels may involve in detection of mechanical forces in brain (Coste et al., 2010, 2012).

1.4 Shear application and signal collection

The common methods of generating shear stress in microfluidic chambers are by using peristaltic pumps, syringe pumps, or by gravity feeding. These methods generate a steady and tunable flow with slow ramping time in the range of seconds. In a recent study on traumatic brain injury, shear stress was controlled by a fast pressure servo (Besch, Suchyna, & Sachs, 2002) that is capable of generating pressure pulses with a rise time of ~1–2 ms in the microfluidic chambers (Besch et al., 2002; Maneshi, Sachs, & Hua, 2015). The piezoelectric valve can produce any desired pulse train to generate shear stress with arbitrary waveforms at high time resolution. The experimental setup and typical pressure characteristics is shown in Fig. 2.

Cell's response to shear stimuli is nonlinear; the viscoelastic nature of cells appears to play a role in the activation of the mechanosensitive Ca^{2+} influx pathways. High-speed stimuli enable the measurements of onset cell responses without interference of other biophysical and biochemical processes, which can help distinguish causality from correlation.

Fluorescence imaging is the most commonly used method to measure cell responses to shear stress. The global and local intracellular Ca^{2+} can

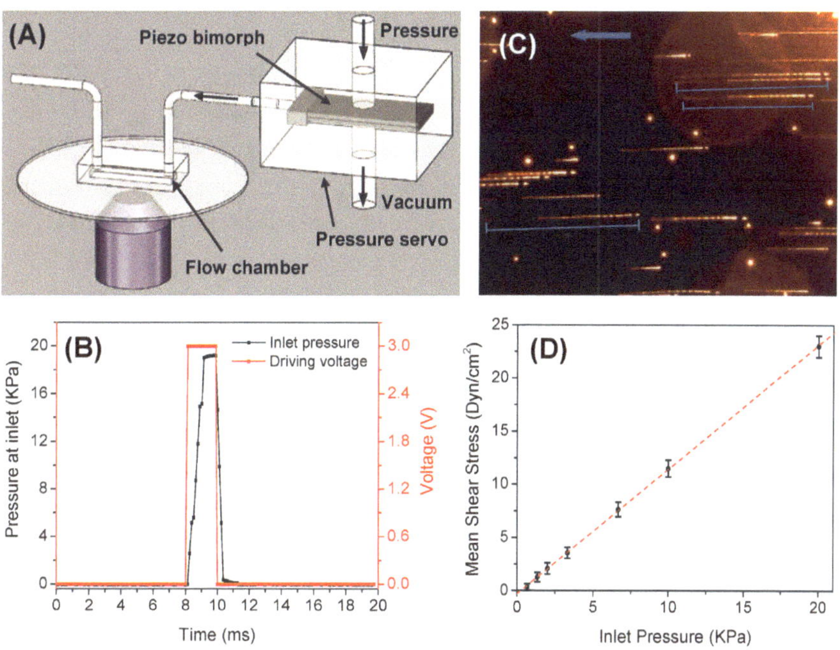

Figure 2 Pressure servo driven microfluidic chamber. (A) Schematic of experimental setup. High-speed pressure servo was connected to a microfluidic chamber to generate controlled fluid shear stimuli. (B) Typical image of trajectory of microbeads subjected to a flow pulse. (C) Comparison of triggering voltage pulse [red] and pressure rise (black) at inlet of the chamber, showing a rise time of 1 ms can be achieved. (D) Calibration of shear stress in the channel.

be measured using fluorescent dyes such as the Fluo family and genetically encoded Ca^{2+} indicator (GECI) such as GCaMP family (Akerboom et al., 2012; Tian, Hires, & Looger, 2012). These probes improve the spatial and time resolution that is particularly useful for localized Ca^{2+} measurements. Many voltage and ligand gated channels are known to be stretch sensitive (Laitko & Morris, 2004; Morris & Juranka, 2007). Activation of these channels can change the membrane potential. Voltage sensitive fluorescence dyes such as ANEPPS-8, which can be ratio-imaged or green fluorescent protein-based probes like Mermaid are used for potential measurements (Tsutsui, Karasawa, Okamura, & Miyawaki, 2008). By placing four electrode arrays in the chamber, the microfluidic chamber can also be used to measure the cell volume in real time (Ateya, Sachs, Gottlieb, Besch, & Hua, 2005; Heo, Sachs, Wang, & Hua, 2012).

1.5 Shear stress calibration

To measure the effects of shear stress on cells, it is imperative to calibrate the generated shear stress in the microfluidic chamber of interest. Several techniques allow calibration of the average shear stress generated by the flow. One of the easiest and most accurate ways of measuring the flow velocity is using fluorescent spherical beads in the fluid. In this technique, fluorescent beads added to the solution are used to track the flow in the chamber. A high-speed camera is used to acquire series of images, and the flow velocity can be determined by the trajectory length for a given frame rate. Alternatively, one can measure displaced volume of fluid during a certain time interval of flow to get the average flow rate.

For pulsatile flow, the beads can be anchored to the surface of the chamber using an elastic linker. The drag force of fluid would then extend the linker by displacing the beads and, by measuring bead movement, the drag force can then be calculated (Ravin et al., 2012).

Based on Poiseuille law, wall shear stress in a rectangular microchannel is calculated by the equation:

$$\tau = \frac{6\mu Q}{wh^2}$$

where Q is the volume flow rate, μ is the viscosity of the fluid, w and h are the width and the height of the chamber, respectively. To measure average wall shear stress, the average fluid velocity is determined.

2. THE SHEAR STRESS RESPONSE

2.1 Shear stress-induced Ca^{2+} influx via mechanosensitive ion channels

Intracellular calcium increase is among the first measurable cell responses to mechanical stimuli (Hua, Gottlieb, Heo, & Sachs, 2010; Pathak et al., 2014; Ranade et al., 2014). Recently Ca^{2+} was shown to flux through cationic MSCs (Coste et al., 2010). Ca^{2+} influx can be triggered by many mechanical stimuli including fluid shear stress and osmotic swelling (Haeberle, Bryan, Vadakkan, Dickinson, & Lumpkin, 2008). Thus, shear stress stimulated Ca^{2+} influx has been widely used as a measure of MSC activity to study force sensitivity, time resolved channel activity, and pharmacological effects. Although a uniform shear stress is applied to the cells, the cells' response can result in a broadly distributed latency and magnitude (Fig. 3A and B), owing to heterogeneous nature of the cells. Using a microfluidic chamber, a large population of cells within a single chamber are measured simultaneously; the large sampling size provides a good statistical analysis (Fig. 3C). In addition, the same shear stress can be applied repeatability. Fig. 3 shows a typical response of HEK293T cells expressing hPIEZO1 to a shear stress pulse of 15 dyn/cm^2 for 10 s. Several important parameters can be extracted from these data, including the level of Ca^{2+} rise, the latency, and the time required to reach the peak value. Together, these characterize the total contribution of initial Ca^{2+} influx through shear sensitive channels and subsequent release of internal stores.

Most MSCs respond to the same shear stress with specific characteristics. The shear stress assay has enough resolution to distinguish them even within the same family of channels. For instance, shear stress of 15 dyn/cm^2 caused 300% Ca^{2+} increase in HEK cells expressing hPIEZO1 and hPIEZO2 (Fig. 4). However, the Ca^{2+} increased immediately in transiently hPIEZO2 expressing cells, while hPizeo1 expressing cells showed a substantial latency of ~5 s (Fig. 4). In contrast, the wild type (WT) HEK293T cells showed a much slower Ca^{2+} rise to less than 80% base level. This suggests PIEZO1 and PIEZO2 channels may use different signaling pathways in response to shear stress, and these channels are shielded from stress in WT HEK cells. The assay could also differentiate between different mutations of a channel. The shear response from cells expressing the double-mutant hPIEZO1 (hPIEZO1-DM) (Bae, Gottlieb, & Sachs, 2013) is very different than the WT hPIEZO1, and the loss of inactivation in hPIEZO1-DM could also

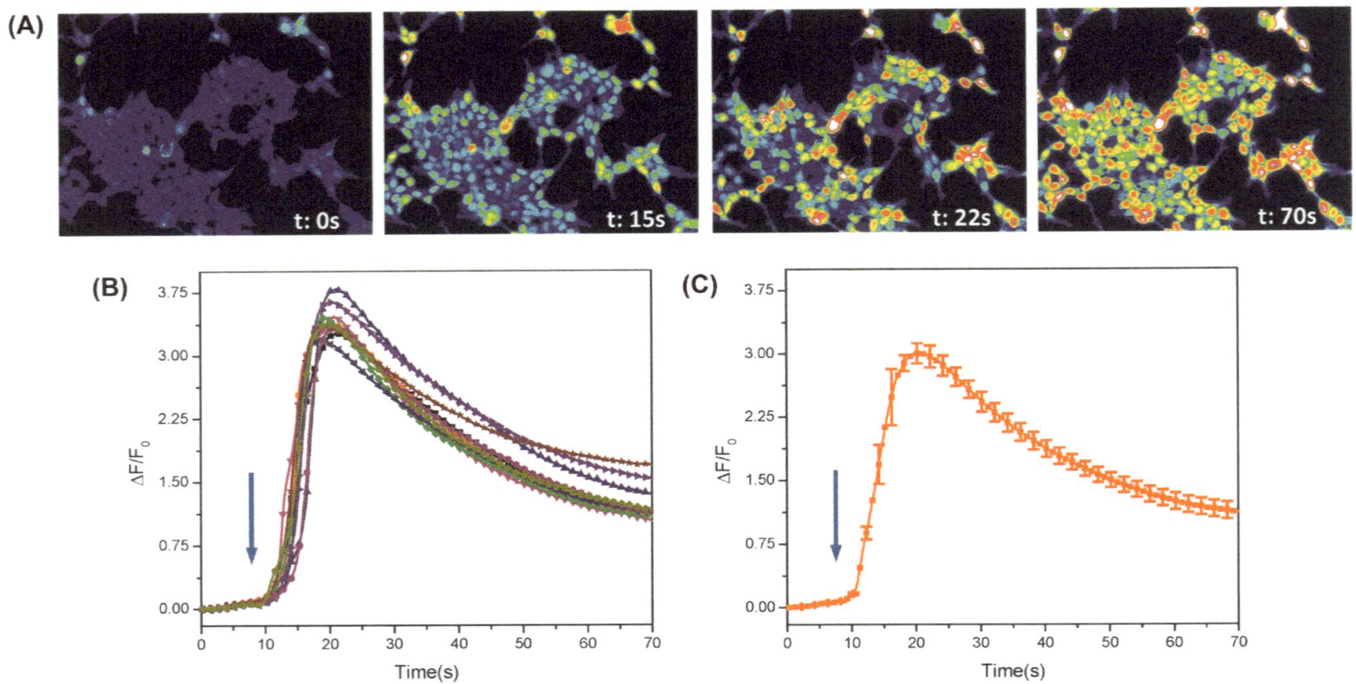

Figure 3 (A) Fluorescence images of Fluo-4 loaded HEK293T cells expressing hPIEZO1 subjected to a narrow shear pulse of 15 dyn/cm^2 for 10 ms. (B) Time-dependent changes of Ca^{2+} intensity from selected cells in the same chamber. (C) Average cell response (from panel B) and standard error for each indicated time.

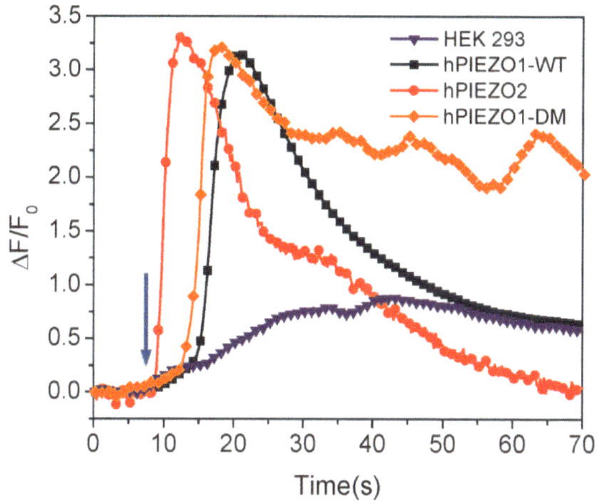

Figure 4 Ca^{2+} response to shear stress in HEK293T cells expressing wild type (WT) and DM hPiezo1 and hPiezo2 compared to WT HEK293T cells. Consistent with previous studies, hPIEZO2 shows a faster response than hPIEZO1. Notice the slow adaptation of Ca^{2+} response in double-mutant PIEZO1 in which channel inactivation is lost.

be detected. The microfluidic shear stress assay provides a quick and robust test of existence of certain MSCs in the cell.

2.2 Determining shear stress thresholds

Recent studies showed that a cell's response to mechanical stimulus is nonlinear, therefore the mechanical thresholds of an MSC influx relies on the features of the stimulus (Chen et al., 2009; Chung et al., 2005; Geddes-Klein, Schiffman, & Meaney, 2006). For brain cells, astrocytes and neurons, rapid shear deformation increases the membrane permeability to Ca^{2+}(Cullen, Vernekar, & LaPlaca, 2011; Maneshi et al., 2015). The response depends not only on the magnitude of the loading, but also on the rate of loading and the repetition rate (Cullen et al., 2011; LaPlaca, Cullen, McLoughlin, & Cargill, 2005; Maneshi et al., 2015). Furthermore, cells respond to repeated subthreshold stimuli with highly nonlinear summation, revealing the presence of an effective viscosity (Maneshi et al., 2015). These studies suggest that subthreshold stimuli exist for an ion channel, and the parameters of the stimulus (intensity, rise time, duration, duty cycle, etc.) affect the open probability.

The shear stress stimulus is controlled with a high speed servo to generate any waveform, such as shear stress of different amplitude, rise time, duration time, and frequency of oscillation. We have analyzed Ca^{2+} sensitivity to a fluid shear pulse by varying its amplitude, width, and rise time. Our data showed that astrocytes reliably responded to a fast shear stimulus of 23 dyn/cm^2 with a transient Ca^{2+} increase but did not respond to a pulse of 11.5 dyn/cm^2, setting an effective threshold of ~11.5 dyn/cm^2 (Fig. 5A). The magnitude of shear is not the only determinant of the stimulus. There are finite relaxation times present in the cells so that reductions in stimulus duration or rise time can affect a subthreshold response. For example, a narrow pulse at low amplitude (10 ms, 11.5 dyn/cm^2) did not trigger Ca^{2+} elevation. However, when the pulse width increased to 1000 ms, there was a clear increase in Ca^{2+} (Fig. 5B). The peak Ca^{2+} response forms a 3-D complex surface relevant to different stimulus parameters (Fig. 5C).

The stimulus rise time, among all parameters, appears critical in activating onset Ca^{2+} responses in astrocytes, probably due to the viscoelastic nature of the cells. The high-speed servo driven microfluidics has unique advantage of studying stimulus ramp time. When the rise time changed from 200 to 2 ms, the peak Ca^{2+} response increased five times (Fig. 5).

2.3 Shear stress assay for pharmacological studies

Most cells have multiple endogenous MSCs that could potentially contribute to Ca^{2+} elevation; pharmacology is a useful initial approach to study ion influx through MSCs. Because known inhibitors are normally insufficient for complete inhibition, a dose dependent study is useful and provides information about the stoichiometry of the inhibitor. When dealing with complex responses, where multiple processes are involved, a cocktail of inhibitors can be used to identify the contributions of individual components.

Microfluidic chips provide advantages of fast and easy solution exchange for drug application and washout. With multiple inlets, either different concentration of a specific drug can be applied for a designated time or multiple drugs can be used sequentially or in parallel. For instance, astrocytes have cationic MSCs that are permeable to Ca^{2+} (Ostrow & Sachs, 2005; Ostrow, Suchyna, & Sachs, 2011; Suchyna et al., 2004) and voltage-gated Ca^{2+} channels that are also sensitive to force. Several MSCs have been identified in astrocytes, they are members of the transient receptor potential (TRP) family including, TRPV4, TRPC1, TRPC4, and TRPC5, where TRPC1 and TRPC5 are located on the cell membrane. Using a membrane-tethered genetically encoded Ca^{2+} indicator (Lck-GCaMP3), Shigetomi, Tong, Kwan, Corey, and Khakh (2012) have observed spotty

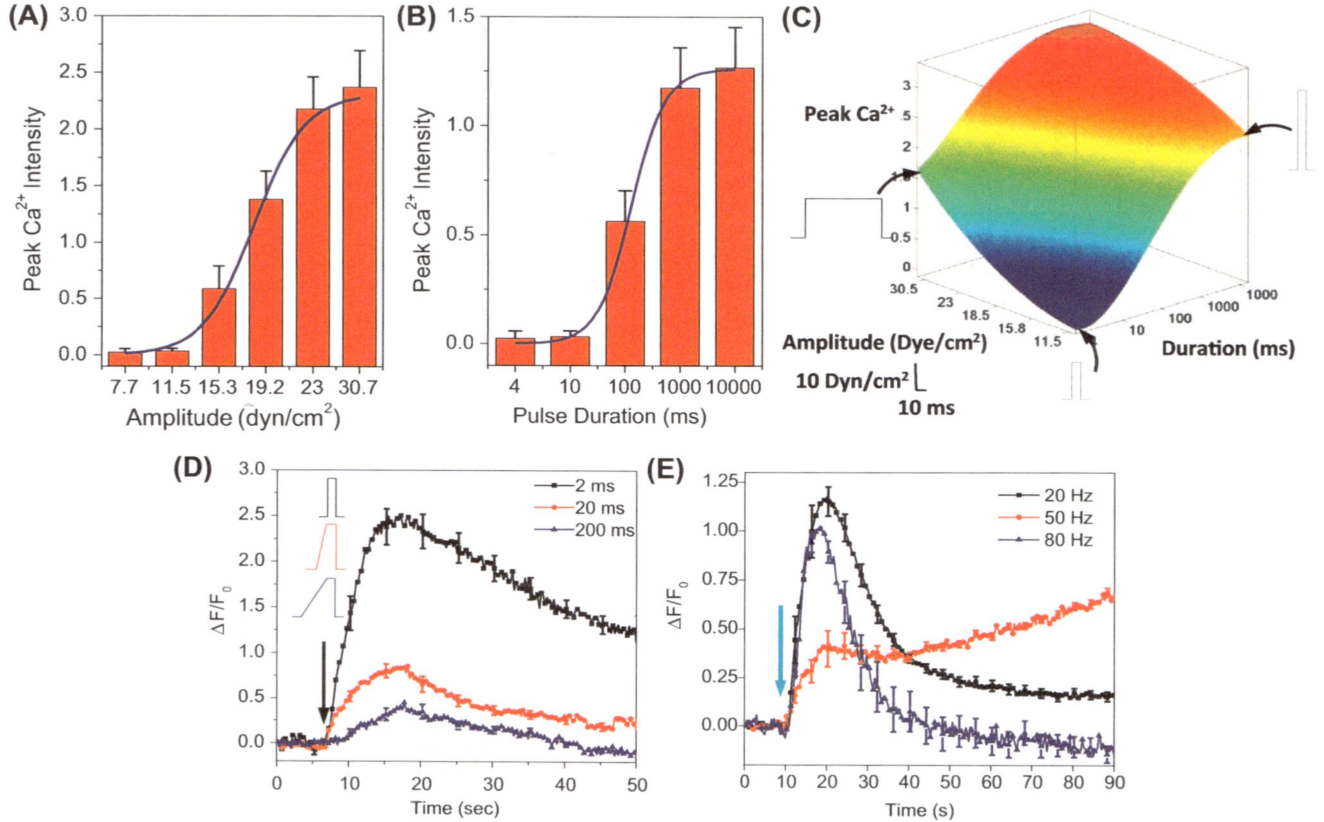

Figure 5 Ca^{2+} response to various shear stress stimuli. (A and B): Peak Ca^{2+} response to different amplitudes (A) and durations (B). (C) Profile of peak Ca^{2+} response to different amplitudes and durations. (D) Ca^{2+} response to shear pulses of different rise time. (E) Ca^{2+} response to sine wave train (10 cycles) of different frequencies, 20, 50, and 80 Hz.

Ca^{2+} signals in cultured astrocytes that were not dependent on Ca^{2+} release from internal stores, but via influx of Ca^{2+} through TRPA1 channels. A servo driven microfluidic chip was used in a series of pharmacological studies to determine the main Ca^{2+} source. The overall Ca^{2+} increase to shear stress has two components: a transient peak that occurs immediately after stimulation that is due to the influx of Ca^{2+} and a second peak (in many cases a shoulder of the first peak) due to the release of Ca^{2+} from intracellular stores (Maneshi et al., 2015). For example, the initial Ca^{2+} increase, due to mechanical stimulation, was eliminated by removal of Ca^{2+} from the media. The response was recovered after Ca^{2+} was reintroduced. Shear stress response generated a transient Ca^{2+} peak after depleting the intracellular Ca^{2+} stores with thapsigargin, supporting a Ca^{2+} influx is required (Maneshi et al., 2015).

It is imperative to know the possible pathways for total Ca^{2+} responses observed by activation of shear stress. One of the common pathways of increase in cytosolic Ca^{2+} is the release from stores. To maintain cells calcium homeostasis, ATP-dependent pumps move calcium from the cytosol into ER and mitochondria calcium stores. One of theories behind mechanical force activated calcium oscillations is by calcium influx—induced calcium release (CICR). This process works as a signal amplifier in sensory cells as well. In this scenario, calcium influx through MSCs would activate pathways that lead to the release of calcium from stores.

We measured the two phases of the CICR for HEK293T cells that stably express hPIEZO1. Cells treated with 5 μM thapsigargin, a known calcium ATPase inhibitor that blocks the Sarco-endoplasmic ATP-dependent calcium pumps (SERCA), causes Ca^{2+} store depletion. Subsequently, cells were stimulated by shear stress to measure the influx of calcium. About 28% of the total calcium was from the depletion of SERCA stores, and nearly 72% of measured calcium increase in HEK293 cells expressing hPIEZO1 was from mechanically activated influx via PIEZO1 channels (Fig. 6)

The shear stress assay can be used to study the mechanical stress that is dependent on downstream biochemical activities. For instance, Voyvodic, Min, and Baker (2012). have used a parallel multichannel system to study human umbilical vein endothelial cells (HUVEC) under shear stress, and they have shown that steady flow have increased the production of endothelial constitutive nitric oxide synthase (eNOS) and formation of actin fibers in cultured HUVEC cells compared to the pulsatile shear stress. Ruiz et al. (2014) used a variation of this approach to measure the toxicity of

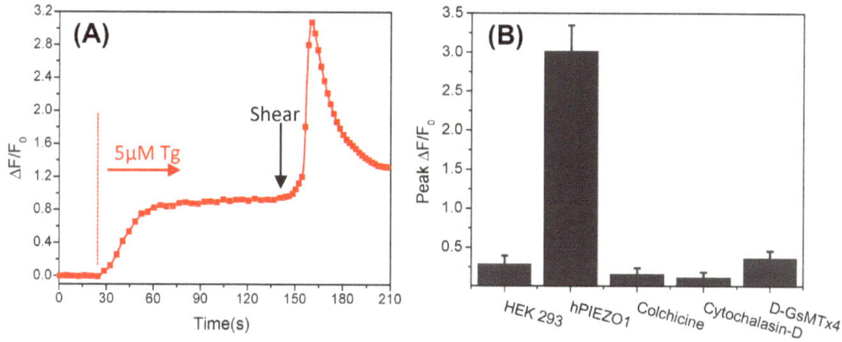

Figure 6 Pharmacological assay. (A) Ca^{2+} response in the presence of Tg (5 μM), showing involvement of Ca^{2+} influx and release of the Ca^{2+} stores. (B) Effect of various drugs on shear stimulated Ca^{2+} elevations.

beta-Amyloids on CNS cultures. In their study, cells in the microfluidic chips were exposed to different aggregated forms of beta-amyloid (Aβ) with CNS drugs, that allows them to assess the integrity of glial cells during pharmacological treatments. Cheung et al. (2009) used shear stress assay to study detachment of cancer cells under flow acceleration in a microchannel and demonstrated that the detachment of cells depends on both flow rate and flow acceleration. The flow rate required for cell detachment is a random variable that can be described by a log-normal distribution. An empirical model was proposed, and Zheng et al. (2012) have used it to study the flow-induced stretch on endothelial involved in blood vessel biomechanics. Cells alignment was correlated with the steady fluid shear stress, cyclic flow, and combination of the two.

3. MECHANICAL COUPLINGS OF CYTOSKELETON TO MECHANOSENSITIVE ION CHANNELS

3.1 Cytochalasin D

Since the cytoskeleton bears the majority of mechanical stress in cell body, it is possible that there is a direct connection between the cytoskeletal proteins and channel activation and kinetics. Thus, the stress from the cell's body transmitted to the membrane affects the open probability of the channel. To probe such links between actin cytoskeleton and Piezo1 channels, cells expressing hPIEZO1 were treated with 5 μM of Cytochalasin D to disrupt actin cytoskeleton. Cells were then stimulated with shear stress. We found that the Ca^{2+} rise in cells treated with Cytochalasin D is less than 10% of control cells, showing that the shear stress response of PIEZO1 requires an intact cytoskeleton.

3.2 Colchicine

We measured the connection between PIEZO channels and microtubules. Microtubules are known to mainly bear compressive stresses in the eukaryotic cells, and they may play a key role in mechanotransductions (Ingber, 2006). To test this, we treated PIEZO1 overexpressing cells with 10 μM of Colchicine depolymerizing the majority of microtubular structures, followed by a shear stress pulse. The Ca^{2+} response was less than 15% of control cells. Like the actin cytoskeleton, microtubes play a role in supporting PIEZO1 activity. This response is consistent with previous work showing that MDX Myofibers (muscular dystrophy), which have acute calcium influx during cell stretch, are inhibited by application of Colchicine (Khairallah et al., 2012).

The loss of the shear stress response by the treatment of either Cytochalasin D or Colchicine resembles the effect observed for whole cell currents where treatment with Cytochalasin D did not generate whole cell currents using a probe to press on the cell (Gottlieb, Bae, & Sachs, 2012). This is not an effect associated with the loss of channel activity because cell-attached patches continued to respond to suction, indicating that the channels were functional.

Whole cell currents generated by cell pressing and shear activation of Piezo channels seem to share a common characteristic in that there is involvement of cytoskeletal elements. We also know that the Piezo channel is sensitive to membrane tension (Cox et al., 2016; Lewis & Grandl, 2015), and this may be a common feature where whole cell currents and shear stress activation activate channels by increasing bilayer stress. Despite this, it is unlikely that the pathway that opens the channels by pressing on cells is common to shear stress activation. This is because whole cell currents have a minimal latency of activation where the channel response is linked to the onset of the stimulus while the response to shear stress requires seconds to develop.

4. SHEAR STRESS AS A HIGH-THROUGHPUT ASSAY FOR MECHANOSENSITIVE ION CHANNEL SPECIFIC DRUGS

Piezo family channels have been found in many types of animal cells, ranging from human neuron stem cells (Pathak et al., 2014) to vascular endothelia (Li et al., 2014; Ranade et al., 2014). To date, the only known specific MSC inhibitor is GsMTx4, a peptide first identified and purified from *Tarantula rosea* (Bowman, Gottlieb, Suchyna, Murphy, & Sachs, 2007; Suchyna et al., 2004). Upon the discovery of PIEZO channels in 2010, it was shown that these channels can be blocked by GsMTx4 (Bae, Sachs, & Gottlieb, 2011). Both enantiomeric forms of GsMTx4 target Piezo1 channels as measured by patch clamp and whole cell recordings (Bae et al.,2011). Most available MSC blockers, such as ruthenium red are nonspecific. Thus, there is constant search for soluble chemical compounds that can directly block specific types of MSCs. Combining different aspects of the microfluidics and shear stress as an assay for MSCs would make this assay a real-time high-throughput assay.

4.1 Inhibition by peptide GsMTx4

To demonstrate the utility of the shear stress response in a microfluidic chamber, we measure the effect of GsMTx4 on the inhibition of Ca^{2+} influx. We cultured HEK293 cells that stably expressed hPIEZO1 in microfluidic chambers. To assist in the surface adhesion and stability during shear stress pulses, the chamber surface was treated with Human Fibronectin. Cells were cultured until they were 75% confluent and cells were loaded with cell permeable Fluo4AM (Maneshi et al., 2015). Various concentrations of the D-form peptide GsMTx4 was applied to designated chambers followed by a shear stress pulse of 15 dyn/cm^2. The fluorescence increase due to the calcium rise was recorded (Fig. 7). The results show that 50% inhibition of channels could be achieved at ~300 nM (Ic_{50}) and that is similar to the concentration derived by whole cell patch clamps (Fig. 8) (Bae et al., 2011). In this study, four fresh cell cultures, each contained ~200 cells, were tested at a given dose. The large number of cells at each dose resulted in a reliable data with minimum error (standard error <0.04). The entire binding curve was produced within hours. In contrast, it would require several weeks to generate same amount of data using patch clamp technique.

4.2 Yoda1 activation

Yoda1 acts as an agonist for both human and mouse Piezo1 (Syeda et al., 2015). Functional studies in cells revealed that Yoda1 affects the activation and the inactivation kinetics of mechanically stimulated responses. Yoda1 characterization in artificial droplet lipid bilayers has shown that Yoda1 activates purified Piezo1 channels in the absence of other cellular components. In cell-attached patch clamp recordings of Piezo1 transiently expressed in

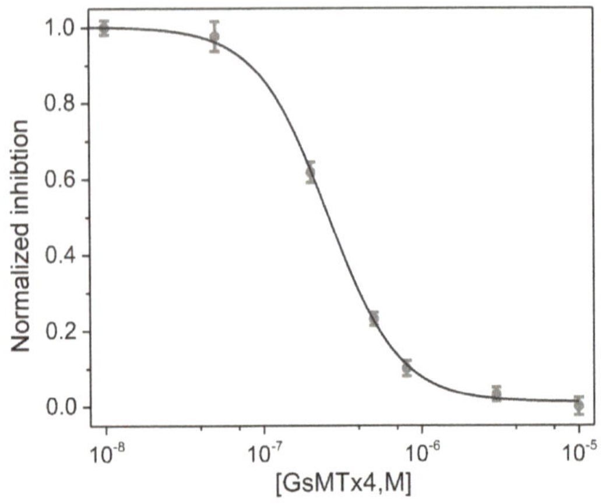

Figure 7 Dose dependent inhibition of shear-activated Ca^{2+} response through hPiezo1 to GsMTx4. A Shear stress pulse of 15 dyn/cm^2 for 5 ms was applied to HEK cells expressing PIEZO1.

293T cells, Yoda1 led to slower channel inactivation kinetics and a shift in the current–pressure relationship, indicating sensitization to membrane stretch-induced activation of Piezo1. This allows Yoda1 to serve as an agonist to study shear-activated response of hPIEZO1 channels.

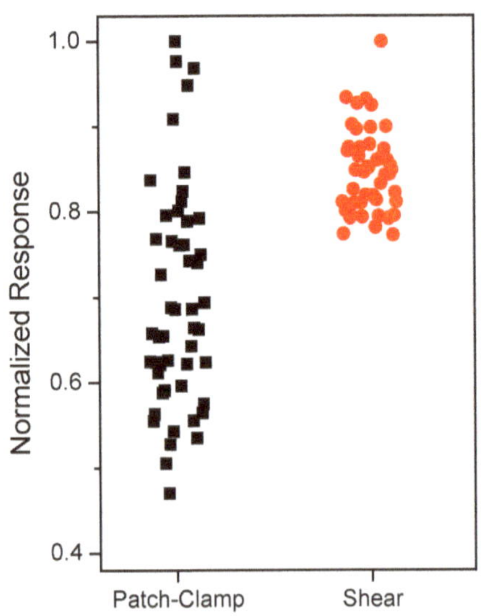

Figure 8 Comparison of cell response to fluid shear stress (red) measured with Ca^{2+} dye to indentation measured with patch-clamp (black), showing the statistical advantage of the shear stress assay. The variance calculated for the shear stress assay is ~8-fold less than the value derived for the patch clamp method.

5. MICROFLUIDIC SHEAR STRESS CHAMBER VERSUS OTHER MECHANICAL ASSAYS

A number of methods exist that can are used to mechanically stimulate cells. They include: pressing on cells with a polished glass pipette (Lamoureux, Altun-Gultekin, Lin, Wagner, & Heidemann, 1997); single or double axial or 3-D stretching or bending of the substrate (Riehl, Park, Kwon, & Lim, 2012); magnetically pulling on beads attached to the cell membrane (Matthews et al., 2010); ultra sound stimulation (Sun, Li, & Meininger, 2012); shaking (Sun et al., 2012) cells; osmotic challenges (Erickson, Northrup, & Guilak, 2003).

For MSCs, the pressure clamp is also widely used to study channels and drugs that affect channel activity at a single or multichannel level. Unlike the microfluidic approach that allows us to generate an inhibition curve within a few hours, single channel activity requires tens of patches to achieve a comparable result that take days. Beyond that, shear stress measurements exerts minimal deformation on the intact cells while the patch clamp method applies considerable tension on the membrane that may also disrupt the cytoskeleton (Sokabe, Sachs, & Jing, 1991).

In a microfluidic chamber, a uniform shear stress is applied to few hundreds of cells, and the activation of MSCs can be measured using Ca^{2+} sensitive or voltage sensitive dyes. The ion channels located on the apical surface of the cells are directly activated by interaction of the fluid and protein channel, a minimal stimulus is required. Moreover, parallel fluid flow chambers fabricated on single chip make an ideal platform for high-throughput drugs testing.

Although every assay would have a number of uncertainties that lead to inconsistency for the assay, one advantage of shear stress assay in microfluidics is that it has more physiological relevance. The cross talk between the channel and the domain in which it is located along with cytoskeletal elements is more reflective of the cellular environment in which the channel exists.

Fig. 8 shows a statistical advantage of the shear stress assay over the whole cell indentation experiments. The data for $N = 50$ cells reduces the error significantly in the measurements. For the two sets of data, the maximum reading on two assays have been normalized to one and the rest of the data in each data set have been normalized to the maximum. The patch data shows a wider distribution with an average of 0.6981 and a variance of 0.016 while the data set from shear stress shows an average of 0.846 and a variance of 0.002.

6. MICROFLUIDIC DEVICES TO MEASURE OTHER PROPERTIES OF PIEZO CHANNELS

Microfluidic devices can be used to measure other properties of Piezo channels. Below we describe some recent advancements in usage of microfluidic devices to study other signaling pathways related to Piezo channels.

6.1 ATP release by red blood cells

Microfluidic chambers have been recently utilized by Cinar et al. (2015) to measure shear-induced ATP release in RBC. There have been previous studies showing the expression of PIEZO1 in RBCs and their possible link to hereditary xerocytosis (Albuisson et al., 2013; Bae, Gnanasambandam, Nicolai, Sachs, & Gottlieb, 2013; Zarychanski et al., 2012). Cinar et. al., cross-section of a microfluidic chamber to apply low shear and high shear on cells, showed that Piezo1 channels regulated mechanotransductive release of ATP from human RBCs by controlling the shear-induced Ca^{2+} influx (Cinar et al., 2015). They have also shown that inhibition of hPIEZO1 channel would significantly decrease ATP release from cells. Cells expressing mutations of hPIEZO1 showed a drastically reduced ATP release by shear stress. In this work, the existence of a threshold Ca^{2+} concentration for triggering ATP release was proposed, and it was shown that membrane ATP stores are released when RBCs experienced shear forces.

6.2 Confinement sensing

In a recent study, parallel signaling pathways of Piezo1 channels and Myosin II in cells motility have been studied using microfluidic chambers as confinements for cell growth (Hung et al., 2016). During the transition from unconfined to confined spaces, intracellular Ca^{2+} level increased and the increase in Ca^{2+} depends on the expression of Piezo1 channels. This increase in Ca^{2+} leads to phosphodiesterase 1—dependent suppression of PKA activity. Simultaneous inhibition of Piezo1 and Myosin II eliminated the effects of confinement on cells. The two pathways appeared work in parallel, and inhibition of either one affected confinement-induced cells stiffness. Both pathways are highly dependent on Actin cytoskeleton and treatment of polymerization inhibiting drugs, for instance, Latrunculin completely abolished the effects due to confinement.

7. PERSPECTIVE AND FUTURE DEVELOPMENT

Numerous studies have demonstrated that shear stress stimuli within a microfluidic chamber provide a state of the art technology for studying MSCs. Use of high speed operation to drive microfluidic chamber will help to separate correlation into cause and effect, which is particularly useful to study mechanosensitive channels, like Piezo. Moreover, it is a useful platform for a high-throughput screen to measure the effects of compounds on MSCs at equilibrium. The method enables sequential and parallel drug applications and provides real-time cell response with high time resolution. Parallel channels with connections to a common mechanical driver could allow the screening of multiple drugs from a common stimulus. All these features would significantly increase throughput of screening specific drugs of MSCs.

REFERENCES

Akerboom, J., Chen, T. W., Wardill, T. J., Tian, L., Marvin, J. S., Mutlu, S., & Looger, L. L. (2012). Optimization of a GCaMP calcium indicator for neural activity imaging. *Journal of Neuroscience, 32*(40), 13819–13840. http://dx.doi.org/10.1523/JNEUROSCI.2601-12.2012.

Albuisson, J., Murthy, S. E., Bandell, M., Coste, B., Louis-dit-Picard, H., Mathur, J., & Syfuss, P.-Y. (2013). Dehydrated hereditary stomatocytosis linked to gain-of-function mutations in mechanically activated PIEZO1 ion channels. *Nature Communications, 4*.

Althaus, M., Bogdan, R., Clauss, W. G., & Fronius, M. (2007). Mechano-sensitivity of epithelial sodium channels (ENaCs): Laminar shear stress increases ion channel open probability. *FASEB Journal, 21*(10), 2389–2399.

Ateya, D. A., Sachs, F., Gottlieb, P. A., Besch, S., & Hua, S. Z. (2005). Volume cytometry: Microfluidic sensor for high-throughput screening in real time. *Analytical Chemistry,* 77(5), 1290–1294.

Bae, C., Gnanasambandam, R., Nicolai, C., Sachs, F., & Gottlieb, P. A. (2013). Xerocytosis is caused by mutations that alter the kinetics of the mechanosensitive channel PIEZO1. *Proceedings of the National Academy of Sciences of the United States of America, 110*(12), E1162–E1168.

Bae, C., Gottlieb, P. A., & Sachs, F. (2013). Human PIEZO1: Removing inactivation. *Biophysical Journal, 105*(4), 880–886. http://dx.doi.org/10.1016/j.bpj.2013.07.019.

Bae, C., Sachs, F., & Gottlieb, P. A. (2011). The mechanosensitive ion channel Piezo1 is inhibited by the peptide GsMTx4. *Biochemistry, 50*(29), 6295–6300. http://dx.doi.org/10.1021/bi200770q.

Besch, S. R., Suchyna, T., & Sachs, F. (2002). High-speed pressure clamp. *Pflugers Archiv, 445*(1), 161–166. http://dx.doi.org/10.1007/s00424-002-0903-0.

Bowman, C. L., Gottlieb, P. A., Suchyna, T. M., Murphy, Y. K., & Sachs, F. (2007). Mechanosensitive ion channels and the peptide inhibitor GsMTx-4: History, properties, mechanisms and pharmacology. *Toxicon, 49*(2), 249–270. http://dx.doi.org/10.1016/j.toxicon.2006.09.030. pii:S0041-0101(06)00356-4.

Chen, Y. C., Smith, D. H., & Meaney, D. F. (2009). In-vitro approaches for studying blast-induced traumatic brain injury. *Journal of Neurotrauma, 26*(6), 861–876. http://dx.doi.org/10.1089/neu.2008.0645.

Cheung, L. S. L., Zheng, X., Stopa, A., Baygents, J. C., Guzman, R., Schroeder, J. A., & Zohar, Y. (2009). Detachment of captured cancer cells under flow acceleration in a bio-functionalized microchannel. *Lab on a Chip, 9*(12), 1721–1731.

Chung, R. S., Staal, J. A., McCormack, G. H., Dickson, T. C., Cozens, M. A., Chuckowree, J. A., & Vickers, J. C. (2005). Mild axonal stretch injury in vitro induces a progressive series of neurofilament alterations ultimately leading to delayed axotomy. *Journal of Neurotrauma, 22*(10), 1081–1091. http://dx.doi.org/10.1089/neu.2005.22.1081.

Cinar, E., Zhou, S., DeCourcey, J., Wang, Y., Waugh, R. E., & Wan, J. (2015). Piezo1 regulates mechanotransductive release of ATP from human RBCs. *Proceedings of the National Academy of Sciences of the United States of America, 112*(38), 11783–11788.

Coste, B., Mathur, J., Schmidt, M., Earley, T. J., Ranade, S., Petrus, M. J., & Patapoutian, A. (2010). Piezo1 and Piezo2 are essential components of distinct mechanically activated cation channels. *Science, 330*(6000), 55–60. http://dx.doi.org/10.1126/science.1193270.

Coste, B., Xiao, B., Santos, J. S., Syeda, R., Grandl, J., Spencer, K. S., & Patapoutian, A. (2012). Piezo proteins are pore-forming subunits of mechanically activated channels. *Nature, 483*(7388), 176–181.

Cox, C. D., Bae, C., Ziegler, L., Hartley, S., Nikolova-Krstevski, V., Rohde, P. R., & Martinac, B. (2016). Removal of the mechanoprotective influence of the cytoskeleton reveals PIEZO1 is gated by bilayer tension. *Nature Communications,* 7, 10366. http://dx.doi.org/10.1038/ncomms10366.

Cullen, D. K., Vernekar, V. N., & LaPlaca, M. C. (2011). Trauma-induced plasmalemma disruptions in three-dimensional neural cultures are dependent on strain modality and rate. *Journal of Neurotrauma, 28*(11), 2219–2233. http://dx.doi.org/10.1089/neu.2011.1841.

Davies, P. F., Spaan, J. A., & Krams, R. (2005). Shear stress biology of the endothelium. *Annals of Biomedical Engineering, 33*(12), 1714–1718. http://dx.doi.org/10.1007/s10439-005-8774-0.

DuFort, C. C., Paszek, M. J., & Weaver, V. M. (2011). Balancing forces: Architectural control of mechanotransduction. *Nature Reviews. Molecular Cell Biology, 12*(5), 308–319. http://dx.doi.org/10.1038/nrm3112.

Erickson, G., Northrup, D., & Guilak, F. (2003). Hypo-osmotic stress induces calcium-dependent actin reorganization in articular chondrocytes. *Osteoarthritis and Cartilage, 11*(3), 187–197.

Fleminger, S., Oliver, D. L., Lovestone, S., Rabe-Hesketh, S., & Giora, A. (2003). Head injury as a risk factor for Alzheimer's disease: The evidence 10 years on; a partial replication. *Journal of Neurology, Neurosurgery, and Psychiatry, 74*(7), 857–862.

Garman, R. H., Jenkins, L. W., Switzer, R. C., 3rd, Bauman, R. A., Tong, L. C., Swauger, P. V., & Kochanek, P. M. (2011). Blast exposure in rats with body shielding is characterized primarily by diffuse axonal injury. *Journal of Neurotrauma, 28*(6), 947–959. http://dx.doi.org/10.1089/neu.2010.1540.

Geddes-Klein, D. M., Schiffman, K. B., & Meaney, D. F. (2006). Mechanisms and consequences of neuronal stretch injury in vitro differ with the model of trauma. *Journal of Neurotrauma, 23*(2), 193–204. http://dx.doi.org/10.1089/neu.2006.23.193.

Gomez, G. A., McLachlan, R. W., & Yap, A. S. (2011). Productive tension: Force-sensing and homeostasis of cell-cell junctions. *Trends Cell Biol, 21*(9), 499–505. http://dx.doi.org/10.1016/j.tcb.2011.05.006.

Gottlieb, P. A., Bae, C., & Sachs, F. (2012). Gating the mechanical channel Piezo1: A comparison between whole-cell and patch recording. *Channels, 6*(4), 282–289.

Guo, Z., Cupples, L. A., Kurz, A., Auerbach, S. H., Volicer, L., Chui, H., & Farrer, L. A. (2000). Head injury and the risk of AD in the MIRAGE study. *Neurology, 54*(6), 1316–1323.

Haeberle, H., Bryan, L. A., Vadakkan, T. J., Dickinson, M. E., & Lumpkin, E. A. (2008). Swelling-activated Ca^{2+} channels trigger Ca^{2+} signals in Merkel cells. *PLoS One, 3*(3), e1750. http://dx.doi.org/10.1371/journal.pone.0001750.

Heo, J., Sachs, F., Wang, J., & Hua, S. Z. (2012). Shear-induced volume decrease in MDCK cells. *Cellular Physiology and Biochemistry, 30.*

Holtzclaw, J. D., Liu, L., Grimm, P. R., & Sansom, S. C. (2010). Shear stress-induced volume decrease in C11-MDCK cells by BK-α/β4. *American Journal of Physiology. Renal Physiology, 299*(3), F507–F516. http://dx.doi.org/10.1152/ajprenal.00222.2010.

Hua, S. Z., Gottlieb, P. A., Heo, J., & Sachs, F. (2010). A mechanosensitive ion channel regulating cell volume. *American Journal of Physiology. Cell Physiology, 298*(6), C1424–C1430.

Hung, W.-C., Yang, J. R., Yankaskas, C. L., Wong, B. S., Wu, P.-H., Pardo-Pastor, C., & Wirtz, D. (2016). Confinement-sensing and signal optimization via Piezo1/PKA and myosin II pathways. *Cell Reports, 15.*

Huveneers, S., & de Rooij, J. (2013). Mechanosensitive systems at the cadherin-F-actin interface. *Journal of Cell Science, 126*(Pt 2), 403–413. http://dx.doi.org/10.1242/jcs.109447.

Ingber, D. E. (2006). Cellular mechanotransduction: Putting all the pieces together again. *FASEB Journal, 20*(7), 811–827. http://dx.doi.org/10.1096/fj.05-5424rev.

Jensen, M. E., Odgaard, E., Christensen, M. H., Praetorius, H. A., & Leipziger, J. (2007). Flow-induced $[Ca^{2+}]i$ increase depends on nucleotide release and subsequent purinergic signaling in the intact nephron. *Journal of the American Society of Nephrology, 18*(7), 2062–2070. http://dx.doi.org/10.1681/ASN.2006070700. pii:ASN.2006070700.

Johnson, V. E., Stewart, W., & Smith, D. H. (2010). Traumatic brain injury and amyloid-beta pathology: A link to Alzheimer's disease? *Nature Reviews. Neuroscience, 11*(5), 361–370. http://dx.doi.org/10.1038/nrn2808.

Khairallah, R. J., Shi, G., Sbrana, F., Prosser, B. L., Borroto, C., Mazaitis, M. J., & Sun, Y. (2012). Microtubules underlie dysfunction in duchenne muscular dystrophy. *Science Signaling, 5*(236).

Kwan, H. Y., Huang, Y., & Yao, X. (2007). Cyclic nucleotides and Ca^{2+} influx pathways in vascular endothelial cells. *Clinical Hemorheology and Microcirculation, 37*(1–2), 63–70.

Laitko, U., & Morris, C. E. (2004). Membrane tension accelerates rate-limiting voltage-dependent activation and slow inactivation steps in a Shaker channel. *Journal of General Physiology, 123*(2), 135–154. http://dx.doi.org/10.1085/jgp.200308965.

Lamoureux, P., Altun-Gultekin, Z. F., Lin, C., Wagner, J. A., & Heidemann, S. R. (1997). Rac is required for growth cone function but not neurite assembly. *Journal of Cell Science, 110*(5), 635–641.

LaPlaca, M. C., Cullen, D. K., McLoughlin, J. J., & Cargill, R. S., 2nd (2005). High rate shear strain of three-dimensional neural cell cultures: A new in vitro traumatic brain injury model. *Journal of Biomechanics, 38*(5), 1093–1105. http://dx.doi.org/10.1016/j.jbiomech.2004.05.032.

Lewis, A. H., & Grandl, J. (2015). Mechanical sensitivity of Piezo1 ion channels can be tuned by cellular membrane tension. *eLife, 4.* http://dx.doi.org/10.7554/eLife.12088.

Li, J., Hou, B., Tumova, S., Muraki, K., Bruns, A., Ludlow, M. J., & Beech, D. J. (2014). Piezo1 integration of vascular architecture with physiological force. *Nature, 515*(7526), 279–282. http://dx.doi.org/10.1038/nature13701.

Liu, W., Morimoto, T., Woda, C., Kleyman, T. R., & Satlin, L. M. (2007). Ca^{2+} dependence of flow-stimulated K secretion in the mammalian cortical collecting duct. *American Journal of Physiology. Renal Physiology, 293*(1), F227–F235.

Liu, W., Xu, S., Woda, C., Kim, P., Weinbaum, S., & Satlin, L. M. (2003). Effect of flow and stretch on the [Ca^{2+}]i response of principal and intercalated cells in cortical collecting duct. *American Journal of Physiology. Renal Physiology, 285*(5), F998–F1012.

Maas, A. I., Stocchetti, N., & Bullock, R. (2008). Moderate and severe traumatic brain injury in adults. *Lancet Neurology, 7*(8), 728–741. http://dx.doi.org/10.1016/S1474-4422(08)70164-9.

Maneshi, M. M., Sachs, F., & Hua, S. Z. (2015). A threshold shear force for calcium influx in an astrocyte model of traumatic brain injury. *Journal of Neurotrauma, 32*(13), 1020–1029. http://dx.doi.org/10.1089/neu.2014.3677.

Matthews, B. D., Thodeti, C. K., Tytell, J. D., Mammoto, A., Overby, D. R., & Ingber, D. E. (2010). Ultra-rapid activation of TRPV4 ion channels by mechanical forces applied to cell surface β1 integrins. *Integrative Biology, 2*(9), 435–442.

Morris, C. E., & Juranka, P. F. (2007). Nav channel mechanosensitivity: Activation and inactivation accelerate reversibly with stretch. *Biophysical Journal, 93*(3), 822–833. http://dx.doi.org/10.1529/biophysj.106.101246.

Mortimer, J. A., French, L. R., Hutton, J. T., & Schuman, L. M. (1985). Head injury as a risk factor for Alzheimer's disease. *Neurology, 35*(2), 264–267.

Mott, R. E., & Helmke, B. P. (2007). Mapping the dynamics of shear stress-induced structural changes in endothelial cells. *American Journal of Physiology. Cell Physiology, 293*(5), C1616–C1626. http://dx.doi.org/10.1152/ajpcell.00457.2006. pii:00457.2006.

Ostrow, L. W., & Sachs, F. (2005). Mechanosensation and endothelin in astrocytes—hypothetical roles in CNS pathophysiology. *Brain Research. Brain Research Reviews, 48*(3), 488–508. http://dx.doi.org/10.1016/j.brainresrev.2004.09.005. pii:S0165-0173(04)00141-9.

Ostrow, L. W., Suchyna, T. M., & Sachs, F. (2011). Stretch induced endothelin-1 secretion by adult rat astrocytes involves calcium influx via stretch-activated ion channels (SACs). *Biochemical and Biophysical Research Communications, 410*(1), 81–86. http://dx.doi.org/10.1016/j.bbrc.2011.05.109. pii:S0006-291X(11)00883-7.

Pathak, M. M., Nourse, J. L., Tran, T., Hwe, J., Arulmoli, J., Le, D. T., … Tombola, F. (2014). Stretch-activated ion channel Piezo1 directs lineage choice in human neural stem cells. *Proceedings of the National Academy of Sciences of the United States of America, 111*(45), 16148–16153. http://dx.doi.org/10.1073/pnas.1409802111.

Praetorius, H. A., & Spring, K. R. (2001). Bending the MDCK cell primary cilium increases intracellular calcium. *Journal of Membrane Biology, 184*(1), 71–79.

Ranade, S. S., Qiu, Z., Woo, S. H., Hur, S. S., Murthy, S. E., Cahalan, S. M., … Patapoutian, A. (2014). Piezo1, a mechanically activated ion channel, is required for vascular development in mice. *Proceedings of the National Academy of Sciences of the United States of America, 111*(28), 10347–10352. http://dx.doi.org/10.1073/pnas.1409233111.

Ravin, R., Blank, P. S., Steinkamp, A., Rappaport, S. M., Ravin, N., Bezrukov, L., … Zimmerberg, J. (2012). Shear forces during blast, not abrupt changes in pressure alone, generate calcium activity in human brain cells. *PLoS One, 7*(6), e39421. http://dx.doi.org/10.1371/journal.pone.0039421.

Resnick, A. (2011). Chronic fluid flow is an environmental modifier of renal epithelial function. *PLoS One, 6*(10), e27058. http://dx.doi.org/10.1371/journal.pone.0027058. pii:PONE-D-11-11412.

Riehl, B. D., Park, J.-H., Kwon, I. K., & Lim, J. Y. (2012). Mechanical stretching for tissue engineering: Two-dimensional and three-dimensional constructs. *Tissue Engineering Part B: Reviews, 18*(4), 288–300.

Rohatgi, R., & Flores, D. (2010). Intratubular hydrodynamic forces influence tubulointerstitial fibrosis in the kidney. *Current Opinion in Nephrology and Hypertension, 19*(1), 65–71. http://dx.doi.org/10.1097/MNH.0b013e32833327f3.

Ruiz, A., Joshi, P., Mastrangelo, R., Francolini, M., Verderio, C., & Matteoli, M. (2014). Testing Abeta toxicity on primary CNS cultures using drug-screening microfluidic chips. *Lab on a Chip, 14*(15), 2860–2866. http://dx.doi.org/10.1039/c4lc00174e.

Sanborn, B., Nie, X., Chen, W., & Weerasooriya, T. (2012). Inertia effects on characterization of dynamic response of brain tissue. *Journal of Biomechanics, 45*(3), 434–439. http://dx.doi.org/10.1016/j.jbiomech.2011.12.017. pii:S0021-9290(11)00785-8.

Satlin, L. M., Sheng, S., Woda, C. B., & Kleyman, T. R. (2001). Epithelial Na(+) channels are regulated by flow. *American Journal of Physiology. Renal Physiology, 280*(6), F1010–F1018.

Schwarz, G., Callewaert, G., Droogmans, G., & Nilius, B. (1992). Shear stress-induced calcium transients in endothelial cells from human umbilical cord veins. *Journal of Physiology, 458*, 527–538.

Shen, J., Luscinskas, F. W., Connolly, A., Dewey, C. F., Jr., & Gimbrone, M. A., Jr. (1992). Fluid shear stress modulates cytosolic free calcium in vascular endothelial cells. *American Journal of Physiology, 262*(2 Pt 1), C384–C390.

Shigetomi, E., Tong, X., Kwan, K. Y., Corey, D. P., & Khakh, B. S. (2012). TRPA1 channels regulate astrocyte resting calcium and inhibitory synapse efficacy through GAT-3. *Nature Neuroscience, 15*(1), 70–80. http://dx.doi.org/10.1038/nn.3000.

Sokabe, M., Sachs, F., & Jing, Z. (1991). Quantitative video microscopy of patch clamped membranes stress, strain, capacitance, and stretch channel activation. *Biophysical Journal, 59*(3), 722.

Suchyna, T. M., Tape, S. E., Koeppe, R. E., 2nd, Andersen, O. S., Sachs, F., & Gottlieb, P. A. (2004). Bilayer-dependent inhibition of mechanosensitive channels by neuroactive peptide enantiomers. *Nature, 430*(6996), 235–240. http://dx.doi.org/10.1038/nature02743.

Sun, Z., Li, Z., & Meininger, G. A. (2012). Mechanotransduction through fibronectin-integrin focal adhesion in microvascular smooth muscle cells: Is calcium essential? *American Journal of Physiology-Heart and Circulatory Physiology, 302*(10), H1965–H1973.

Syeda, R., Xu, J., Dubin, A. E., Coste, B., Mathur, J., Huynh, T., … Engels, I. H. (2015). Chemical activation of the mechanotransduction channel Piezo1. *eLife, 4*, e07369.

Tian, L., Hires, S. A., & Looger, L. L. (2012). Imaging neuronal activity with genetically encoded calcium indicators. *Cold Spring Harbor Protocols, 2012*(6), 647–656. http://dx.doi.org/10.1101/pdb.top069609.

Tsutsui, H., Karasawa, S., Okamura, Y., & Miyawaki, A. (2008). Improving membrane voltage measurements using FRET with new fluorescent proteins. *Nature Methods, 5*(8), 683–685. http://dx.doi.org/10.1038/nmeth.1235.

Vogel, V., & Sheetz, M. (2006). Local force and geometry sensing regulate cell functions. *Nature Reviews. Molecular Cell Biology,* 7(4), 265–275. http://dx.doi.org/10.1038/nrm1890.

Voyvodic, P. L., Min, D., & Baker, A. B. (2012). A multichannel dampened flow system for studies on shear stress-mediated mechanotransduction. *Lab on a Chip, 12*(18), 3322–3330. http://dx.doi.org/10.1039/c2lc40526a.

Weinbaum, S., Duan, Y., Satlin, L. M., Wang, T., & Weinstein, A. M. (2010). Mechanotransduction in the renal tubule. *American Journal of Physiology. Renal Physiology, 299*(6), F1220–F1236. http://dx.doi.org/10.1152/ajprenal.00453.2010. pii: ajprenal.00453.2010.

Woda, C. B., Bragin, A., Kleyman, T. R., & Satlin, L. M. (2001). Flow-dependent K^+ secretion in the cortical collecting duct is mediated by a maxi-K channel. *American Journal of Physiology. Renal Physiology, 49*(5), F786–F793.

Zarychanski, R., Schulz, V. P., Houston, B. L., Maksimova, Y., Houston, D. S., Smith, B., … Gallagher, P. G. (2012). Mutations in the mechanotransduction protein PIEZO1 are associated with hereditary xerocytosis. *Blood, 120*(9), 1908–1915.

Zheng, W., Jiang, B., Wang, D., Zhang, W., Wang, Z., & Jiang, X. (2012). A microfluidic flow-stretch chip for investigating blood vessel biomechanics. *Lab on a Chip, 12*(18), 3441–3450. http://dx.doi.org/10.1039/c2lc40173h.

FSC
www.fsc.org
MIX
Papier | Fördert
gute Waldnutzung
FSC® C083411

Kolibri 360 GmbH
Ferdinand-Jühlke-Straße 7
99095 Erfurt, Deutschland
produktsicherheit@kolibri360.de